DIGITAL SIGNAL PROCESSING FOR HIGH-SPEED OPTICAL COMMUNICATION

DIGITAL SIGNAL PROCESSING FOR HIGH-SPEED OPTICAL COMMUNICATION

Jianjun Yu
ZTE (TX) Inc, USA

Xinying Li
Georgia Institute of Technology, USA

Junwen Zhang
ZTE (TX) Inc, USA

World Scientific

NEW JERSEY · LONDON · SINGAPORE · BEIJING · SHANGHAI · HONG KONG · TAIPEI · CHENNAI · TOKYO

Published by

World Scientific Publishing Co. Pte. Ltd.
5 Toh Tuck Link, Singapore 596224
USA office: 27 Warren Street, Suite 401-402, Hackensack, NJ 07601
UK office: 57 Shelton Street, Covent Garden, London WC2H 9HE

Library of Congress Cataloging-in-Publication Data
Names: Yu, Jianjun, (Optical engineer), author.
Title: Digital signal processing for high-speed optical communication / Jianjun Yu
　　(ZTE (TX) Inc, USA), Xinying Li (Georgia Institute of Technology, USA),
　　Junwen Zhang (ZTE (TX) Inc, USA).
Description: [Hackensack] New Jersey : World Scientific, [2018] |
　　Includes bibliographical references.
Identifiers: LCCN 2017044681 | ISBN 9789813233973 (hc : alk. paper)
Subjects: LCSH: Optical fiber communication. | Signal processing--Digital techniques.
Classification: LCC TK5103.592.O68 Y84 2018 | DDC 621.382/7--dc23
LC record available at https://lccn.loc.gov/2017044681

British Library Cataloguing-in-Publication Data
A catalogue record for this book is available from the British Library.

Copyright © 2018 by World Scientific Publishing Co. Pte. Ltd.

All rights reserved. This book, or parts thereof, may not be reproduced in any form or by any means, electronic or mechanical, including photocopying, recording or any information storage and retrieval system now known or to be invented, without written permission from the publisher.

For photocopying of material in this volume, please pay a copying fee through the Copyright Clearance Center, Inc., 222 Rosewood Drive, Danvers, MA 01923, USA. In this case permission to photocopy is not required from the publisher.

For any available supplementary material, please visit
http://www.worldscientific.com/worldscibooks/10.1142/10818#t=suppl

Desk Editors: V. Vishnu Mohan/Amanda Yun

Typeset by Stallion Press
Email: enquiries@stallionpress.com

Printed in Singapore

Preface

With the increase of the data throughput at an exponential rate, there is an urgent need for the ultra-large bandwidth. Over 90 percent of the communication capacity is transmitted by the fiber-optic. The traditional technique of intensity-modulation and direct-detection employed by fiber-optic communication has the advantages of a simple architecture and a low cost, but it also has the disadvantages of a low spectral efficiency and a short transmission distance. Moreover, it is difficult to realize long distance transmission of over 40-Gb/s/channel signals. The technique of the coherent optical communication based on digital signal processing (DSP) can realize large-capacity high-spectral-efficiency high-receiver-sensitivity data transmission. The employed DSP algorithms by coherent optical communication not only can pre-equalize the transmitter signals, but also can effectively compensate for various kinds of linear and nonlinear impairments caused by components and fiber-optic transmission links. Therefore, the technique of the coherent optical communication based on DSP is a very attractive and hot research topic.

This book summarizes various kinds of advanced DSP techniques reported in recent years, which aim at various kinds of application scenarios, including single-carrier long-haul/large-capacity optical coherent transmission systems, short-haul access networks, direct-detection/coherent-detection optical orthogonal-frequency-division-multiplexing (OFDM) systems, free-space optical communication, optical-wireless integration systems, and so on. The principle of these advanced DSP techniques is introduced in detail in this book.

The author, Dr. Jianjun Yu, is the Fellow of the Optical Society of America (OSA), and he has published over 600 high-quality academic papers and possesses over 40 authorized US patents.

The author, Dr. Xinying, now works as a Postdoctoral Fellow at Georgia Institute of Technology in USA, and she has published over 170 academic papers in top international journals and conferences in the field of optical communication.

The author, Dr. Junwen Zhang, now is a Senior Engineer at ZTE (TX) Inc. in USA, and he has published over 170 academic papers in top international journals and conferences in the field of optical communication.

We would like to thank Prof. Nan Chi, Dr. Fan Li and Dr. Yuanquan Wang for their help in finishing this book writing.

Jianjun Yu
Xinying Li
Junwen Zhang

September 15, 2017

About the Authors

Jianjun Yu is the department head of Wireline Technologies and Vice President of ZTE TX Inc. and Special Professor at Fudan University since 2010. Prior to that Dr. Yu was a senior technical staff member and project leader at the NEC Laboratory America. He has also worked on the technical staff at Bell labs, Lucent Technologies and Agere Systems, and has served on the research faculty at the Georgia Institute of Technology and Denmark Technical University. He holds a PhD and an MD in telecommunications from the Beijing University of Posts and Telecommunications. Jianjun is an Optical Society of America (OSA) Fellow and a senior Member of IEEE, IEEE Photonics Society, and serves as Associate Editor of *IEEE Photonics Journal, IEEE/OSA Journal of Lightwave Technology, OSA/IEEE Journal of Optical Communication and Networking,* and *OSA Journal of Optical Communication.* He served as the Editor-in-Chief for *Recent Patents on Engineering* from 2008 to 2016. Dr. Yu is an expert in high-speed optical communication systems and networks. He has authored/co-authored 600 peer-reviewed technical journal and international conference papers as well as 60 US patents.

Xinying Li is a Postdoctoral Fellow specializing in fiber-wireless integration systems and networks. She is currently at the School of Electrical and Computer Engineering, Georgia Institute of Technology, USA, working for the National Science Foundation, USA, on fiber wireless integration and networking for next generation heterogeneous mobile data communications. She holds a PhD and a BSc in Communication Science and Engineering, from Fudan University, China. Previously, she worked for the International Program of Shanghai Science and Technology Association, China, from 2012 to 2014. Xinying is a Member of IEEE, IEEE Photonics Society, and Optical Society of America (OSA), Associate Editor of *Recent Patents on Engineering and Digital Communications and Networks*, and an active reviewer of the *Journal of Lightwave Technology, Photonics Journal, Journal of Optical Communications and Networking, Optics Letters*, and *Optics Express*. She has also authored/co-authored 173 peer-reviewed technical journal and international conference papers as well as four US patents.

Junwen Zhang received his Ph.D. degree in optical communication and signal processing from Fudan University, Shanghai, China in 2014. During 2012 and 2014, he was a visiting student in Georgia Institute of Technology, and worked at the National Science Foundation (NSF) Research Center on Optical Wireless Applications (COWA), where he did research on 100G/400G/1T high speed transmission systems and digital signal processing. In 2015, he joined Georgia Tech as a Post-doc Research Fellow and served as a 5G Research Lab manager in the NSF-sponsored Research Center of Fiber-wireless Integration and Networking (FiWIN), doing research on high-speed optical-wireless integration Access Network, and novel Fronthaul and Backhaul networking technologies for next generation mobile networks. Dr Zhang joined ZTE Tx Inc. in 2016 as a Senior Research Engineer, working on high-speed optical access networks for fixed (NG-PON) and mobile services. His research interests cover many aspects of fiber and wireless communication systems and networks, including high-speed transmission, advanced modulation formats, digital signal processing, radio-over-fiber, fiber-wireless integration, mobile fronthaul/backhaul, etc. Dr Zhang has published more than 160 peer-reviewed papers in IEEE/OSA/SPIE journals and conferences, and has achieved several record-breaking transmission results in 400G optical transport systems. He has also co-authored eight US patent applications (three granted, five pending). Given his outstanding work, he received the Marconi Society Paul Baran Young Scholar Award in 2016. He was also the recipient of IEEE Photonics Society Graduate Student Fellowship in 2013, the Wang Daheng Optics Award in 2013, and the China Scholarship Council Fellowship in 2012.

Contents

Preface v

About the Authors vii

Chapter 1. Digital Signal Processing for Optical Coherent Long-Haul Transmission System   1

Chapter 2. Advanced DSP for Super-Nyquist Transmission System   25

Chapter 3. Advanced DSP for Short-Haul and Access Network   51

Chapter 4. DSP for Direct-Detection OFDM System   77

Chapter 5. DFT-Spread for OFDM Based on Coherent Detection   107

Chapter 6. Digital Signal Processing for Dual/Quad Subcarrier OFDM Coherent Detection   133

Chapter 7. DSP for MIMO OFDM Signal   163

Chapter 8. DSP Implementation in OFDM Signal Systems   175

Chapter 9. Advanced DSP for Free-Space Optical Communication   197

Chapter 10. DSP Precoding for Photonic Vector Signal Generation   229

Index   255

Chapter 1

Digital Signal Processing for Optical Coherent Long-Haul Transmission System

1.1. Introduction

Coherent optical communication based on digital signal processing (DSP) for long-haul transmission system has been proved to be a great success in the last decade [1–20]. However, coherent optical communication itself is not a new technology, people have started the research over 40 years ago, and the first coherent detection experimental demonstration was reported in 1970s [1–5]. The early stage of research is not accepted by the industry due to the complexity of phase, frequency and polarization tracking. After a long-time stagnation during the time of intensity modulation and direct detection (IM/DD), digital coherent optical communication based on DSP has again revolutionized the optical communication for high-speed, high-capacity and long-distance transmission [21–37].

With the development and maturation of high-speed digital-to-analog convertor (DAC), analog-to-digital converter (ADC) and application-specific integrated circuit (ASIC), DSP for high-speed optical signal becomes possible, which moves the complexity of phase, frequency and polarization tracking into the digital domain [9–20]. Therefore, it simplifies the reception of advanced modulation formats (i.e., quadrature phase shift keying (QPSK), 16QAM, 64QAM) and also enables the major electrical and optical impairments (bandwidth limitation, chromatic dispersion, polarization mode dispersion, fiber nonlinear impairments) being processed and compensation in the digital domain at the transmitter or receiver side. Therefore, the DSP-based coherent optical communication system has

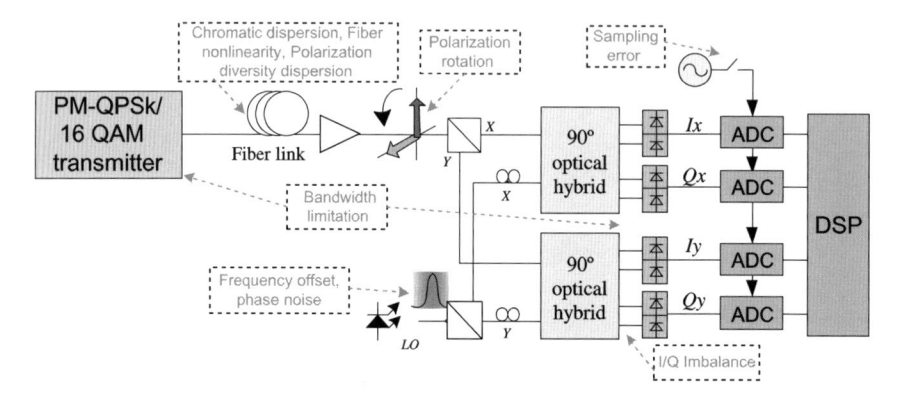

Fig. 1.1. The typical digital coherent system and the impairments from the system, which are needed to be handled using the advanced DSP.

become an active research topic, which is a promising technology for future high spectrum efficiency and high-speed transmission.

The chapter will introduce the recent progress of DSP for high-speed and long-haul optical coherent transmission system. Figure 1.1 shows the typical digital coherent system and the impairments from the system, which are needed to be handled using the advanced DSP. Section 1.2 will introduce the classic DSP algorithms used in the coherent system, then Section 1.3 will discuss the digital pre-equalization for bandwidth limited system. Advanced DSP for fiber nonlinearity compensation is introduced in Section 1.4. Finally, Section 1.5 summarizes this chapter.

1.2. DSP for Digital Coherent System at Receiver Side

1.2.1. The basic DSP blocks

The typical digital coherent receiver with DSP for signal equalization and recovery are shown in Fig. 1.2 [2, 3]. The optical signal is coherently detected by a polarization and phase diverse hybrid with a local oscillator, which converts the optical signals into electrical signals with both in-phase and quadrature (I/Q) signals in X and Y polarizations after balanced photo-detectors. The electrical signals after PDs are then sampled and digitized by the ADCs and then are processed by the DSP as shown in Fig. 1.2. Generally, the DSP at the receiver side can be divided into several subsystems or subunits, and each of them is applied to handle one specific impairment in the transmission link and transponders.

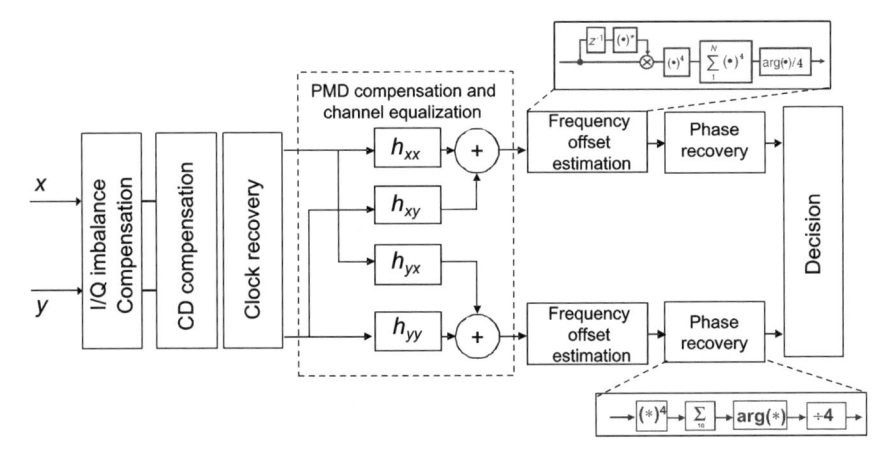

Fig. 1.2. Basic DSP blocks at the receiver side for coherent optical communication [2, 3].

Ideally, the I and Q signals should be orthogonal to each other. However, in the practical system, the I/Q signals are not orthogonal to each other due to the imbalance between these two components. These imbalances can be caused by bias drift, device defect of modulators and PDs or power differences after drivers. In order to get the signals recovered correctly, I/Q imbalance compensation should be applied first to the digitized signals. This process can be carried out by using the Gram–Schmidt Orthogonalization Process (GSOP) algorithm [4]. The GSOP creates a set of mutually orthogonal vectors, taking the first vector as a reference against which all subsequent vectors are orthogonalized [2]. The GSOP algorithm is an effective method for converting the unbalanced or non-orthogonal data samples to the balanced and orthogonal data samples.

1.2.2. *Chromatic dispersion linear channel equalization*

After that, the signals are then processed by the linear signal process to electrically compensate the chromatic dispersion (CD). The channel response of a fiber with CD can be represented as a linear function. CD compensation can be realized at the receiver side in frequency domain or time domain (Fig. 1.3) [2, 3].

The digital CD compensation can be processed in either time domain based on finite impulse response (FIR) or frequency domain based on the transfer function. In time domain, the required filter coefficients for FIR can

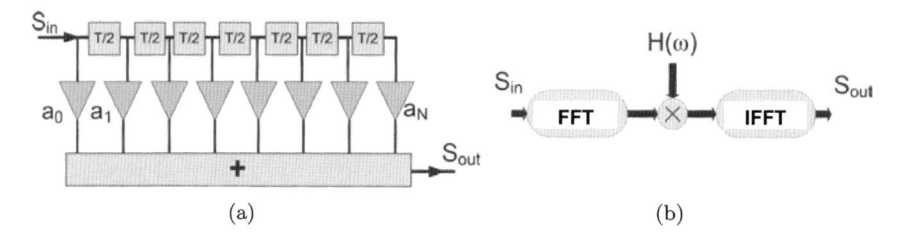

Fig. 1.3. (a) Time-domain FIR structure of CD compensation and (b) frequency-domain CD compensation based on the FFT and IFFT processes.

be obtained by the fiber CD transfer function using either the frequency-domain truncation method or the time-domain truncation method [3]. The FIR coefficient of each tap a_k can be calculated by the following equations:

$$a_k = \sqrt{\frac{jcT^2}{D\lambda^2 z}} \exp\left(-j\frac{\pi c T^2}{D\lambda^2 z}k^2\right), \quad -\left\lfloor\frac{N}{2}\right\rfloor \leq k \leq \left\lfloor\frac{N}{2}\right\rfloor \quad (1.1)$$

$$N = 2 \times \left\lfloor\frac{|D|\lambda^2 z}{2cT^2}\right\rfloor + 1 \quad (1.2)$$

In the above equations, D is the dispersion coefficient, λ is the wavelength and z is the fiber length, c is the speed of light, T is the symbol period, k is the order of taps, and N is the total number of FIR taps.

For short distance, time-domain method shows less complexity, however, for long-haul transmission (over 1000 km), frequency-domain equalization (FDE) based on FFT shows less computation complexity:

$$H(w) = \exp\left(j\frac{D\lambda^2 z}{4\pi c}\omega^2\right) \quad (1.3)$$

The frequency-domain channel response is shown in Eq. (1.3). When doing the frequency-domain CD compensation, one needs to transfer the data into frequency domain by using the FFT process. After multiplying the transfer function, the data can be recovered by the IFFT process.

1.2.3. Clock recovery

The other impairments, such as the timing offset caused by the ADC can also be compensated in the digital domain, known as the clock-recovery algorithms [5–7]. Several classical timing phase estimation and recovery schemes can be successfully applied in the coherent optical communication

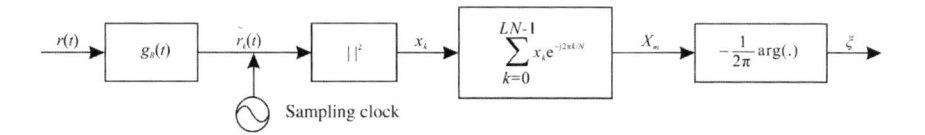

Fig. 1.4. Square-timing recovery based on feedforward structure.

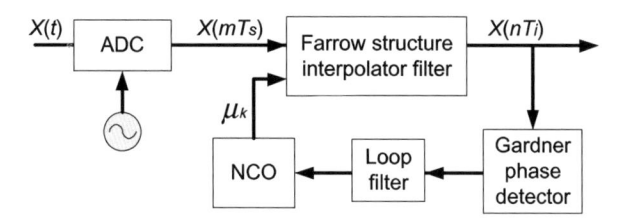

Fig. 1.5. Gardner-timing recovery method based on feedback loop with error update.

system, such as the square-timing recovery [5], Gardner-timing recovery [6] and Godard scheme [7].

Generally, the square-timing recovery is a feedforward method as shown in Fig. 1.4, and it requires at least four samples per symbol to find the right clock information. Therefore, it cannot track the clock offset and has large computation complexity.

Another method, Gardner-timing recovery widely used in the digital coherent system, only requires two samples per symbol. On the other hand, it is feedback-type algorithm (Fig. 1.5).

Assume the kth symbol with the I and Q components as $\{y_I(k)\}$, $\{y_Q(k)\}$ and $\{y_I(k-1/2)\}$, $\{y_Q(k-1/2)\}$. Then, according to the Gardner time recovery algorithm [6], the error function for the clock update is based on the following equation:

$$\mu_k = y_I(k-1/2)[y_I(k) - y_I(k-1)] \\ + y_Q(k-1/2)[y_Q(k) - y_Q(k-1)] \quad (1.4)$$

1.2.4. *Polarization demultiplexing and PMD compensation*

As analyzed above, there are two linear impairments that would affect the polarization multiplexed (PM) signals. To obtain the data for each polarization, one needs to do polarization demultiplexing and polarization

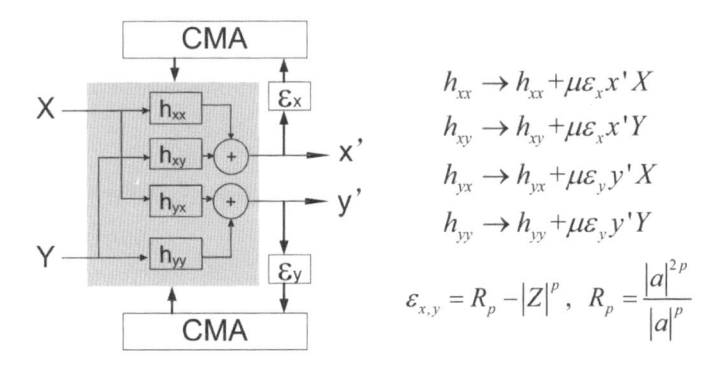

Fig. 1.6. The four adaptive butterfly equalizers based on CMA and its updating functions.

mode dispersion (PMD) compensation. After that, the following four adaptive butterfly equalizers are used for polarization demultiplexing, channel equalization and also PMD compensation.

Figure 1.6 shows the structure and the principle of signal recovery by the adaptive and butterfly equalizers based on constant modulus algorithm (CMA) and also the filter coefficients updating equations. The cost function and error function are based on the modules of the output and input data. The linear distortions caused by optical filtering effects in the link are also equalized by the time-domain FIR filters. Generally, T/2-spaced FIR filters are used in the DSP and the filter coefficients are adaptively updated by using the CMA [8–10], cascaded multi-modulus algorithm (CMMA) [11] or decision-directed least-mean square (DD-LMS) algorithm [12], depending on the modulation formats used in the system. For QPSK signals, the CMA-based scheme is mostly used in the previous works, and CMMA and DD-LMS are required for higher modulation formats for further performance improvements, i.e., 16QAM and 64QAM. Results in [12] indicate that the CMMA and DD-LMS show better performances for high-level modulation formats.

1.2.5. *Frequency-offset estimation and phase recovery*

Finally, the carrier recovery is realized by two-step DSP sub-units, frequency-offset estimation (FOE) and phase recovery. Since most phase recovery methods require the zero frequency offset, the FOE is always carried out before the phase recovery. For FOE, there are two main algorithms for FOE, the first is based on the differential phase-based methods [13, 14], the second is based on the spectral methods using the FFT spectra analysis [15]. The former is also known as mth-power Viterbi–Viterbi algorithm for QPSK signals [13, 14]. However, for high-level modulation formats, partitioning is required to improve the accuracy. The latter, on the other hand, is a universal method for any modulation format [15].

Generally, there are two major schemes for phase recovery. One is called feedforward method, known as mth-power Viterbi–Viterbi algorithm, blind-phase search (BPS) method, and the other one is decision-direct feedbackward method, such as DD-maximum-likelihood (ML) method [16, 17]. Viterbi–Viterbi algorithm method shows small computation complexity, however, the performances for high-level modulation formats are not good. Similar to the FOE, partitioning is also required [16]. BPS is a universal method for any modulation format, however, to achieve better performances, the computation complexity increases with the increase of modulation formats [17]. Several discussions were conducted on complexity reduction using the hybrid method based on BPS, ML, and Viterbi–Viterbi algorithm [16].

1.2.6. *Output results of different blocks*

As a demonstration and conclusion, Fig. 1.7 shows the simulation results of 32GBaud signals (QPSK, 8QAM and 16QAM) after 1000-km transmission

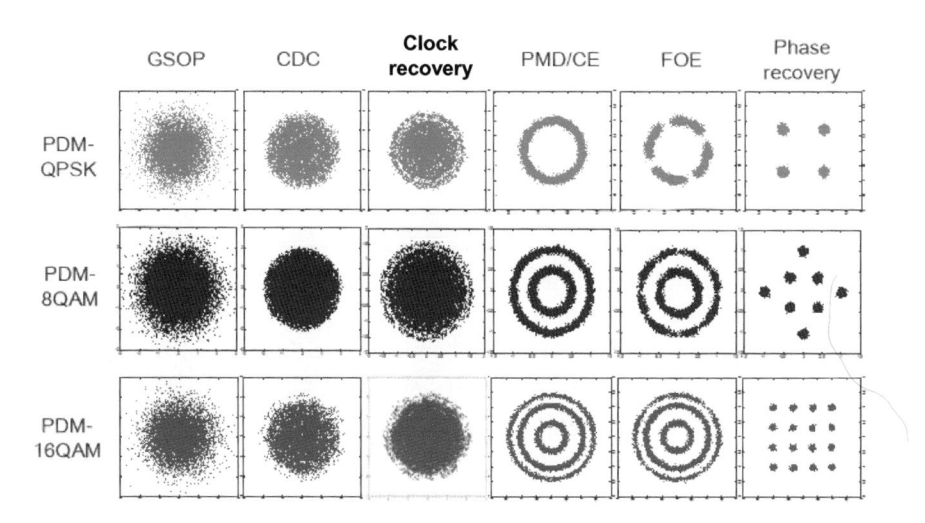

Fig. 1.7. The constellation of different signals after 1000-km transmission with DSP.

with basic DSP subunits. Based on the DPS subunits, we can observe the constellation recovery process after all these algorithms. In the above simulation demonstration, the QPSK signals are processed based on CMA equalization, while 8QAM and 16QAM are based on CMMA and DD-LMS. For FOE and phase recovery, QPSK signal is processed by fourth power Viterbi–Viterbi algorithm, while 8QAM and 16QAM are processed by frequency-domain FOE and BPS method.

1.3. DSP at Transmitter Side for Pre-equalizations

With the advent of high-speed DAC, signal generation based on DAC becomes an attractive method due to the simple configuration and flexible signal generation capability, and it has been attracting a great deal of interest in recent years for the transmission of 100G and beyond [25–32]. DAC for signal generation allows the software-defined optics (SDO) with arbitrary waveform generation, which can be used for signal switch software in different modulation formats [25–30]. On the other hand, it also allows the DSP at the transmitter side (Tx) with pre-compensation or pre-equalizations [26–35]. One of the first electronic pre-equalization technologies for CD was demonstrated in 5120-km transmission for a 10 Gb/s Differential Phase Shift Keying (DPSK) system [31]. In order to achieve high-speed signal generation, the industrial research communities have taken great efforts to increase the bandwidth and sample rate of DAC.

However, the 3-dB analog bandwidth of state-of-the-art DACs is still much less than half of its sample rate, which means that the generated signals suffer the distortions caused by the bandwidth limitation, especially for generated signals with high baud rate. Meanwhile, when it operates at high baud rate, other optoelectronic devices, such as the electrical drivers and modulators which work beyond their specified bandwidth can further suppress the signal spectrum. Due to such cascade bandwidth narrowing effect, the system performance is seriously degraded by inter-symbol interference (ISI), noise and inter-channel crosstalk enhancement [26–31]. Electrical domain pre-equalization for the bandwidth-limitation impairments, which is a well-known technique in optical communication, has been widely utilized in recent publications [26–36]. The first pre-equalization of the filtering penalty for the 43 Gb/s optical DQPSK signals is reported in [32] using DAC. In the previous works [26–31], zero-forcing FDEs are carried out to pre-equalize the linear band-limiting effects. The inverse transfer function of DAC and other optoelectronic devices is measured by calculating the fast Fourier transform (FFT) of both transmitted and received binary data using a known training signal sequence.

1.3.1. The principle of digital pre-equalizations

Figure 1.8 shows the principle of the proposed pre-equalization by leveraging the estimated inverse channel given by the receiver-side adaptive equalizer, which comes with two stages [35]. The first stage is the channel estimation, where a data $X(t)$ without pre-equalization is transmitted and passes through the channel $H(t)$. The channel is band-limited with a channel response of $H(t)$, which causes the signal distortion with ISI due to the narrow filtering effect. In a coherent optical system, such a channel response $H(t)$ represents an end-to-end transfer function taking the analog bandwidths of the DAC, the driver, the modulator at the transmitters, and that of the photo-detector and the ADC at the receivers into consideration. Again, it is worth noting that we use the channel here describing the transfer function of the transmitter hardware, but not the outside plant of the transmitter.

Assuming the noise $n(t)$ is additive white Gaussian noise (AWGN), then the received signal can be expressed as

$$Y(t) = X(t) * H(t) + n(t) \quad (1.5)$$

The received signal $Y(t)$ suffers distortions with ISI caused by bandwidth limitation. As analyzed in [35], to compensate the channel distortion, we can employ a linear filter with adjustable taps to equalize the

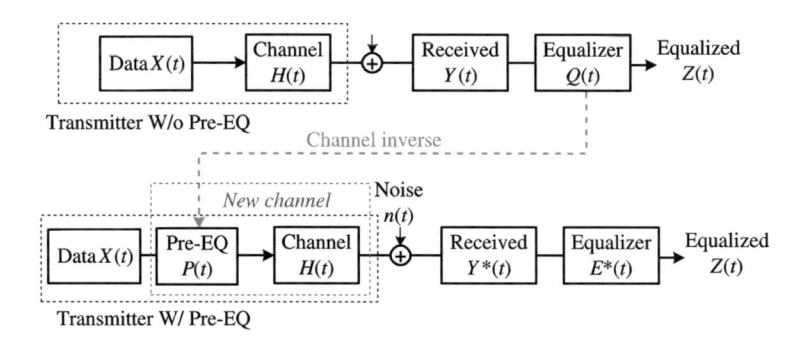

Fig. 1.8. The principles of digital pre-equalizations.

received signal

$$Z(t) = Y(t) * Q(t) = X(t) * H(t) * Q(t) + n(t) * Q(t) \qquad (1.6)$$

where $Z(t)$ is the equalized signal. $Q(t)$ is the impulse response of an adaptive linear equalizer for channel equalization. $Q(t)$ can be implemented using either zero-forcing or minimum mean square error (MMSE) criterions. In practice, MSE-criterion equalizers are better and more widely used since the zero-forcing equalizers may result in noise amplification. Here, we choose $Q(t)$ as an MSE-criterion-based equalizer using stochastic gradient algorithms, such as CMA and LMS [35]. These two algorithms are widely adopted in optical coherent communication system. Since CMA- and LMS-based equalizers take both ISI and noise into account in the filter tap update, the final goal is to find the optimal filter $Q(t)$ having minimum MSE. In our case, we can minimize the noise in order to obtain the exact channel response. Assuming that the noise $n(t)$ is small and negligible, we can remove the noise part in Eq. (1.6) and have

$$Z(t) \cong X(t) * H(t) * Q(t) \qquad (1.7)$$

Therefore, it is clear that we have $Q(t) = H(t)^{-1}$ if $Z(t) = X(t)$. In this case, $H(t) * Q(t) = 1$. It shows that the equalizer is the channel inverse when the noise is small. More specifically, in practice, the ISI is limited to a finite number of samples in real channels. Therefore, the channel equalizer is approximated by a finite duration impulse response with symbol-spaced or fractionally spaced taps.

Considering the common CMA or DD-LMS equalizer in coherent optical communication system, in which four $T/2$-spaced FIRs are used as the channel equalizer with T-spaced update and detector loop. The error function approaches zero during the convergence, therefore, ISI can approach zero at the T sampling points when noise is negligible.

Since only T sampling points are calculated during the updating and convergence, the output symbols of equalizer $Q(t)$ after convergence (ISI is 0 at the T sampling points) can be expressed as

$$Z(t) = X(kT) * X_N(t) \qquad (1.8)$$

where $X_N(t)$ is a Nyquist pulse-shaping criterion filter, and comparing Eqs. (1.7) and (1.8), we have

$$Q(t) \cong H(t)^{-1} * X_N(t) \qquad (1.9)$$

In frequency domain, the response of the equalizer $Q(t)$ can be expressed as

$$Q(f) \cong 1/H(f), \quad |f| < 1/2T \tag{1.10}$$

Here, the $H(f)$ is the frequency response of the bandwidth-limited channel. It shows that the frequency response of the equalizer is the inverse of the channel response $H(f)$ within the Nyquist bandwidth when the noise is negligible. In this way, for the $T/2$-spaced DD-LMS with T-spaced detection and update loop, the channel response within Nyquist bandwidth can be estimated, otherwise, it approaches 1. Therefore, a time domain pre-equalization method can be employed based on the receiver-side adaptive equalizer. For optical coherent transmission system, one can simply record the FIR tap coefficients as the output of those commonly used linear equalizers (such as CMA, CMMA and DD-LMS equalizers) and feedback that information to the transmitter for pre-equalization.

Note that the above analysis excludes the factor of noise, which, however, is represented in real systems. On the other hand, as analyzed in [35], the performance of channel estimation by adaptive filter is significantly affected by the filter tap length and also the OSNR in channel. Therefore, considering all the factors, the channel inverse calculated by adaptive filter taps is a function of channel response, noise level, tap length, which can be expressed as

$$Q(f) = F[H(f), N_0, L] \tag{1.11}$$

where N_0 is the AWGN power spectrum density and L is the tap length. These factors should be considered in practical implementation.

As analyzed in [35], the benefit of the pre-equalization can be proved by the smaller MSE of the recovered signal compared with post-equalization-only case in symbol detection systems. Assuming filter tap is long enough and the step size is small enough, one can obtain an optimal filter tap, which has the minimum MSE based on the MMSE criterion algorithms as

$$Q(f)_{\text{MMSE}} = 1/(N_0 + H(f)), \quad |f| < 1/2T \tag{1.12}$$

The minimum MSE achievable by a linear equalizer using the above optimal filter is given by

$$\text{MSE}_{\text{min\_post-eq}} = T \int_{-f/2}^{f/2} N_0/[N_0 + H(f)] df \tag{1.13}$$

We can see that the minimum MSE is determined by the noise power and also the channel response. The minimum MSE can be very large even with linear equalizations when the $H(f)$ is small, which means the bandwidth of channel is significantly limited. For the pre-equalization case, assuming that the channel response is exactly estimated as Eq. (1.10), the bandwidth limitation can be fully compensated. In this way, the new channel response including the pre-equalization is $H_{\text{pre}}(f) = Q(f)H(f) = 1$. Therefore, the minimum MSE can be given by

$$\text{MSE}_{\text{min\_Pre-eq}} = T \int_{-f/2}^{f/2} N_0/[N_0 + 1] df \qquad (1.14)$$

Therefore, the minimum MSE of pre-equalization case is smaller than that of post-equalization-only case when the channel is bandwidth-limited with narrow filtering effect. Larger gain can be obtained using the pre-equalization for narrower filtering bandwidth. Equation (1.14) gives a qualitative analysis of the factors that affect the estimation results. The estimated channel response is determined by the OSNR, tap length and bandwidth. The required OSNR or tap length can be different under different BER tolerance and different channel responses. One should adjust and optimize these factors for practical use.

1.3.2. *FFT and IFFT-based digital pre-equalizations*

Figure 1.9 depicts the principle of frequency-domain pre-equalization method proposed in our previous works [26–28]. Compared to binary-phase-shift-keying (BPSK) signal, the high-level signal suffers much more uncontrolled nonlinear effects due to the imperfection of the DAC, electrical amplifier (EA), I/Q MOD, optical filter and ADC. As a result, they calculate the transfer function of BPSK, which is then used to pre-distort the high-level signal in the AWG. A known binary sequence with a word length of $2^{15} - 1$ is used for the generation of BPSK. The continuous wavelength (CW) lightwave (CW1), generated by an external cavity laser (ECL), is used as both the signal source and the local oscillator (LO) source in a self-homodyne coherent detection. The LPF adopted here is used to suppress out-of-band noise of AWG before BPSK signal generation. The optical BPSK signal passes through a wavelength selective switch (WSS) before coherent detection. A real-time scope is used to capture the detected electrical signal, which is used to calculate the transfer function of transmitter with Nyquist band in the frequency domain. Then the transfer function

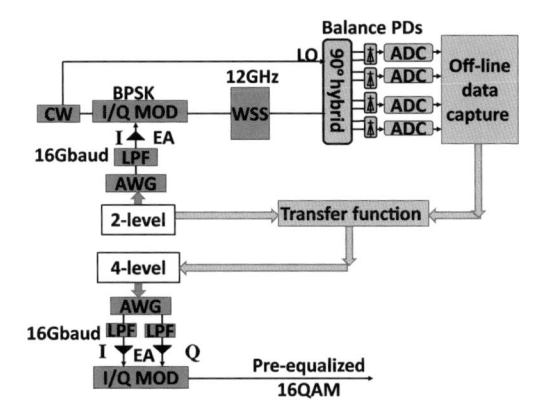

Digital Signal Processing for Optical Coherent Long-Haul Transmission System 13

Fig. 1.9. The principle of pre-equalization. AWG: arbitrary waveform generator, LPF: low pass filter, EA: electrical amplifier, WSS: wavelength selective switch, CW: continuous lightwave, LO: local oscillator, ADC: analog-to-digital converter, I: in-phase, Q: quadrature.

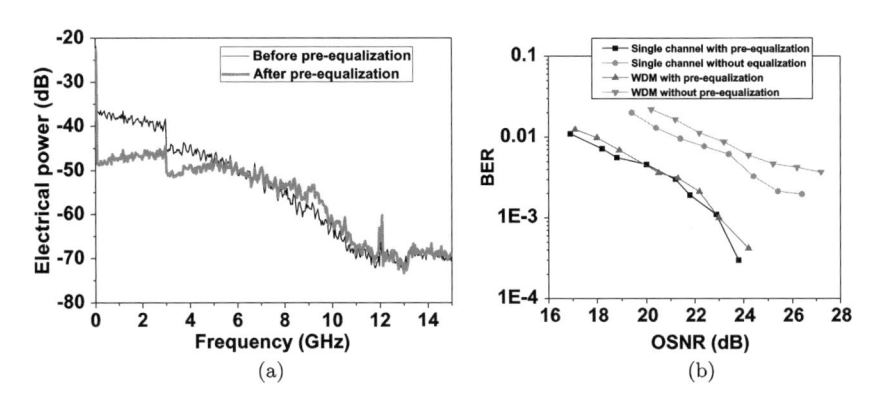

Fig. 1.10. (a) The electrical spectrum for the N-WDM PDM 16-QAM signal with and without pre-equalization. (b) The BER as a function of OSNR for four different cases after passing through 10-GHz WSS.

is used to pre-equalize the 4-level signal. The 4-level signal with a word length of $2^{15} - 1$ is used to generate the optical 16QAM via IQ modulation. It is noted that the pre-equalization for the I and Q outputs of AWG show similar performance and therefore, we simply choose the I output of AWG to implement pre-equalization in our experiment.

Figure 1.10(a) shows the electrical spectrum for the 16GBaud N-WDM PDM 16-QAM signal with and without pre-equalization [26–28]. Clearly, it can be seen that some certain high frequency components, lost due to aggressive spectral filtering, are recovered with pre-equalization. Figure 1.10(b) shows the BER performance, as a function of optical signal-to-noise ratio (OSNR), for four different cases. The scope has a bandwidth of 9 GHz and a sample rate of 50 Gsa/s. These four different cases include the N-WDM PDM-16QAM transmission and the single-channel PDM-16QAM transmission, each with and without pre-equalization. We can clearly find from Fig. 1.14 that the single-channel case has a rather better BER than the N-WDM case without pre-equalization, while in the case of pre-equalization, the two cases almost have similar BER. It is because the adoption of pre-equalization can effectively pre-compensate transmitter impairments as well as reduce the effect of nonlinear propagation impairments and laser phase noise. As a result, adopting pre-equalization improves the BER performance of the N-WDM.

1.3.3. *Time-domain digital pre-equalization*

From the perspective of system implementation, such approach based on the FFT/IFFT process may not be easily utilized in current 100G or 400G systems since an additional DSP block at the receiver needs to be developed to deal with the channel estimation. Furthermore, to remove the fluctuation caused by signal and noise randomness, more than 100 measurements are required to be carried out for averaging [28–30]. Also, a strict time-domain synchronization is also required in [30]. On the other hand, to increase the measurement accuracy in the high frequency region, the spectrum of the De Bruijn BPSK signal needs special pre-emphasized process [29].

Alternatively, a time-domain pre-equalization method can be a good solution [33–35]. In most band-limited systems, adaptive equalizers are used for ISI equalization at the receiver side (post-equalization). Theoretically, in a zero-ISI system, the receiver-side linear adaptive equalizers approach the channel inverse within the sampling rate of equalization taps to compensate the bandwidth limitation [37]. In fact, this channel inverse is also effective for pre-equalization. It is worth noting that we use the channel here describing the transfer function of the transmitter hardware, but not the outside plant of the transmitter. The linear equalizers used in the receiver side is a good tool for channel estimation, which has been employed in digital cable television (CATV) and wireless transmission system for pre-equalizations [33, 34]. Significant signal BER gain can be obtained using this method. Numerical results also show that the pre-equalization outperforms post-equalization-only in band-limited system with narrow-band filtering impairments [35]. In optical coherent communication system, such as constant modulus algorithm (CMA), multi-modulus algorithm (MMA) or decision-directed least-mean-squares (DD-LMS), these adaptive equalizers' transfer functions are naturally modeled with the inverse Jones matrices of the channel. However, when there is no PMD, the frequency response of these adaptive equalizers is just the inverse transfer functions of the channel. With this feature in mind, one can simply get the inverse of channel transfer function for pre-equalization. Since the adaptive equalizers are blind to the data pattern, different from the scheme used in [27–30], there is no need to do data pattern alignment. Only clock recovery is needed for symbol synchronization. Compared to the prior arts, the proposed method, featuring no additional DSP, no precise symbol alignment, is advantageous for the system implementation. Using this scheme, recently we have demonstrated the improved performance for DAC generated signals [33]. We also realized a band-limited 480-Gb/s dual-carrier PDM-8QAM transmission and have demonstrated the BER performance improvements by experimental results in [34].

Figure 1.11 shows the principle of channel estimation for the proposed DPEQ at the transmitter in [35]. The linear pre-equalization is based on the receiver-side blind and adaptive equalizers for channel estimation. We first use the DAC to generate the mQAM data without pre-equalization for channel estimation. Since the bandwidth limitation impairment is mainly caused by the DAC, the electrical drivers, the modulator, the receiver-side PDs and the ADC, only single-polarization signal is used to avoid the polarization crosstalk. One CW lightwave external-cavity laser is used as both the signal source and the local oscillator for the self-homodyne coherent detection. In this case, the traditional post-equalization methods

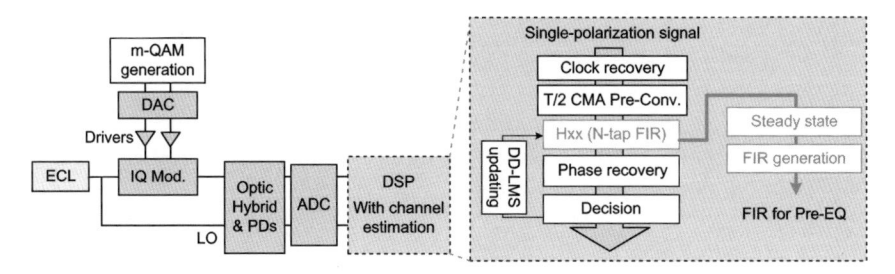

Fig. 1.11. The principle of channel estimation for the adaptive pre-equalization based on DD-LMS.

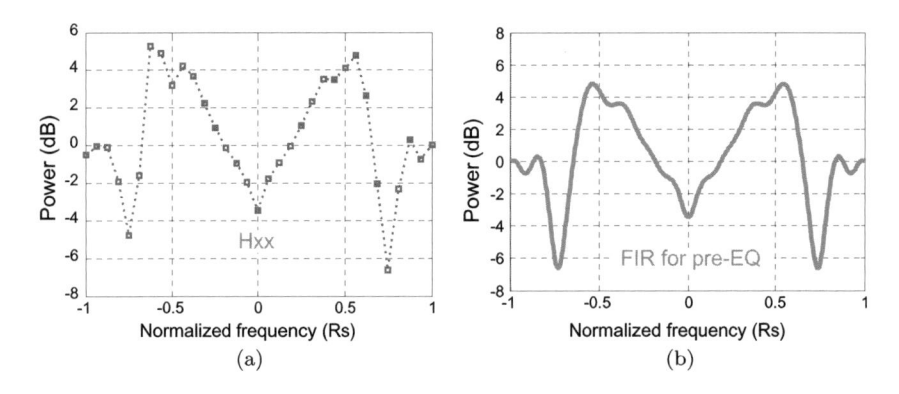

Fig. 1.12. The frequency response of (a) Hxx and (b) regenerated FIR.

for polarization demultiplexing, e.g., CMA and DD-LMS, are actually the channel equalizers for the bandwidth limitation impairment, which can be used for channel estimation. The amplitude frequency response of these equalizers is the inverse transfer function of the channel. The DD-LMS loop, which is after CMA for pre-convergence, consists of four complex-valued, N-tap, finite-impulse-response (FIR) filters for equalization. After convergence, these FIR filters achieved the steady state. After normalization and frequency symmetrization, the time-domain FIR for pre-equalization can be regenerated.

The results of channel estimation and pre-equalization are shown in Fig. 1.12. Since the bandwidth limitation impairment is mainly caused by the DAC, the electrical drivers, the modulator and the ADC, only single-polarization signal is used to avoid the polarization crosstalk. Self-homodyne coherent detection is applied using the same CW lightwave (ECL1) as both the signal source and the LO source. Figures 1.12(a) and 1.12(b) show the frequency response of 33-taps FIR filter Hxx in DD-LMS and the regenerated FIR for DPEQ, which indicates the inverse transfer functions of the commercial 64-GSa/s DAC.

Figure 1.13 shows the back-to-back (BTB) BER performance versus the OSNR for single carrier 40-Gbaud PM-QPSK, PM-8QAM and PM-16QAM signal with and without pre-equalization, respectively. Here, 33-taps FIR filters are used. We can see that about 3.5dB, 2.5dB and 1.5dB OSNR improvement can be obtained at the BER of 1×10^{-3} by DPEQ based on DD-LMS method for the 40-Gbaud PM-QPSK/8QAM/16QAM signals, respectively. As a comparison, we also plot the theoretical BER curves

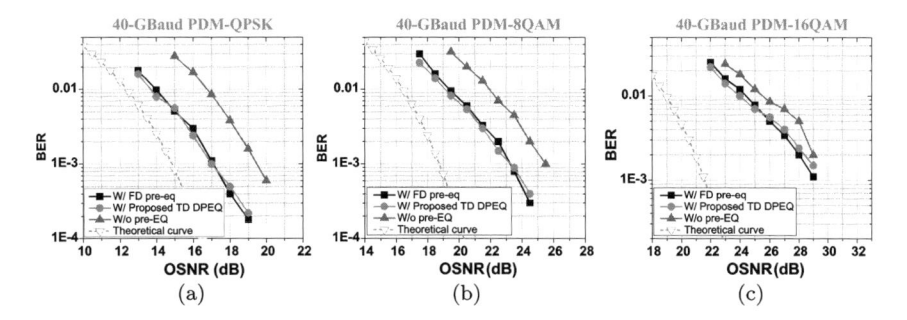

Fig. 1.13. The BTB BER results versus the OSNR with and without pre-equalization for (a) 40-GBaud PDM-QPSK, (b) 40-GBaud PDM-8QAM, and (c) 40GBaud PDM-16QAM signals.

in these figures. The BER results of signals with a previously reported frequency-domain pre-equalization (FD Pre-eq) method [35] are also plotted. The result of the proposed DPEQ has a performance similar to that of the frequency-domain method, since both schemes use the signals with very high OSNR for channel estimation. On the other hand, we also observe the OSNR penalties compared with the theoretical performances. About 1.5-, 2.5- and 4.5-dB OSNR penalties are observed for signals with pre-equalization compared with the theoretical curves. Since higher modulation formats require higher OSNR and they are more sensitive to the ISI, the pre-equalization is less effective for the high modulation formats.

1.4. Receiver-Side Fiber Nonlinear Compensation

Nonlinear compensation (NLC) by using DSP has become an attractive research topic these years for long-haul high-speed coherent transmission system [18–24]. As shown in Fig. 1.14, digital backward propagation (DBP), based on split-step Fourier method (SSFM) by backward solving the nonlinear Schrodinger equation (NLSE), has been proved as an effective way to compensate for the nonlinear effect including self-phase modulation (SPM), cross-phase modulation (XPM) and four-wave mixing (FWM) [18–20]. In the previous works [18–20], the DBP method has been theoretically and experimentally demonstrated for single channel polarization division multiplexing (PDM) system by using an improved NLSE. For the WDM system with inter-channel nonlinear effects, it is believed that a fully reconstructed multi-channel signal shows better NLC performance [21–23]. In [19, 20], they have shown the NLC for the PDM WDM system by DBP method using

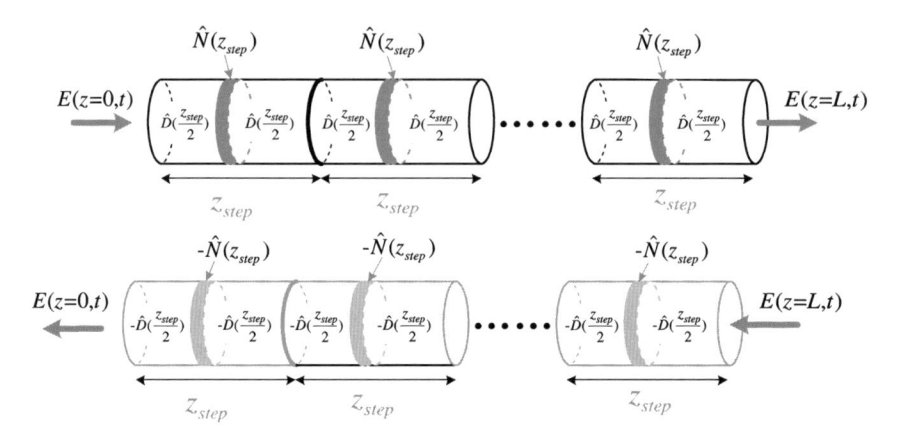

Fig. 1.14. The principle of the fiber nonlinear compensation based on digital backward propagation (DBP) method.

multi-channel signals. However, the above DBP techniques are implemented with constant step-size SSFM, where the step size is equal in each SSFM computation and the performance of NLC is significantly dependent on the computational step size or step number for each span fiber [18–20, 36, 37]. Thus, reducing the number of DBP calculation steps per fiber span is an effective way to reduce the computations of the algorithm. On the other hand, people have also proposed the non-constant step size in forward simulation to enhance the accuracy in the estimation of signal distortions compared with constant step-size distribution [37]. It has also been demonstrated that a logarithmic non-constant step-size distribution can reduce the step number while keeping the same performance in individual intra-channel NLC [37].

The fiber nonlinear impairment can be compensated by digital backward propagation (DBP) method based on the solving of nonlinear Schrodinger (NLS) Eqs. [18–21]. In our case, an improved DBP method for polarization multiplexed WDM system is used, which can be realized by solving Manakov function as [19, 20, 36, 37]

$$\frac{\partial E_{x,y}}{\partial z} = -\frac{\alpha}{2}E_{x,y} + \frac{\beta_2}{2}\frac{\partial^2 E_{x,y}}{\partial t^2} - \frac{\beta_3}{6}\frac{\partial^3 E_{x,y}}{\partial t^3}$$
$$+ i\frac{8}{9}\gamma(|E_{x,y}|^2 + |E_{y,x}|^2)E_{x,y} \qquad (1.15)$$

where $E_{x,y}$ is the multi-channel optical field of X- or Y-polarization signal, β_i is the in-order dispersion, α is the fiber loss, γ is the nonlinear parameter and z is the step fiber length. By using split-step method (SSM), we can compensate the linear fiber CD and nonlinear impairment by backward solving the above-mentioned function. During each step length z, we first compensate the linear CD and fiber loss for the $z/2$ fiber length in the frequency domain. Then, we calculate and compensate the nonlinear phase shift in the time domain. Finally, the CD and loss of the other half step length $z/2$ is compensated again in the frequency domain.

Assume the fiber length per span is L and the number of steps of DBP calculation per span is N. As shown in Fig. 1.14, for the regular DBP method based on SSFM, a constant step size z is used and the step size is equal in each SSFM computation as $z(n) = L/N$ [36]. In this way, the performance of NLC is significantly dependent on the computational step size or step number for each span fiber. Thus, reducing the number of DBP calculation steps per fiber span is an effective way to reduce the computations of the algorithm. However, when we consider the fiber attenuation, the power is a nonlinear distribution. The power is larger at the beginning of the propagation with stronger nonlinear impairments, which becomes much smaller at the end as shown in Fig. 1.15. Thus, a smaller step size is needed for the larger power to enhance the accuracy in the estimation of signal distortions. Alternatively, non-constant logarithmic step-size distribution, where the step size decreases as power increases, has also been proposed in the forward propagation simulation to avoid the overestimated production of spurious FWM peaks by constant step-size methods. Figure 1.15 shows the principle of NLC based on constant and non-constant logarithmic step size proposed in [37].

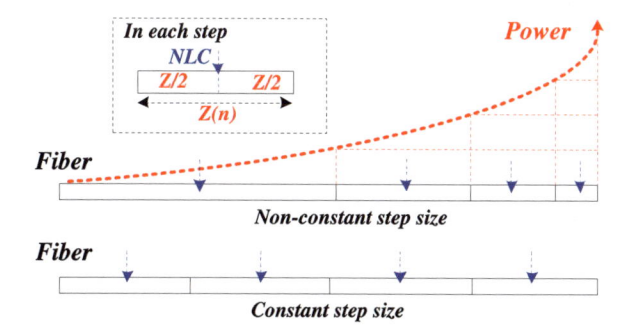

Fig. 1.15. The principle of NLC based on constant and non-constant logarithmic step size. The red-dashed curve shows the power distribution in the fiber span; the blue arrows show the nonlinear operation in each step [37].

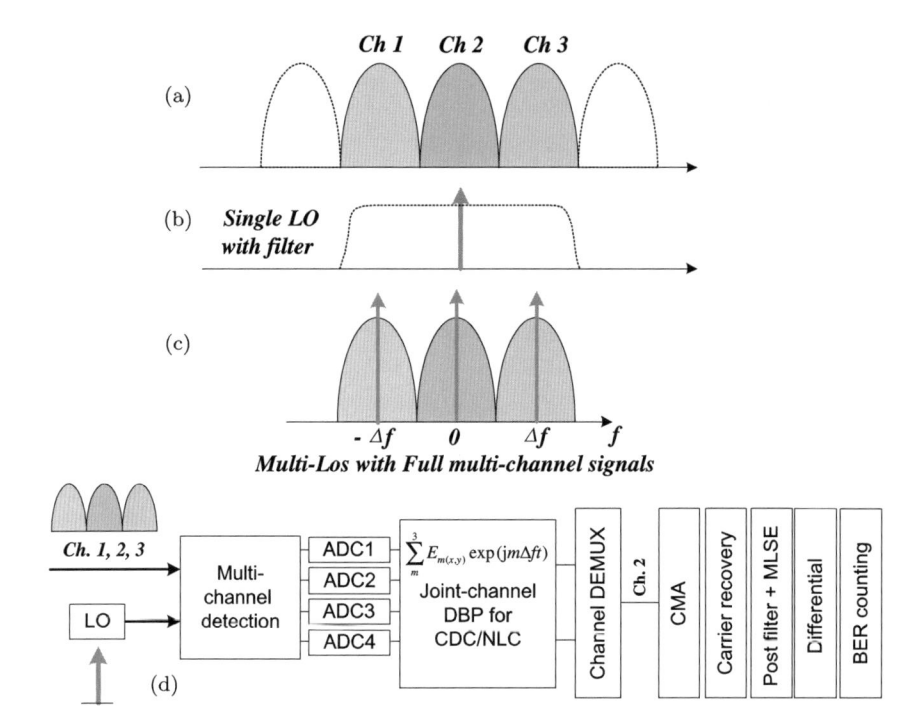

Fig. 1.16. (a) The Nyquist WDM multi-channel signals; (b) the principle of multi-channel detection for joint-channel NLC; (c) the principle of independent detection with synchronized sampling of each sub-channel; and (d) the block diagrams of proposed JC-NLC and CDC based on the DBP method, channel DEMUX and following equalizations.

Figures 1.16(a)–(c) show the Nyquist WDM multi-channel signals and (b) shows the principle of multi-channel detection for joint-channel (JC) NLC [37]. For the NWDM system, the channel spacing is equal to the baud rate with a high spectral efficiency. The inter-channel nonlinear impairments such as XPM and FWM are stronger for these sub-channels. However, based on the broadband multi-channel detection, it enables the compensation of inter-channel effects such as XPM and FWM. Figure 1.16(d) shows the block diagrams of the proposed multi-channel detection receiver and DSP with JC-NLC and CD compensation (CDC), channel demultiplexing and following equalizations. After NLC and CDC, the three-channel signals are demultiplexed in the electrical domain and down-converted to baseband by frequency shifting. Then each subchannel is processed by subsequent DSP, including polarization demultiplexing based on CMA, FOE and phase recovery as analyzed in Section 1.2. A detailed joint channel and the modified logarithmic step-size distribution NLC can be found in [37], where the results show large improvement using the above scheme in Fig. 1.16.

1.5. Summary

In this chapter, we introduce the recent progress of DSP for high-speed and long-haul optical coherent transmission system. A typical digital coherent system transfers the impairments from the system into digital domain, which equalizes and compensates those impairments by using advanced DSP. In Section 1.2, the classic DSP algorithms used in the coherent system are introduced, and then the digital pre-equalization for bandwidth limited system are discussed in Section 1.3. Finally, advanced DSP for fiber-nonlinearity impairment compensation is introduced in Section 1.4.

References

[1] T. Okoshi and K. Kikuchi, *Coherent Optical Fiber Communications* (Springer, 1988).
[2] S. J. Savory, Digital filters for coherent optical receivers, *Optics Express*, **16**(2) (2008) 804–817.
[3] X. Zhou and J. Yu, Digital signal processing for coherent optical communication, *IEEE WOCC* (2009).
[4] S. Haykin, *Adaptive Filter Theory* (Prentice-Hall, Englewood Cliffs, NJ, 2001).
[5] M. Oerder and H. Meyr, Digital filter and square timing recovery, *IEEE Trans. Commun.* **36**(5) (1988) 605–612.
[6] F. Gardner, A BPSK/QPSK timing-error detector for sampled receivers, *IEEE Trans. Commun.* **34**(5) (1986) 423–429.
[7] D. Godard, Passband timing recovery in an all-digital modem receiver, *IEEE Trans. Commun.* **26**(5) (1978) 517–523.
[8] G. Proakis, Carrier and symbol synchronization, Chapter 6, in *Digital Communications*, 4th edn. (McGraw-Hill, 2001).
[9] D. N. Godard, Self-recovering equalization and carrier tracking in two-dimensional data communication systems, *IEEE Trans. Commun.* **28** (1980) 1867–1875.
[10] C. R. Johnson, P. Schniter, T. J. Endres, J. D. Behm, D. R. Brown and R. A. Casas, Blind equalization using the constant modulus criterion: A review, *Proc. IEEE* **86** (1998) 1927–1950.
[11] X. Zhou, J. Yu and P. D. Magill, Cascaded two-modulus algorithm for blind polarization de-multiplexing of 114-Gb/s PDM-8-QAM optical signals, *OFC 2009*, paper OWG3 (2009).

[12] X. Zhou, L. E. Nelson, P. Magill, R. Isaac, B. Zhu, D. W. Peckham, P. I. Borel and K. Carlson, High spectral efficiency 400 Gb/s transmission using PDM time-domain hybrid 32–64 QAM and training-assisted carrier recovery, *J. Lightwave Technol.* **31** (2013) 999–1005.

[13] A. J. Viterbi and A. M. Viterbi, Nonlinear estimation of PSK modulated carrier phase with application to burst digital transmission, *IEEE Trans. Inf. Theory* **IT-29**(4) (1983) 543–551.

[14] I. Fatadin and S. J. Savory, Compensation of frequency offset for 16-QAM optical coherent systems using QPSK partitioning, *IEEE Photon. Technol. Lett.* **23**(17) (2011) 1246–1248.

[15] T. Nakagawa, Frequency-domain signal processing for chromatic dispersion equalization and carrier frequency offset estimation in optical coherent receivers, in *Advanced Photonics Congress*, OSA Technical Digest (online) (Optical Society of America, 2012), paper SpTh1B.4.

[16] Y. Gao *et al.*, Low-complexity two-stage carrier phase estimation for 16-QAM systems using QPSK partitioning and maximum likelihood detection, Los Angeles, CA (2011), pp. 1–3.

[17] T. Pfau, S. Hoffmann and R. Noe, Hardware-efficient coherent digital receiver concept with feedforward carrier recovery for M-QAM constellations, *J. Lightwave Technol.* **27**(8) (2009) 989–999.

[18] K.-P. Ho and J. M. Kahn, Electronic compensation technique to mitigate nonlinear phase noise, *J. Lightwave Technol.* **22**(3) (2004) 779–783.

[19] E. Ip and J. M. Kahn, Compensation of dispersion and nonlinear impairments using digital backpropagation, *J. Lightwave Technol.* **26** (2008) 3416–3425.

[20] E. Ip and J. M. Kahn, Fiber impairment compensation using coherent detection and digital signal processing, *J. Lightwave Technol.* **28**(4) (2010) 502–519.

[21] X. Li, X. Chen, G. Goldfarb, E. F. Mateo, I. Kim, F. Yaman and G. Li, Electronic post-compensation of WDM transmission impairments using coherent detection and digital signal processing, *Opt. Express* **16**(2) (2008) 880–888.

[22] S. Zhang, M. Huang, F. Yaman, E. Mateo, D. Qian, Y. Zhang, L. Xu, Y. Shao, I. Djordjevic, T. Wang, Y. Inada, T. Inoue, T. Ogata and Y. Aoki, 40 × 117.6 Gb/s PDM-16QAM OFDM Transmission over 10,181 km with Soft-Decision LDPC Coding and Nonlinearity Compensation, in *National Fiber Optic Engineers Conference*, OSA Technical Digest (Optical Society of America, 2012), paper PDP5C.4.

[23] J. Cai, H. Zhang, H. G. Batshon, M. Mazurczyk, O. Sinkin, Y. Sun, A. Pilipetskii and D. Foursa, Transmission over 9,100 km with a capacity of 49.3 Tb/s using variable spectral efficiency 16 QAM based coded modulation, in *Proc. OFC 2014*, Postdeadline Papers, paper Th5B.4.

[24] G. Bosco, V. Curri, A. Carena, P. Poggiolini and F. Forghieri, On the performance of Nyquist-WDM terabit superchannels based on PM-BPSK, PM-QPSK, PM-8QAM or PM-16QAM subcarriers, *J. Lightwave Technol.* **29** (2011) 53–61.

[25] Y. Gao, J. C. Cartledge, A. S. Karar, S. S.-H. Yam, M. O'Sullivan, C. Laperle, A. Borowiec and K. Roberts, Reducing the complexity of perturbation based nonlinearity pre-compensation using symmetric EDC and pulse shaping, *Opt. Express* **22** (2014) 1209–1219.

[26] J. Wang, C. Xie and Z. Pan, Generation of spectrally efficient Nyquist-WDM QPSK signals using digital FIR or FDE filters at transmitters, *J. Lightwave Technol.* **30** (2012) 3679–3686.

[27] Z. Dong, X. Li, J. Yu and N. Chi, 6 × 144-Gb/s Nyquist-WDM PDM-64QAM generation and transmission on a 12-GHz WDM Grid equipped with Nyquist-band pre-equalization, *J. Lightwave Technol.* **30** (2012) 3687–3692.

[28] Z. Dong, X. Li, J. Yu and N. Chi, 6×128-Gb/s Nyquist-WDM PDM-16QAM generation and transmission over 1200-km SMF-28 with SE of 7.47 b/s/Hz, *J. Lightwave Technol.* **30** (2012) 4000–4005.

[29] X. Zhou, J. Yu, M.-F. Huang, Y. Shao, T. Wang, L. Nelson, P. Magill, M. Birk, P. I. Borel, D.W. Peckham, R. Lingle and B. Zhu, 64-Tb/s, 8 b/s/Hz, PDM-36QAM transmission over 320 km using both pre- and post-transmission digital signal processing, *J. Lightwave Technol.* **29** (2011) 571–577.

[30] X. Zhou, L. E. Nelson, P. Magill, B. Zhu and D. W. Peckham, 8x450-Gb/s, 50-GHz-spaced, PDM-32QAM transmission over 400 km and one 50 GHz-grid ROADM, in *Proc. OFC 2011*, paper PDPB3.

[31] D. McGhan, C. Laperle, A. Savchenko, C. Li, G. Mak and M. O'Sullivan, 5120 km RZ-DPSK transmission over G652 fiber at 10 Gb/s with no optical dispersion compensation, in *Proc. OFC 2005*, PDP27.

[32] T. Sugihara, T. Kobayashi, Y. Konishi, S. Hirano, K. Tsutsumi, K. Yamagishi, T. Ichikawa, S. Inoue, K. Kubo, Y. Takahashi, K. Goto, T. Fujimori, K. Uto, T. Yoshida, K. Sawada, S. Kametani, H. Bessho, T. Inoue, K. Koguchi, K. Shimizu and T. Mizuochi, 43 Gb/s DQPSK pre-equalization employing 6-bit, 43 GS/s DAC integrated LSI for cascaded ROADM filtering, in *Proc. OFC 2010*, paper PDPB6.

[33] J. Zhang and H. Chien, A novel adaptive digital pre-equalization scheme for bandwidth limited optical coherent system with DAC for signal generation, in *Proc. OFC 2014*, paper W3K.4.

[34] J. Zhang, H. Chien, Z. Dong and J. Xiao, Transmission of 480-Gb/s dual-carrier PM-8QAM over 2550km SMF-28 using adaptive pre-equalization, in *Proc. OFC 2014*, paper Th4F.6.

[35] J. Zhang, J. Yu, N. Chi, and H.-C. Chien, Time-domain digital pre-equalization for band-limited signals based on receiver-side adaptive equalizers, *Opt. Express* **22** (2014) 20515–20529.

[36] J. Zhang, J. Yu, N. Chi, Z. Dong and X. Li, Nonlinear compensation and crosstalk suppression for 4 × 160.8Gb/s WDM PDM-QPSK signal with heterodyne detection, *Opt. Express* **21** (2013) 9230–9237.

[37] J. Zhang, X. Li and Z. Dong, Digital nonlinear compensation based on the modified logarithmic step size, *J. Lightwave Technol.* **31** (2013) 3546–3555.

Chapter 2

Advanced DSP for Super-Nyquist Transmission System

2.1. Introduction

Bandwidth demand of Internet and private line service continues to grow at around 30% to 50% per year driven by more and more video streaming and proliferation of cloud computing, social media and mobile data delivery [1–4]. This trend combined with the requirement of cost reduction per bit per Hz directly leads to the need for higher speed underlying optical transmission interfaces. At present, the 100-Gb/s long-haul systems, whether in development or in deployment, are all based on single-carrier (SC) polarization division multiplexed quadrature phase shift keying (PDM-QPSK) modulation format associated with coherent detection and digital signal processing (DSP) [5, 6]. On the other hand, the technology options for transmission going beyond 100G are being intensively studied and are multidimensional, which comes in three major categories: increasing the symbol rate [7], the number of subcarriers [8] and the modulation levels [9, 10].

Based on the polarization multiplexing quadrature phase-shift keying (PM-QPSK) modulation format, there are two mainstream multicarrier multiplexing and transmission proposals for next-generation terabit optical transport: no-guard-interval coherent optical orthogonal frequency division multiplexing (NGI-CO-OFDM) [11–13] and Nyquist wavelength division multiplexing (N-WDM) [14–16]. It was shown that, in principle, both techniques exhibit the same sensitivity and spectral efficiency (SE). However, in practice, there are several physical limitations for their implementation leading to suboptimal performance: NGI-CO-OFDM requires

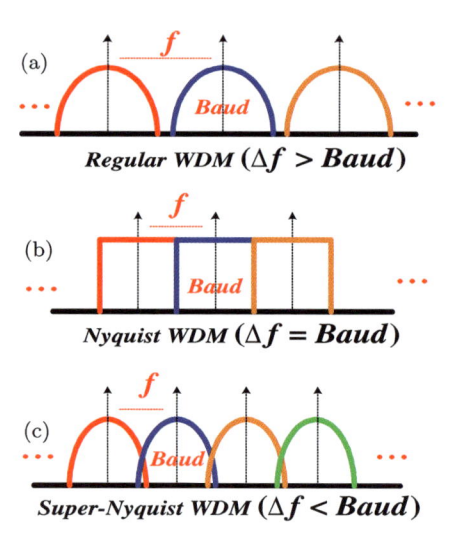

Fig. 2.1. The comparison of different channel multiplexing schemes.

analog-to-digital converters (ADC) with much larger bandwidth and higher sampling rate at the receiver. We have demonstrated NGI-CO-OFDM PM-QPSK superchannel transmission over 3200 km SMF-28 [17]. On the other hand, Nyquist pulse is generated to achieve the Nyquist limit of SE for a given baud rate. But the operating rate is limited by the speed of digital-to-analog converter (DAC) [16].

Figure 2.1 shows the comparison of different channel multiplexing schemes, including the regular WDM (symbol bandwidth < channel spacing), the Nyquist-WDM (symbol bandwidth = channel spacing) and the super-Nyquist WDM or faster-than-Nyquist WDM signals (symbol bandwidth < channel spacing). The regular WDM scheme with guard-bands between the channels has no inter-channel crosstalk and no inter-symbol interference (ISI). However, this scheme has lowest spectrum efficiency (SE) due to the guard-bands. Nyquist-WDM, utilizing the time-domain orthogonal pluses, has the subcarriers spectrally shaped so that their occupancy is close or equal to the Nyquist limit for zero inter-channel crosstalk and ISI-free transmission. However, when considering the forward error correction (FEC) overhead, the transmission of 100G or 400G channels on existing optical line systems based on an ITU grid presents a difficult challenge due to the limited bandwidth available for each channel [23]. The excess bandwidth causes severe crosstalk [8]. On the other hand, the super-Nyquist

WDM with narrowband optical filtering or electrical pre-filtering can be used with the symbol bandwidth less than the channel spacing [20, 25–29]. In previous works [20, 25–30], a simple Gaussian optical narrowband filtering was used for super-Nyquist filtering. However, due to the filtering effect, the system performance was seriously degraded by noise and inter-channel crosstalk enhancement and ISI. Therefore, additional processing is needed for higher SE to counter the noise and crosstalk enhancement and ISI in order to maintain the reasonable long-haul transmission distance with FEC overhead.

This chapter introduces the recently reported advanced DSP for super-Nyquist transmission systems. It is organized as follows. Section 2.2 shows the theory and principles of super-Nyquist WDM systems. Section 2.3 introduces three mainstream DSP algorithms for super-Nyquist channels. Section 2.4 introduces the DSP for super-Nyquist signal generation at the transmitter side. Finally, Section 2.5 summarizes this chapter.

2.2. Theory and Principles of Super-Nyquist Channel

Figure 2.2 shows the principle of super-Nyquist filtering 9-QAM-like signal or quadrature duobinary (QDB) spectral shaping for PM-QPSK signal [40, 41]. Generally, the ideal NWDM transmission with Nyquist pulse requires DAC at a high sampling rate to implement raised cosine (RC) pulse shaping, which unfortunately is not commercially available yet, and a simpler and more realistic way which has been widely used in previous works [38, 39] is the use of fourth-order super-Gaussian narrowband filtering, such as wavelength selective switch (WSS) for the optical Nyquist filtering as shown in Fig. 2.14. For PM-QPSK signals with symbol rate of

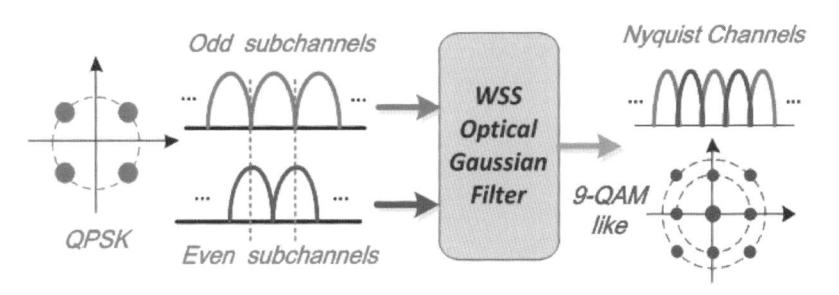

Fig. 2.2. The principle of super-Nyquist filtering WDM channels of 9-QAM-like signal generation by optical Gaussian filtering from QPSK signals.

Rs, we use a waveshaper or WSS with 3-dB pass-band bandwidth of Rs or less for Nyquist spectral shaping. After Nyquist filtering, the 4-point QPSK becomes a 9-QAM-like signal in constellation due to the filtering effect. The Nyquist filtered signal significantly narrows the spectrum compared with QPSK, and the spectral side lobes are also greatly suppressed, which provides a practical means for achieving transmission of the Nyquist limit of SE for a given baud rate.

Partial response system gives the same high SE at a cost of optical signal-to-noise ratio (OSNR) penalty due to its severe noise and crosstalk. Recently, QDB has been proposed to approach SE of 4 bit/s/Hz [18–24]. A simpler and more realistic means is spectrum shaping by using the optical narrow filter or the WSS as shown in Fig. 2.1 [20–24]. Instead of adopting DAC with large bandwidth and high sampling speed at the transmitter, sharp optical filtering is used to perform aggressive spectral shaping and multiplexing function in order to obtain N-WDM. However, due to the strong filtering effect, the system performance is seriously degraded by the strong ISI, severe noise enhancement and inter-channel crosstalk [24]. The noise in high frequency components of the signal spectrum and the inter-channel crosstalk are both enhanced after linear equalization algorithm, such as conventional constant modulus algorithm (CMA).

2.3. Advanced DSP for Super-Nyquist Signal Recovery

Three successful DSP solutions have been reported for super-Nyquist signal recovery [19–26]. One simple way to suppress the noise and crosstalk is to use a post-filter with two taps (i.e., one symbol delay and add) in DSP, while applying CMA with constant modulus equalizations (CMEQ), which performs the function of duobinary shaping, is proposed in [20–24] to release strong filtering limitation. Then, 1-bit maximum likelihood sequence estimation (MLSE) is used before taking a decision to equalize ISI impairment [19].

Alternatively, from the view of constellation point, the effect of the optical Nyquist filtering also turns the 4-point QPSK to 9-point QDB signals. The constant modulus algorithm (CMA) is a kind of blind equalization algorithm, which tries to adaptively balance the noise and ISI, and minimize the mean squared error (MSE) of the signal. The CMA does not work well when ISI is too large [25]. Novel DSP schemes for this optical Nyquist filtering 9-QAM-like signals are reported based on multi-modulus equalization

(MMEQ) without post-filter, which directly recovers the Nyquist filtered QPSK to a 9-QAM-like signal based on the cascaded multi-modulus algorithm (CMMA) [25] and decision-directed least radius distance (DD-LRD) algorithm [26].

2.3.1. CMEQ-based DSP with post-filter

As discussed in the introduction, the real implementation of the optical filter is far away from the requirement in the ideal condition of Nyquist spectral shaping. To avoid the need for high bandwidth and sampling speed DAC at Tx, typically, the optical multiplexer with narrowband optical filtering function is used to perform aggressive spectrum shaping and multiplexing function to obtain Nyquist (symbol bandwidth = channel spacing) or faster-than-Nyquist-WDM signals (symbol bandwidth < channel spacing). However, the spectral shaping induced crosstalk for the intra- and inter-channels will largely limit the transmission performance, which means that innovative signal processing techniques are needed at the receiver side to realize high SE while countering ISI.

The idea of partial response signaling, also called duobinary signaling or correlative coding [18–20], is to introduce a controlled amount of ISI into the signal rather than trying to eliminate it completely. This can be compensated for at the receiver, thereby achieving the ideal symbol-rate packing of two symbols per Hertz but without the requirements of unrealizable filters from Nyquist theorem. Therefore, the multi-symbol optimal detection schemes, like MAP and MLSE, are necessary to take advantage of the symbol correlation contained in the received partial response signals. Successful transmission of $198 \times 100\,\text{G}$ bandwidth-constrained PDM RZ-QPSK channels with an SE of $4\,\text{bit/s/Hz}$ over long-haul ultra-large-effective-area (ULEA) fiber has been reported recently [27]. The challenge is that the number of states and transitions grows exponentially with the memory length, for instance, the adopted MLSE length of 10 means 4^{10} states and 4^{11} transitions in lane-dependent PDM-QPSK signals [8], which significantly imposes the computational complexity in practical implementation. On the other hand, the noise in high-frequency components of the signal spectrum is enhanced after conventional linear impairment equalization algorithm (like conventional CMA) in bandwidth-constraint optical coherent system [18, 19]. One linear electrical delay-and-add digital filter is a simple way to achieve partial response while mitigating the enhanced noise [18]. The MLSE algorithm is still employed to realize symbol decoding

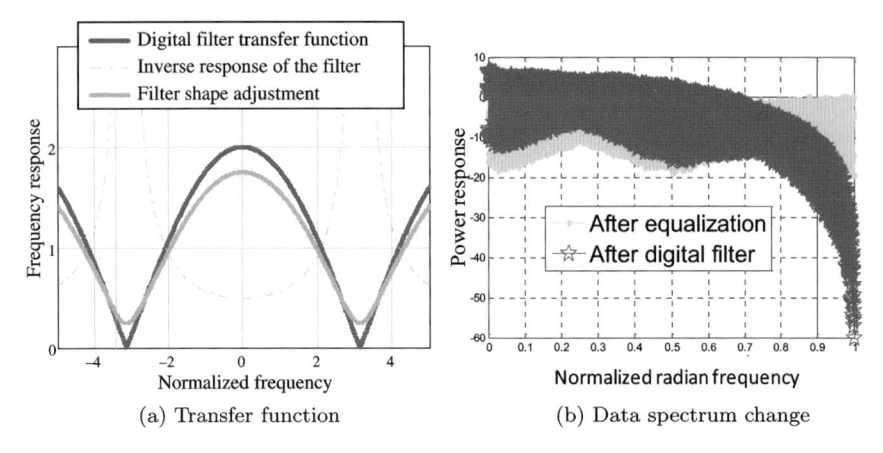

Fig. 2.3. Transfer profile of digital filter and data spectrum after equalization and filtering.

and optimal detection but with the significant reduction of memory length. The transfer function of the digital filter is shown in Fig. 2.3(a). The filtering function is performed after the carrier phase estimation in the conventional DSP flow of the coherent receiver, the filtering impact on the high-frequency components is shown in Fig. 2.3(b). It is noted that the second-tap coefficient of the filter is adjusted to optimize the overall performance by following the MLSE detection algorithm. After the filtering, the enhanced noise and ISI are suppressed.

From the view of the constellation point, the effect of the digital filter turns the 4-point QPSK to 9-point QDP signals. The evolution of this transformation is illustrated in Fig. 2.4. As a result of the delay-and-add effect, the 2-ASK in-phase and quadrature components disappear and independently change into two 3-ASK symbol series. The generation mechanism of 9-QAM signals can be considered as superposition from two 3-ASK vectors on a complex plane. The size of constellation points represents the relative number of points generated after the digital filter.

Figure 2.5 shows the simple way for Nyquist signal processing based on CMEQ with post-filter [18–22]. The received signal is first recovered to QPSK, and then converted to the 9-QAM-like signal using the delay-and-add post-filter to suppress the noise. Before the decision, 1-bit MLSE is used to equalize ISI impairment. The sampled signals are first resampled and then applied by CD compensation and clock recovery. After a $T/2$-spaced time-domain finite impulse response (FIR) filter for the

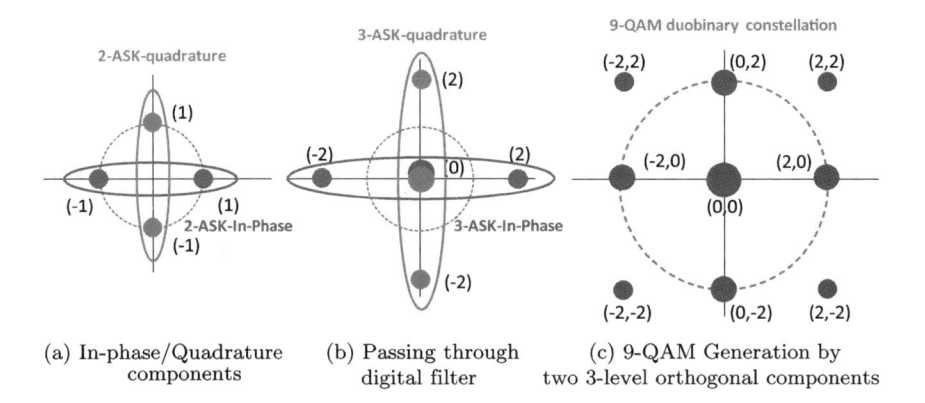

Fig. 2.4. An illustration of 9-QAM signal generation by digital filter.

(a) In-phase/Quadrature components
(b) Passing through digital filter
(c) 9-QAM Generation by two 3-level orthogonal components

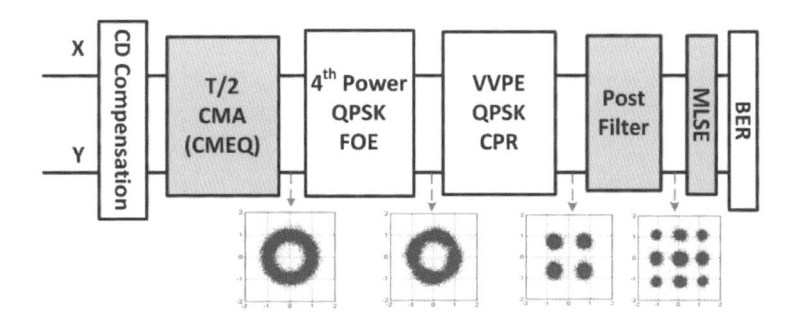

Fig. 2.5. The main DSP blocks based on CMEQ with post-filter.

electronic dispersion compensation (EDC, T is the sampling time period), the polarization recovery by using classic CMA is done with 21-tap, $T/2$-spaced adaptive FIR filters, and the carrier recovery including frequency offset estimation by fast Fourier transform method and carrier phase recovery (CPR) by fourth-power Viterbi–Viterbi algorithm are performed. It is then additionally equipped with the filter and followed by use of MLSE with a memory length of 1 bit.

2.3.2. MMEQ-based DSP for super-Nyquist signal

Consider a QDB spectrum shaped PM-QPSK system where the received signal is sampled and processed in a DSP after CD and possibly nonlinearity compensation, timing recovery. Since the constellation of QDB signal is 9-QAM like, and with three moduli, with the MMEQ is more compatible. Further results show that the MMEQ scheme has improved filtering tolerance with better performance on noise and crosstalk suppression. Here, we present two multi-modulus processing schemes, the CMMA and decision-directed least radius distance (DD-LRD) algorithm. These key multi-modulus DSP algorithms include polarization demultiplexing based on CMMA, frequency offset estimation (FOE) and CPR based on multi-modulus QPSK partitioning and the DD-LRD for 9-QAM-like constellations.

2.3.2.1. CMMA with modified carrier recovery scheme

Figure 2.6 shows the main DSP blocks for MMEQ based on CMMA and modified carrier recovery scheme [25]. In the MMEQ scheme, we handle the Nyquist signal as 9-QAM-like signal with three moduli; thus, CMMA is used with a modified carrier phase recovery scheme to obtain 9-QAM-like signals directly. For QDB spectrum shaped PM-QPSK, the classic CMA is not well compatible [18]. This is because 9-point signal does not present constant symbol amplitude. It not only leads to extra noise after equalization, but also causes a problem with filter tap frequency response. Thus, we use CMMA proposed and used in PM 8-QAM and PM 16-QAM systems [30, 31] with good modulus decision performances for blind polarization demultiplexing.

The principle of CMMA for QDB spectrum shaped PDM-QPSK signals is shown in Fig. 2.7. It also has a four butterfly-configured adaptive digital equalizers [32, 33]. Here, $\varepsilon_{x,y}$ is the feedback signal error for filter tap updating. The corresponding filter tap weight updating equalizations are

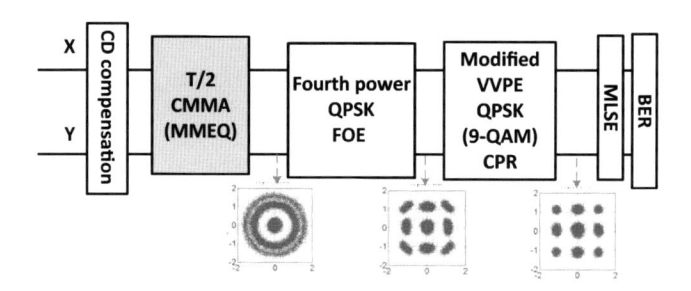

Fig. 2.6. The main DSP blocks of CMMA-based MMEQ with modified carrier recovery scheme.

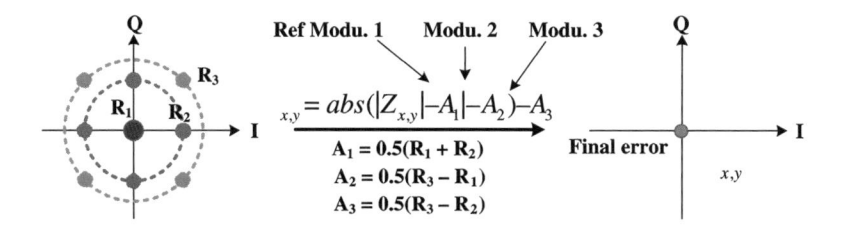

Fig. 2.7. The principle of CMMA for QDB spectrum shaped PDM-QPSK.

given as follows:

$$\begin{aligned} h_{xx}(k) &\to h_{xx}(k) + \mu\varepsilon_x(i)e_x(i)\hat{x}(i-k) \\ h_{xy}(k) &\to h_{xy}(k) + \mu\varepsilon_x(i)e_x(i)\hat{y}(i-k) \\ h_{yx}(k) &\to h_{yx}(k) + \mu\varepsilon_y(i)e_y(i)\hat{x}(i-k) \\ h_{yy}(k) &\to h_{yy}(k) + \mu\varepsilon_y(i)e_y(i)\hat{y}(i-k) \end{aligned} \quad (2.1)$$

and $e_{x,y}(i)$ for QDB 9-point signal is given by

$$e_{x,y}(i) = \text{sign}(||Z_{x,y}(i)| - A_1| - A_2) \cdot \text{sign}(|Z_{x,y}(i)| - A_1) \cdot \text{sign}(Z_{x,y}(i)) \quad (2.2)$$

Here, $\hat{x}$ and $\hat{y}$ denote the complex conjugates of received signals x and y, respectively; sign(x) is the sign function, and μ is the convergence parameter. By introducing three reference circles, A_1–A_3, the final error can approach zero for ideal QDB signal as shown in 8QAM signals. R_1 and R_2 are the radii of the three-moduli QDB PDM-QPSK signal and $Z_{x,y}$ is the output of the equalizer. As a result, it is clear that the regular CMA error signal will not approach zero even for an ideal 9-point signal.

The partition scheme has been presented in [34] for FOE in a 16-QAM coherent system. In this way, the regular m-power algorithm can also be used for FOE for the 9-point QDB spectrum shaped signal with partitioning. On the other hand, for polarization multiplexed coherent system, the same transmitter and local oscillator (LO) are used for the two polarization signals. In this way, both polarization signals are affected by the same frequency offset. To address this issue, we propose a joint-polarization QPSK partitioning algorithm for FOE.

Figure 2.8 shows the principle of QPSK partition and rotation for the 9-point QDB spectrum shaped signal. The processing consists of three steps. Since the amplitude of R1 symbols is zero, the frequency offset and carrier phase noise have no impact on the R1 symbols. Thus, we only need to use the R2 and R3 symbols for frequency offset estimation and phase recovery.

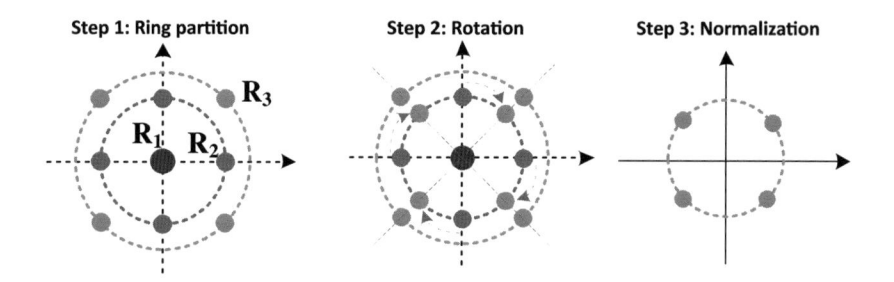

Fig. 2.8. The principle of QPSK partition and rotation.

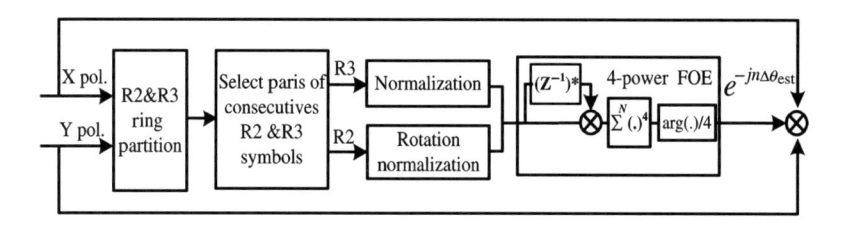

Fig. 2.9. The block diagram for joint-polarization QPSK partitioning FOE.

The input symbols are first ring partitioned by their amplitudes with three values. Then, only the middle ring symbols in R2 are constellation rotated with $-\pi/4$. Finally, the symbols in the middle and outside two rings, R2 and R3, are normalized according to their amplitudes. In this way, the middle and outside two rings are combined to one ring with QPSK 4-point constellation.

Figure 2.9 shows the block diagram for the proposed joint-polarization QPSK partitioning algorithm for FOE. After polarization demultiplexing by CMMA, the incoming X and Y pol. symbols are first portioned into three groups with different circle radii. Here, only pairs of consecutive R1 and R2 symbols are selected for estimation to reduce the complexity of the algorithm. Afterwards, R2 symbols are first rotated with $-\pi/4$ angle and then normalized. However, R1 symbols are only normalized. After that, the two groups can combine together with a "QPSK"-like constellation. In this way, the fourth-power frequency estimation for QPSK can operate now. For N pairs of R1 and R2 symbols, the phase angle estimation caused by frequency offset is

$$\Delta\theta_{\text{est}} = 2\pi\Delta f_{\text{est}}T_s = \frac{1}{4}\arg\sum_{1}^{N}(S_{k+1} \cdot S_k^*)^4 \qquad (2.3)$$

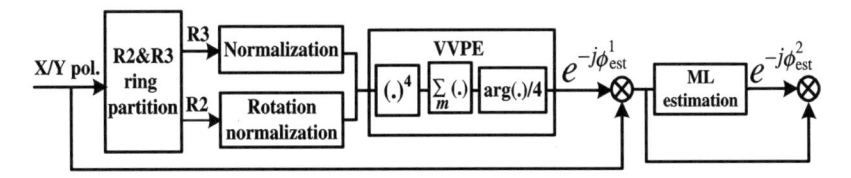

Fig. 2.10. Block diagram for two-stage phase estimation based on QPSK partitioning/ML.

where S_k is the combined normalized symbols of R1 and R2 groups, T_s is the symbol duration and Δf_{est} is the estimated frequency offset. Then the frequency offset can be compensated by $e^{-jn\Delta\theta_{\text{est}}}$ for the nth received symbol in both polarizations. The frequency offset Δf_{est} can be estimated within $[-1/(8T_s), +1/(8T_s)]$ for the proposed fourth-power operation.

As analyzed above, the QPSK partitioning scheme can also be used in the CPR for QDB spectrum shaped signals. On the other hand, maximum likelihood algorithm shows good improvement and low complexity for 16 QAM phase estimation [35]. In this way, we propose a two-stage phase recovery for the 9-point QDB spectrum shaped signal based on QPSK Partition/ML as shown in Fig. 2.10.

The principle of R2 and R3 ring partition, rotation and normalization is the same as shown in Fig. 2.9. We also only need R2 and R3 symbols for phase recovery here. In practice, the partition steps for FOE and phase recovery can become one. After that, the symbols in R2 are first rotated with $-\pi/4$ angle and then normalized, while R1 symbols are only normalized and then combined with R2 symbols. Here, all the combined symbols can be used in groups and m is the number of symbols in each group. In this way, the first stage phase ϕ_{est}^1 can be eliminated by Viterbi and Viterbi phase estimation (VVPE) as

$$\phi_{\text{est}}^1 = \left(\sum_m S_k^4\right)\bigg/4 \tag{2.4}$$

and the second stage of phase estimation ϕ_{est}^2 based on ML [10] as

$$h = \sum_m S_k \cdot D_k^* \tag{2.5}$$

$$\phi_{\text{est}}^2 = \tan^{-1}(\text{Im}[h]/\text{Re}[h]) \tag{2.6}$$

Here, D_k is the decision of S_k after the first stage phase recovery. The second stage phase recovery is implemented before the final output.

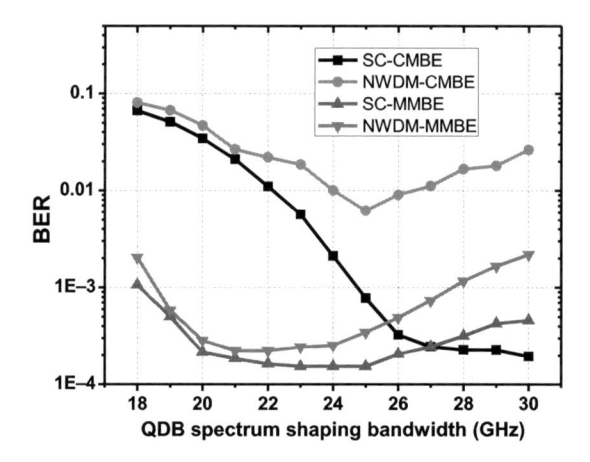

Fig. 2.11. The simulation results of BER performance varying with QDB spectrum shaping bandwidth for different DSP schemes.

Figure 2.11 shows the simulation results of back-to-back (BTB) bit error ratio (BER) performance varying with QDB spectrum shaping bandwidth for the conventional CMAs of CMBE, and our proposed multi-modulus algorithms of MMBE for both SC and NWDM. Here, we keep the OSNR at 16 dB. For NWDM channels, the channel spacing is set at 25 GHz. It shows that the proposed MMBE scheme has better tolerance for strong QDB spectrum shaping and also crosstalk from other channels. For the SC case, the MMBE scheme shows better BER performance when the QDB spectrum shaping bandwidth is less than 27 GHz. On the other hand, the MMBE scheme shows significant BER improvement over conventional CMBE for NWDM case. The conventional CMBE for NWDM has the worst performance due to the crosstalk and strong shaping. It also shows that the optimal QDB spectrum shaping bandwidth for our proposed MMBE scheme is from 23 to 25 GHz for SC and 21 to 23 for NWDM. It is worth noting that our scheme has two additional steps compared with regular QPSK signals for the proposed 9-point QDB, including the QPSK partitioning and rotation. These two additional steps require additional computation. However, the QDB spectrum shaped signals have a higher spectrum efficiency (nearly doubled) and tolerance to channel crosstalk with respect to QPSK signals. Thus, we believe that there is a trade-off, when considering the spectrum efficiency and real-time implementation complexity.

2.3.2.2. Decision-directed least radius distance (DD-LRD) algorithm

Figure 2.12 shows the major DSP functional blocks based on DD-LRD [26]. A novel cascaded, 9-QAM-based, highly filtering-tolerant MMEQ is used for polarization demultiplexing, robust filtering compensation, and other channel distortion mitigation. First, a 17-tap T/2-spaced CMA equalizer is used to perform the pre-equalization. The output of this CMA equalizer is used for the initial frequency-domain frequency offset estimation and compensation [37]. Then, a 17-tap T/2-spaced 2 × 2 equalizer as the second stage equalization based on decision-directed least radius distance (DD-LRD) algorithm is used for polarization demultiplexing. The carrier frequency and phase recovery are performed within the DD-LRD loop. The frequency offset is also estimated and compensated using a frequency-domain method. The phase recovery is realized by novel decision-directed blind phase search (BPS) method [38] within a small phase-varying range: the initial phase is recovered by the last symbol but then refined using the BPS over a nonlinear distributed phase range. Such a two-stage algorithm can effectively mitigate cyclic phase slipping. Before the calculation of the BER, the 1-bit MLSE based on Viterbi algorithm is utilized for symbol decoding and detection to eliminate the ISI impact.

Again, we investigate the 28GBuad PM-QPSK signal for the performance of DD-LRD-based MMEQ in simulation. The following optical bandpass filter of fourth-order Gaussian type is utilized to shape the spectrum of QPSK signal. The 3-dB filter bandwidth is emulated from 22 to 30 GHz in simulation, so that the spectrum is significantly compressed to approach

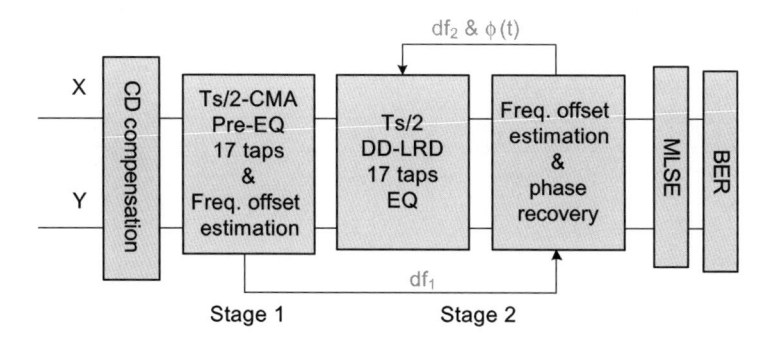

Fig. 2.12. An illustration of major DSP blocks of MMEQ based on DD-LRD.

Nyquist bandwidth. The ASE noise is added before optical homodyne coherent detection. The OSNR is 30 dB which is defined in a 0.1-nm noise bandwidth. We ignore the influence of carrier phase drift in the simulation. Continuous wave (CW) laser sources at the transmitter and for local oscillator (LO) at the coherent receiver are both of 0-Hz linewidth. The signal is finally sampled at twice the baud rate.

The sampled signal is divided into two groups: one is the signal sampled at time T, i.e., the timing phase is 0, and the other is the signal sampled at time $T/2$, i.e., the timing phase is π offset. Figure 2.13 illustrates two groups of samples. We define T samples as S_T and $T/2$ samples as $S_{T/2}$. The constellations of the received signal with the optical Gaussian filter of 28 GHz and 24 GHz are shown in Figs. 2.14(a) and 2.14(b) respectively. We observe that T samples are like 4-QAM (blue dots) and $T/2$ samples are like 9-QAM (red dots). Obvious ISI occurs after fourth-order Gaussian filter. We can see that each constellation point becomes a square-like distribution. Comparing constellations of the sampled signals with 28- and 24-GHz

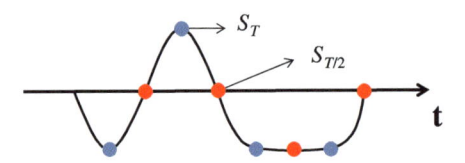

Fig. 2.13. Signal is sampled at the time of T (blue sample) and $T/2$ (red sample).

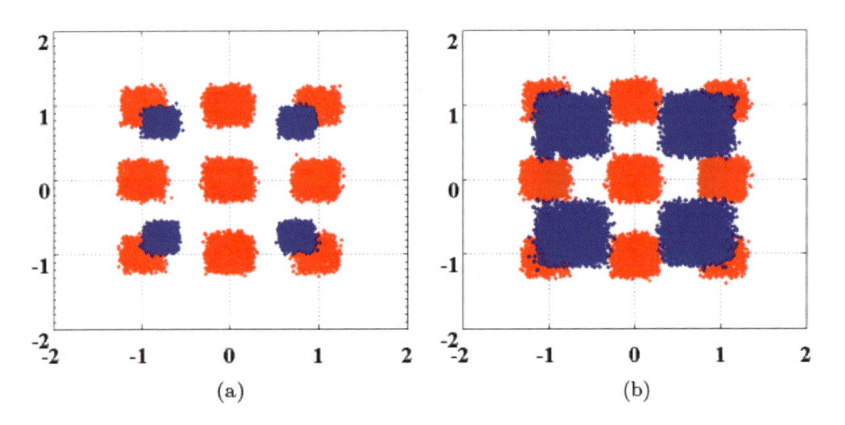

Fig. 2.14. Constellations of the sampled signals with the filter of (a) 28 GHz and (b) 24 GHz. The blue and red dots represent the signals sampled at the time of T and $T/2$ respectively.

filtering, T samples have much larger ISI (blue dots in Fig. 2.14(b)) when 24-GHz filter is applied, while the $T/2$ samples keep nearly unchanged (red dots in Fig. 2.14). In that case, we can recover the 9-QAM-like $S_{T/2}$ samples instead of conventional processing to S_T samples in the presence of strong filtering. We expect the new concept of data recovery would have better performance.

Assuming S_T is known, $S_{T/2}$ can be estimated approximately by linear interpolation which is expressed as Eq. (1.7)

$$S_{T/2} = \frac{S_T(k) + S_T(k+1)}{2} \quad (k=1,2,3,\ldots) \tag{2.7}$$

It is quite interesting that $S_{T/2}$ is to a certain extent duobinary shaped signal which will share the nature of narrow bandwidth of the duobinary coded signal. Therefore, $T/2$ samples have larger tight filtering tolerance and lower ISI thanks to the partial response system. However, a little higher OSNR is required since multi-level detection must be applied to the 9-QAM signal. The signal can be finally detected with MLSD algorithm which makes use of the inherent inter-symbol memory and minimizes the number of errors by selecting the most probable trellis path [18].

In order to decrease the signal ISI, an adaptive 9-tap FIR filter can be applied. The CMA is normally used to blindly update the FIR tap weights for the QPSK signal which has constant modulus. But CMA does not work well enough when big ISI exists. A post-filter with two taps (i.e., one symbol delay and add), which performs the function of duobinary shaping, is proposed to further release strong filtering limitation. Figure 2.15(a) shows the T samples suffering from severe ISI. CMA (Fig. 2.15(b)) and post-filter (Fig. 2.15(c)) can reduce the ISI to a certain degree.

In our scheme, we propose to recover the 9-QAM $T/2$ samples directly. CMA is not well compatible for this signal because 9-QAM does not present constant symbol amplitude and the error signal cannot approach zero. We propose to use the DD-LRD algorithm [36], which is used for 16QAM signal and much more accurate than CMA, to update the filter tap weights. The error function is given by

$$e(n) = y(n)(|\hat{d}(n)|^2 - |y(n)|^2) \tag{2.8}$$

where $y(n)$ is the equalized signal, and $\hat{d}(n)$ is the decided symbol. And the filter tap weight updating function is given by

$$w(n) = w(n-1) + \mu e(n) x(n)^* \tag{2.9}$$

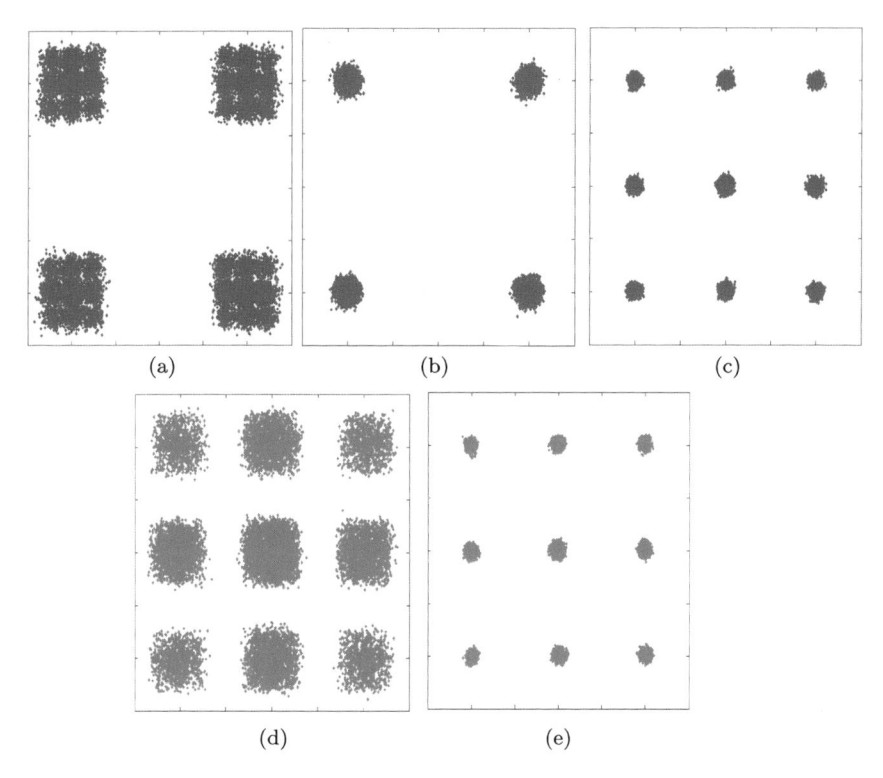

Fig. 2.15. Constellations of (a) T samples, (b) CMA processing, (c) CMA + post-filter processing, (d) $T/2$ samples, and (e) DD-LRD processing.

where $w(n)$ is the adaptive FIR filter, and μ is the convergence parameter. The DD-LRD has superior tolerance to ISI because of its phase independence manner. Besides, it has rapid convergence speed which is robust to time varying situation. Figure 2.15(d) shows the $T/2$ samples suffering from severe ISI. LRD algorithm presents excellent performance to restrain the ISI as shown in Fig. 2.15(e).

Figure 2.16 shows measured MSE, varying with different DSP algorithms. The hollow curves present the processing to T samples of the QPSK constellation. 1 dB MSE improvement is obtained using CMA compared to T sampling signal and an additional 0.5 dB is gained using post-filter with 26-GHz optical filtering. However, tighter filter bandwidth below 26 GHz induces even more serious ISI. MSE degrades rapidly with narrower bandwidth and the benefit of ISI compression from CMA decreases to less than 0.5 dB when the filter bandwidth is 22 GHz. The solid curves present

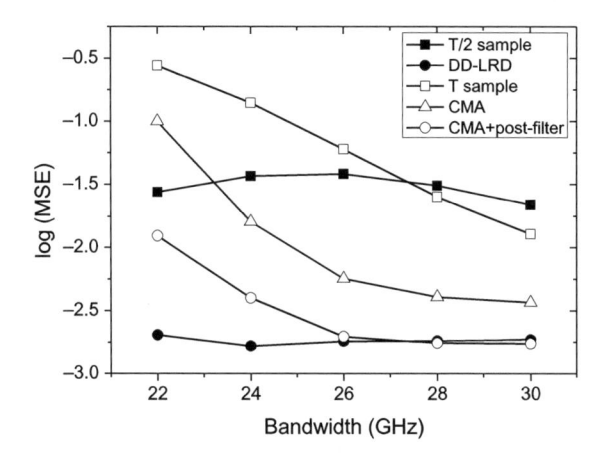

Fig. 2.16. Measured MSE of the signals after T and $T/2$ sampling, DD-LRD, CMA, and CMA + post-filter processing.

the processing to $T/2$ samples of the 9-QAM constellation. 1.4 dB MSE improvement is obtained using DD-LRD compared to $T/2$ sampling signal. It is found that the MSE changes very little as the filter bandwidth varies from 22 to 30 GHz. The performance is quite comparable with larger than 28 GHz filtering between DD-LRD and CMA + post-filter algorithms. So, it is more robust to strong filtering to recover 9-QAM signal on $T/2$ samples than QPSK on the T sample.

Lots of experimental results of the MMEQ for high-speed data transmission system have proved the success of 9-QAM-based super-Nyquist channel with high-SE and long-haul transmission distance, for example, the 400G system based on dual-carrier 60GBaud [39], SC 110GBaud in 100 GHz-grid [40–42] and with ROADMs in the fiber link, and QPSK with baud rate up to 128.8GBaud in 100 GHz-grid [43–45] transmission over long-distance for fiber. All these experimental results show the advantages of MMEQ for super-Nyquist channels in future 400G or beyond transmission systems.

2.4. DSP for Super-Nyquist 9-QAM (QDB) Signal Generation

Figure 2.17 shows the principle of the proposed novel DAC-based super-Nyquist 9-QAM signal generation compared with regular Nyquist QPSK signal. For regular Nyquist filtering, a raised cosine (RC) or square root raised cosine (SRRC) filter with roll-off factor of 0 is used for Nyquist pulse

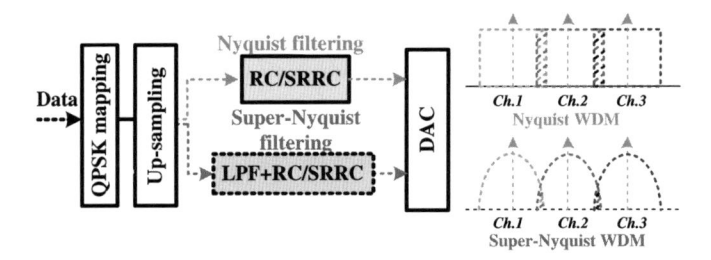

Fig. 2.17. The principle of DAC-based Nyquist and super-Nyquist 9-QAM signal generation and the crosstalk impairments in SN-WDM.

generation. However, as shown in Fig. 2.17, when the channel spacing is less than the baud rate, the excess bandwidth causes severe crosstalk. In order to realize the super-Nyquist transmission, additional low pass filter (LPF) is added for super-Nyquist pulse generation. In this way, the signal spectrum is further suppressed to reduce the channel crosstalk.

In our case, the LPF can be simply realized by the quadrature duobinary (QDB) delay and add filter, of which the transfer function in *z-transform* is given by

$$H_{\text{QDB}}(Z) = 1 + Z^{-1} \tag{2.10}$$

It can be simply implemented by a 2-tap FIR digital filter with good performance and turn the QPSK to the 9-QAM signal. The super-Nyquist digital filter in the time domain by cascading the QDB and SRRC filters is

$$h_{\text{SN}}(t) = h_{\text{QDB}}(t) \otimes h_{\text{srrc}}(t) \tag{2.11}$$

where the $h_{\text{srrc}}(t)$ is the typical time-domain impulse response of SRRC filter as described in [46, 47]. The $h_{\text{QDB}}(t)$ is the impulse response of QDB filter H_{QDB} as described in Eq. (2.10). When the roll-off factor is zero, the SRRC filter has the same pulse and frequency response with RC. Therefore, the Nyquist filtering can also be implemented with RC filter here.

Figures 2.18(a) and 2.18(d) show the time-domain impulse response of the regular Nyquist filter based on SRRC and the super-Nyquist filter based on cascading QDB and SRRC filters, respectively. Here, the roll-off factor of SRRC is set at 0. Less oscillations and faster convergence can be observed for the super-Nyquist digital filter compared with the Nyquist filter. Figures 2.18(b) and 2.18(e) show eye diagrams of the generated Nyquist QPSK 2-level baseband signal and the super-Nyquist 9-QAM 3-level baseband signal, respectively. The electrical power spectra of

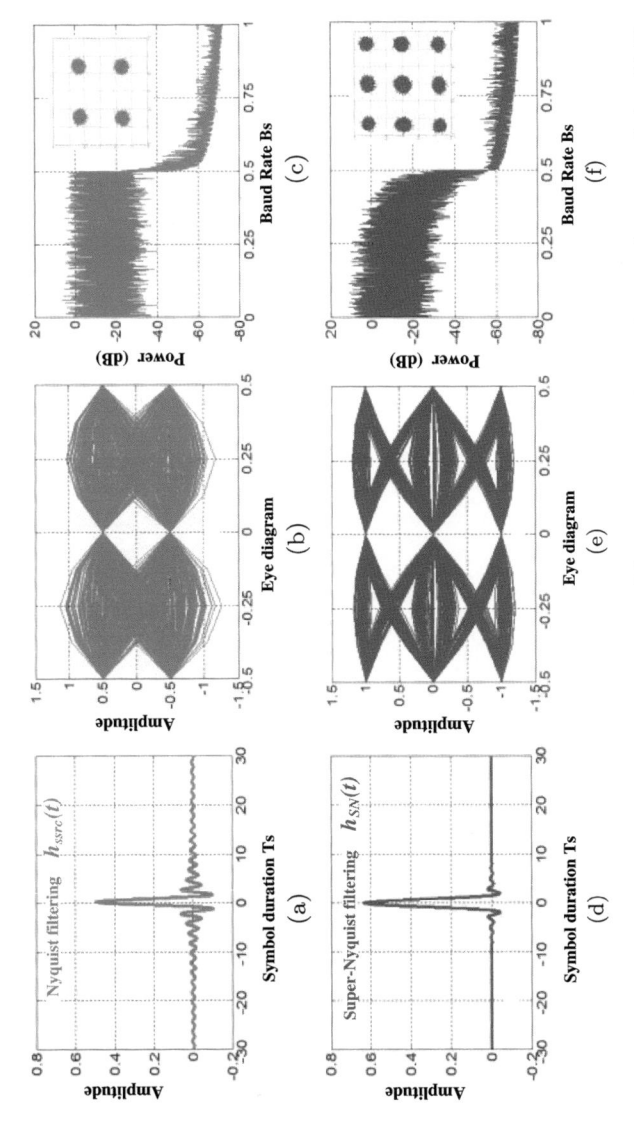

Fig. 2.18. The impulse response of the (a) Nyquist filter and (d) super-Nyquist signal. The FFT spectrum of the generated (c) Nyquist QPSK signal and (f) super-Nyquist 9-QAM signal. The eye diagrams of the (b) Nyquist QPSK signal and (e) super-Nyquist 9-QAM signal.

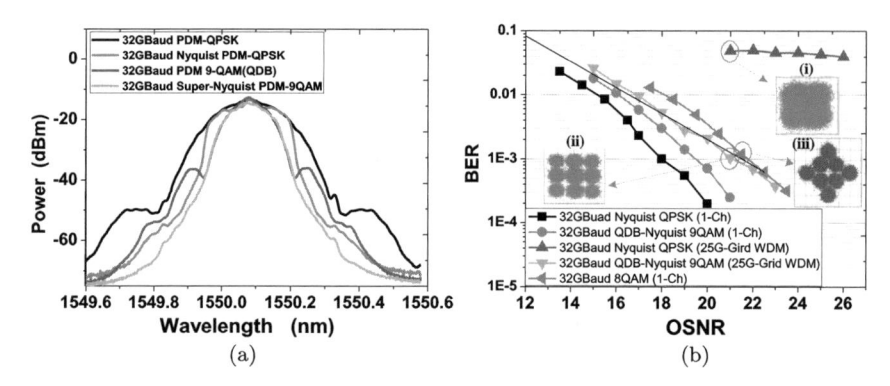

Fig. 2.19. (a) The optical spectrum of single channel 32-Gbaud PDM-QPSK, Nyquist PDM-QPSK, PDM-9QAM and super-Nyquist PDM-9-QAM signal. (b) The BTB BER performance of 32-GBaud Nyquist-QPSK, super-Nyquist 9-QAM and 32GBaud 8QAM signals versus OSNR in single channel case and WDM case.

Nyquist QPSK and super-Nyquist 9-QAM signals are shown in Figs. 2.18(c) and 2.18(f), respectively. We can see that the power spectrum of the super-Nyquist signal is significantly suppressed compared with the Nyquist signal, and the spectral side lobes are also greatly suppressed. The 3-dB bandwidth is less than 0.5 baud rate. Insets (i) and (ii) show the constellations of Nyquist QPSK and super-Nyquist 9-QAM signal, respectively. In this way, a super-Nyquist signal is generated by the Nyquist filtering of the QDB 9-QAM signal.

Figure 2.19(a) shows the optical spectrum of generated single channel (SC) 32-GBaud PDM-QPSK, Nyquist PDM-QPSK, PDM-9QAM (with QDB filter only) and proposed super-Nyquist PDM-9-QAM signal. We can see that the PDM-QPSK signal without any operation occupies the largest bandwidth [47]. The bandwidth of Nyquist PDM-QPSK is equal to the baud rate, which is 32 GHz and exceeds the 25-GHz carrier spacing. The 9-QAM signal with only QDB filer has two side-lobes. Our proposed super-Nyquist PDM-9-QAM signal shows the narrowest bandwidth compared with other three types of signals. The 3-dB bandwidth is less than 0.5 baud rate. Figure 2.19(b) shows the BTB BER performance of 32-GBaud Nyquist PDM-QPSK and super-Nyquist PDM-9 QAM signals versus the optical signal-to-noise ratio (OSNR) in the single channel case and 25-GHz grid WDM case [47]. In the single channel case, the Nyquist signal shows the best BER performance. Due to the narrow digital QDB filtering, the super-Nyquist 9-QAM signal shows about 1.5-dB OSNR penalty at BER of 1×10^{-3} compared with SC Nyquist QPSK. However, for 25-GHz grid WDM case, the 32-GBaud Nyquist signal cannot be recovered due to large crosstalk from adjacent channels. There is an error floor at 4×10^{-2} for Nyquist QPSK signal even under large OSNR condition. Only about 1.5-dB OSNR penalty is observed for the 32GBaud super-Nyquist signal in the 25-GHz grid WDM case compared with the SC case. Insets (i) and (ii) show the constellations of Nyquist QPSK and super-Nyquist 9-QAM signal in WDM case with OSNR at 21 dB, respectively. We also measure the BER of SC 32GBaud PDM-8QAM using the same experiment setup. It shows about 2.5-dB OSNR penalty compared with the QDB 9-QAM signal. Thus, 9-QAM signal has better BER performance compared with 8-QAM. We believe there are two reasons: first, the minimum Euclidean distance of the QDB 9-QAM signal is larger than that of the 8-QAM signal; second, a multi-symbol equalization and decision based on MLSE is used for 9-QAM signal.

2.5. Summary

In this chapter, we introduce the recent progress of DSP for high-speed and long-haul optical coherent transmission system. A typical digital coherent system transfers the impairments from the system into digital domain, which equalize and compensate those impairments by using advanced DSP. In Section 2.2, the classic DSP algorithms used in the coherent system are introduced, and then the digital pre-equalization for bandwidth limited system are discussed in Section 2.3. Finally, advanced DSP for super-Nyquist signal generation and recovery is introduced in Section 2.4.

References

[1] B. Swanson and G. Gilder, Estimating the Exaflood — The impact of video and rich media on the internet — A zettabyte by 2015? Discovery Institute (January 29, 2008).
[2] R. Tkach, Scaling optical communications for the next decade and beyond, *Bell Labs Technon. J.* **14**(4) (2010) 3–9.
[3] R. J. Essiambre and R. Tkach, Capacity trends and limits of optical communication networks, *Proc. IEEE* **100**(4) (2012) 1–21.
[4] C. V. N. Index, Forecast and methodology, 2010–2015, White Paper, Cisco Systems, San Jose, CA (June 1, 2011) [Online]. Available: http://www.cisco.com/en/US/solutions/collateral/ns341/ns525/ns537/ns705/ns827/white_paper_c11-481360_ns827_Networking_Solutions_White_Paper.html.
[5] K. Roberts, M. O'Sullivan, K.-T. Wu, H. Sun, A. Awadalla, D. J. Krause and C. Laperle, Performance of dual-polarization QPSK for optical transport systems, *J. Lightwave Technol.* **27**(16) (2009) 3546–3559.

[6] J. Renaudier, O. Bertran-Pardo, H. Mardoyan, P. Tran, G. Charlet, S. Bigo, M. Lefrancois, B. Lavigne, J.-L. Auge, L. Piriou and O. Courtois, Performance comparison of 40G and 100G coherent PDM-QPSK for upgrading dispersion managed legacy systems, presented at the Opt. Fiber Commun. Conf./Nat. Fiber Opt. Eng. Conf., San Diego, CA (2009), paper NWD5.

[7] C. Xie, G. Raybon and P. J. Winzer, Transmission of mixed 224-Gb/s and 112-Gb/s PDM-QPSK at 50-GHz channel spacing over 1200-km dispersion-managed LEAF® spans and three ROADMs, *J. Lightwave Technol.* **30**(4) (2012) 547–552.

[8] J.-X. Cai, C. R.Davidson, A. Lucero, H. Zhang, D. G. Foursa, O. V. Sinkin, W. W. Patterson, A. N. Pilipetskii, G. Mohs and N. S. Bergano, 20 Tbit/s transmission over 6860 km with sub-Nyquist channel spacing, *J. Lightwave Technol.* **30**(4) (2012) 651–657.

[9] X. Zhou, J. Yu, M.-F. Huang, Y. Shao, T. Wang, L. Nelson, P. Magill, M. Birk, P. I. Borel, D. W. Peckham, R. Lingle and B. Zhu, 64-Tb/s, 8 b/s/Hz, PDM-36QAM transmission over 320 km using both pre- and post-transmission digital signal processing, *J. Lightwave Technol.* **29**(4) (2011) 571–577.

[10] A. H. Gnauck, P. J. Winzer, A. Konczykowska, F. Jorge, J.-Y. Dupuy, M. Riet, G. Charlet, B. Zhu and D. W. Peckham, Generation and transmission of 21.4-Gbaud PDM 64-QAM using a novel high-power DAC driving a single I/Q modulator, *J. Lightwave Technol.* **30**(4) (2012) 532–536.

[11] S. Chandrasekhar and X. Liu, Experimental investigation on the performance of closely spaced multi-carrier PDM-QPSK with digital coherent detection, *Opt. Express* **17**(24) (2009) 21350–21361.

[12] B. Zhu, X. Liu, S. Chandrasekhar, D. W. Peckham and R. Lingle, Jr., Ultra-long-haul transmission of 1.2-Tb/s multicarrier No-Guard-Interval CO-OFDM superchannel using ultra-large-area fiber, *IEEE Photon. Technol. Lett.* **22**(11) (2010) 826–828.

[13] A. Sano, E. Yamada, H. Masuda, E. Yamazaki, T. Kobayashi, E. Yoshida, Y. Miyamoto, R. Kudo, K. Ishihara and Y. Takatori, No-Guard-Interval Coherent Optical OFDM for 100-Gb/s long-haul WDM transmission, *J. Lightwave Technol.* **27**(16) (2009) 3705–3713.

[14] G. Bosco, A. Carena, V. Curri, P. Poggiolini and F. Forghieri, Performance limits of Nyquist-WDM and CO-OFDM in high-speed PM-QPSK systems, *IEEE Photon. Technol. Lett.* **22**(15) (2010) 1129–1131.

[15] J. Yu, X. Zhou, M. F. Huang, D. Qian, L. Xu and P. N. Ji, Transmission of hybrid 112 and 44 Gb/s PolMux-QPSK in 25 GHz channel spacing over 1600 km SSMF employing digital coherent detection and EDFA-Only amplification, presented at the Opt. Fiber Commun. Conf./Nat. Fiber Opt. Eng. Conf., San Diego, CA (2009), paper OThR3.

[16] M. Yan, Z. Tao, W. Yan, L. Li, T. Hoshida and J. C. Rasmussen, Experimental comparison of No-Guard-Interval-OFDM and Nyquist-WDM superchannels, presented at the Opt. Fiber Commun. Conf./Nat. Fiber Opt. Eng. Conf., Los Angeles, CA (2012), paper OTh1B.2.

[17] J. Yu, Z. Dong and N. Chi, 1.96Tb/s (21 × 100Gb/s) OFDM optical signal generation and transmission over 3200-km fiber, *IEEE Photon. Technol. Lett.* **23**(15) (2011) 1061–1063.

[18] J. Li, Z. Tao, H. Zhang, W. Yan, T. Hoshida and J. C. Rasmussen, Spectrally efficient quadrature duobinary coherent systems with symbol-rate digital signal processing, *J. Lightwave Technol.* **29**(8) (2011) 1098–1104.

[19] J. Li, E. Tipsuwannakul, T. Eriksson, M. Karlsson and P. A. Andrekson, Approaching Nyquist limit in WDM systems by low-complexity receiver-side duobinary shaping, *J. Lightwave Technol.* **30** (2012) 1664–1676.

[20] Z. Jia, J. Yu, H. Chien, Z. Dong and D. Di Huo, Field transmission of 100 G and beyond: Multiple baud rates and mixed line rates using Nyquist-WDM technology, *J. Lightwave Technol.* **30** (2012) 3793–3804.

[21] Z. Dong, J. Yu, Z. Jia, H. Chien, X. Li and G. Chang, 7 × 224 Gb/s/ch Nyquist-WDM transmission over 1600-km SMF-28 using PDM-CSRZ-QPSK modulation, *IEEE Photon. Technol. Lett.* **24**(13) (2012).

[22] J. Yu, Z. Dong, H. Chien, Z. Jia, X. Li, D. Huo, M. Gunkel, P. Wagner, H. Mayer and A. Schippel, Transmission of 200 G PDM-CSRZ-QPSK and PDM-16 QAM With a SE of 4 b/s/Hz, *J. Lightwave Technol.* **31** (2013) 515–522.

[23] H. Chien, J. Yu, Z. Jia, Z. Dong and X. Xiao, Performance assessment of noise-suppressed Nyquist-WDM for terabit superchannel transmission, *J. Lightwave Technol.* **30** (2012) 3965–3971.

[24] J. Yu, Z. Dong, H.-C. Chien, Z. Jia, D. Huo, H. Yi, M. Li, Z. Ren, N. Lu, L. Xie, K. Liu, X. Zhang, Y. Xia, Y. Cai, M. Gunkel, P. Wagner, H. Mayer and A. Schippel, Field trial Nyquist-WDM transmission of 8 × 216.4Gb/s PDM-CSRZ-QPSK exceeding 4b/s/Hz spectral efficiency, presented at the Opt. Fiber Commun. Conf./Nat. Fiber Opt. Eng. Conf., Los Angeles, CA (2012), paper PDP5D.3.

[25] J. Zhang, J. Yu, N. Chi, Z. Dong, J. Yu, X. Li, L. Tao and Y. Shao, Multi-modulus blind equalizations for coherent quadrature duobinary spectrum shaped PM-QPSK digital signal processing, *J. Lightwave Technol.* **31** (2013) 1073–1078.

[26] B. Huang, J. Zhang, J. Yu, Z. Dong, X. Li, H. Ou, N. Chi and W. Liu, Robust 9-QAM digital recovery for spectrum shaped coherent QPSK signal, *Opt. Express* **21** (2013) 7216–7221.

[27] J.-X. Cai, Y. Cai, C. R. Davidson, D. G. Foursa, A. J. Lucero, O. V. Sinkin, W. W. Patterson, A. N. Pilipetskii, G. Mohs and N. S. Bergano, Transmission of 96 × 100-Gb/s bandwidth-constrained PDM-RZ-QPSK channels with 300% spectral efficiency over 10610 km and 400%spectral efficiency over 4370 km, *J. Lightwave Technol.* **29**(4) (2011) 491–498.

[28] J. Yu, Z. Dong, H.-C. Chien, Z. Jia, D. Huo, H. Yi, M. Li, Z. Ren, N. Lu, L. Xie, K. Liu, X. Zhang, Y. Xia, Y. Cai, M. Gunkel, P. Wagner, H. Mayer and A. Schippel, Field trial Nyquist-WDM transmission of 8 × 216.4 Gb/s PDM-CSRZ-QPSK exceeding 4b/s/Hz spectral efficiency, presented at the

Opt. Fiber Commun. Conf./Nat. Fiber Opt. Eng. Conf., Los Angeles, CA (2012), paper PDP5D.3.

[29] R. Dischler, A. Klekamp, F. Buchali, W. Idler, E. Lach, A. Schippel, M. Schneiders, S. Vorbeck and R.-P. Braun, Transmission of 3 × 253-Gb/s OFDM-superchannels over 764 km field deployed single mode fibers, presented at the Opt. Fiber Commun. Conf./Nat. Fiber Opt. Eng. Conf., San Diego, CA (2010), paper PDPD2.

[30] X. Zhou, J. Yu and P. D. Magill, Cascaded two-modulus algorithm for blind polarization de-multiplexing of 114-Gb/s PDM-8-QAM optical signals, presented at the OFC2009, San Diego, CA (March 2009), paper OWG3.

[31] X. Zhou and J. Yu, Multi-level, multi-dimensional coding for high-speed and high-spectral-efficiency optical transmission, *J. Lightwave Technol.* **27** (2009) 3641–3653.

[32] S. J. Savory, G. Gavioli, R. I. Killey and P. Bayvel, Electronic compensation of chromatic dispersion using a digital coherent receiver, *Opt. Express* **15**(5) (2007) 2120–2126.

[33] S. J. Savory, Digital filters for coherent optical receivers, *Opt. Express* **16**(2) (2008) 804–817.

[34] I. Fatadin and S. J. Savory, Compensation of frequency offset for 16-QAM optical coherent systems using QPSK partitioning, *IEEE Photon. Technol. Lett.* **23**(17) (2011).

[35] Y. Gao, A. P. T. Lau, C. Lu, Y. Li, J. Wu, K. Xu, W. Li and J. Lin, Low-complexity two-stage carrier phase estimation for 16-QAM systems using QPSK partitioning and maximum likelihood detection, in *Proc. OFC/NFOEC*, Los Angeles, CA (2011), paper OMJ6.

[36] X. Xu, B. Chatelain and D. V. Plant, Decision directed least radius distance algorithm for blind equalization in a dual-polarization 16-QAM system, *Proc. OFC2012*, Los Angeles, CA (2012), paper OM2H.

[37] M. Selmi, Y. Jaouen and P. Ciblat, Accurate digital frequency offset estimator for coherent PolMux QAM transmission systems, *Proc. ECOC2009*, Vienna, Austria, (2009), paper P3.08.

[38] T. Pfau, S. Hoffmann and R. Noe, Hardware-efficient coherent digital receiver concept with feedforward carrier recovery for M-QAM constellations, *J. Lightwave Technol.* **27** (2009) 989–999.

[39] J. Yu, J. Zhang, Z. Dong, Z. Jia, H.-C. Chien, Y. Cai, X. Xiao and X. Li, Transmission of 8 × 480-Gb/s super-Nyquist-filtering 9-QAM-like signal at 100 GHz-grid over 5000-km SMF-28 and twenty-five 100 GHz-grid ROADMs, *Opt. Express* **21** (2013) 15686–15691.

[40] J. Zhang, Z. Dong, H. Chien, Z. Jia, Y. Xia and Y. Chen, Transmission of 20 × 440-Gb/s super-Nyquist-filtered signals over 3600 km based on single-carrier 110-GBaud PDM QPSK with 100-GHz Grid, in *Proc. OFC 2014*, paper Th5B.3.

[41] J. Zhang, J. Yu, Z. Dong, H.-C. Chien and Z. Jia, 10 × 440-Gb/s super-Nyquist-filtered signal transmission over 3000-km fiber and 10 cascaded ROADMs with 100-GHz grid based on single-carrier ETDM 110-GBaud QPSK, *Proc. ECOC 2014*, paper P. 5.16.

[42] J. Zhang, J. Yu, Z. Jia and H.-C. Chien, 400 G Transmission of super-Nyquist-filtered signal based on single-carrier 110-GBaud PDM QPSK with 100-GHz grid, *J. Lightwave Technol.* **32** (2014) 3239–3246.

[43] J. Zhang and J. Yu, Generation and transmission of high symbol rate single carrier electronically time-division multiplexing signals, *IEEE Photon. J.* **8**(2) (2016) 1–6.

[44] J. Zhang, J. Yu, B. Zhu, F. Li, H.-C. Chien, Z. Jia, Yi Cai, X. Li, X. Xiao, Y. Fang and Y. Wang, Transmission of single-carrier 400G signals (515.2-Gb/s) based on 128.8-GBaud PDM QPSK over 10,130- and 6,078-km terrestrial fiber links, *Opt. Express* **23**(13) (2015) 16540–16545.

[45] J. Yu, J. Zhang, Z. Jia, X. Li, H.-C. Chien, Y. Cai, F. Li, Y. Wang and X. Xiao, Transmission of 8 × 128.8 Gbaud Single-Carrier PDM-QPSK Signal over 2800 km SMF-28 with EDFA-only, *ECOC 2015*, #ID-243 (2015).

[46] J. Qi, B. Mao, N. Gonzalez, L. N. Binh and N. Stojanovic, Generation of 28GBaud and 32GBaud PDM-Nyquist-QPSK by a DAC with 11.3GHz analog bandwidth, *Proc. OFC 2013*, paper OTh1F.1.

[47] J. Zhang, J. Yu and N. Chi, Generation and transmission of 512-Gb/s quad-carrier digital super-Nyquist spectral shaped signal, *Opt. Express* **21** (2013) 31212–31217.

Chapter 3

Advanced DSP for Short-Haul and Access Network

3.1. Introduction

Digital signal processing (DSP) has been proved to be a successful technology recently in high speed and high spectrum-efficiency optical short-haul and access network, which enables high performances based on digital equalizations and compensations [1–16]. Based on different scenarios, DSP in optical communications systems is different to solve different problems. For high-speed optical transmission system based on high baud rate signals, the ISI caused by the bandwidth limitation of available device and components is the major problem [17–25]. In addition, with the development and maturation of high-speed digital-to-analog convertor (DAC), analog-to-digital converter (ADC) and application specific integrated circuit (ASIC), DSP for high spectrum-efficiency (SE) signal becomes possible, which moves the complexity of phase, frequency and polarization tracking into the digital domain [9–24]. It simplifies the reception of advanced modulation formats (i.e., QPSK, 16QAM, 64QAM) and also enables the major electrical and optical impairments (bandwidth limitation, chromatic dispersion, polarization mode dispersion, fiber nonlinear impairments) being processed and compensation in the digital domain, at the transmitter or receiver side [21–24]. Therefore, the advanced DSP for short haul and access network has become an active research topic, which is a promising technology for future high spectrum efficiency and high speed.

For short-haul network, the data rate of both line-side and client-side per port has been pushed to 400GE [1–9]. There are several solutions to increase the data rate to 400 Gb/s, including increasing the number of channels with wavelength-division multiplexing (WDM), using higher baud rate signals or increasing the modulation format orders [4–8]. Considering the implementation complexity, cost and power consumption, one should try to reduce the number of lanes as much as possible to achieve the highest economically feasible component density for optical interfaces [18–25]. Therefore, it leaves us with the last two ways: higher baud rate and high-order modulation (HOM) formats. On the other hand, due to the simple setup, low-cost and doubled spectrum efficiency (SE) compared with non-return-to-zero (NRZ) signal, the 4-level pulse amplitude modulation (PAM-4) has attracted a lot of research interest in both short-range optical communication (∼a few kilometers) and metro/regional networks (∼a few hundred kilometers) [4–6]. To further improve the data rate per lane, the baud rate of the PAM-4 signal has been increased year by year. Recently, 56 GBaud PAM-4 in intensity modulation and direct-detection (IM/DD) system has been demonstrated with 112-Gb/s line rate as the highest baud rate [1, 2]. To achieve 400GE, one solution is to increase the lane number, i.e., four-lane 112-Gb/s PAM-4 for 400GBase [4, 5], or two-lane 224-Gb/s polarization-division multiplexing (PDM) PAM-4 with multiple-input multiple-output (MIMO) receivers [6]. A 2 × 4 PDM/WDM monolithic VCSEL array for 400-Gb/s is also demonstrated adopting 25 GBaud PAM-4 for metro networks based on coherent detection [7–9]. Consequently, using these PAM-4 signals with baud rate less than 56 GBaud, it requires more than two lanes for 400GE solutions, even considering the polarization diversity. We believe continuously increasing the baud rate to achieve 400GE on a single-optical carrier PAM-4 for short-range optical communication is an attractive, promising and cost-effective solution for the next generation high-speed networks. On the other hand, the employment of PDM makes the direct-detection receiver also complex [5, 6] a better solution may be coherent detection with better performances and the similar hardware complexity, especially for metro and regional networks [5, 6].

The same trend can be found in the access network. To meet these capacity needs, access networks are moving from the classic spectral inefficient non-return-to-zero (NRZ) time-division multiplexing (TDM), to more advanced modulation formats [10–18]. Different techniques have been used for advanced modulation in short-range communication, such as orthogonal frequency division multiplexing (OFDM) [14, 15], and carrierless amplitude/phase (CAP) modulation [16–25] based on 16QAM, 64QAM and even higher order modulation. Using a higher level modulation format, which is a natural approach for high SE, leads to higher implementation penalty, higher receiver sensitivity requirement, and higher requirement for equalizations. DSP, again, shows its unequal advantages for handling these problems.

In this chapter, we introduce the advanced DSP at the transmitter side and the receiver side for signal pre-equalization and post-equalization in short haul and access optical communication network. A novel DSP-based digital and optical pre-equalization scheme has been proposed for bandwidth-limited high speed short-distance communication system, which is based on the feedback of receiver-side adaptive equalizers, such as least-mean-square (LMS) algorithm and constant or multi-modulus algorithms (CMA, MMA) [21-23].

This chapter is organized as follows. In Section 3.2, we show the idea of the advanced DSP for single-band CAP for short-haul data center interconnections by using high-order QAMs [21, 22]. Multi-band CAP is introduced in Section 3.3 for optical access networks [23]. The novel DSP-based digital and optical pre-equalization scheme is discussed and introduced in Section 3.3 with PAM-4 signals [25]. Finally, we summarize the whole chapter in Section 3.4.

3.2. Sing-Band CAP in Short-Haul Access Network

Considering the cost and complexity, intensity modulation and direction detection (IM/DD) with high-order modulation formats is a more practical and attractive method [1-12]. Using IM/DD, different techniques have been proposed, such as the quadrature amplitude modulation (QAM) subcarrier modulation (SCM) [1, 2], pulse amplitude modulation (PAM) [3], the discrete multi-tone (DMT) or orthogonal frequency division multiplexing (OFDM) [4, 5], and the CAP modulation [6-14].

Overall, the CAP architecture based on IM/DD has been demonstrated to be less complex and with good performance, which allows relatively high data using low cost optical components such as directly modulated laser (DML) and vertical cavity surface emitting laser (VCSEL) and optical and electrical components of limited bandwidth [6-14]. No electrical complex-to-real-value conversion, complex mixer, radio frequency (RF) source or optical in-phase/quadrature (I/Q) modulator are required for CAP, compared with QAM-SCM [1, 2] and OFDM [4, 5]. Neither does it require the discrete Fourier transform (DFT) that is utilized in OFDM signal generation and demodulation [12]. In [6], authors have proved that

CAP has great potential of being with high power efficiency with low cost compared with PAM or OFDM for short-range optical transmission. Higher spectrum efficiency can be obtained for CAP compared with the ordinary full-band QAM-SCM and PAM in [1–3]. A number of optical communication systems based on CAP have been demonstrated recently [7–14]. In [10], multi-band CAP-16QAM has been proposed to extend the bandwidth for high speed short-range data transmission. In [14], systems based on CAP-16QAM and CAP-64QAM are proposed but the bit rate is only 2 Gb/s and 2.1 Gb/s, respectively. In [12], the digital equalizer based on cascaded multi-modulus algorithm (CMMA) has been proposed for CAP-16QAM ISI equalization with good performance. However, the higher level modulation format CAP system, such as CAP-64QAM with tens of Gb/s data rate has not been demonstrated. The performance of digital equalization for CAP-64QAM with tens of Gb/s has also not been investigated.

3.2.1. *The principles of single-band CAP*

Figure 3.1 shows the schematic diagrams of transmitter and receiver based on CAP m-QAM for fiber-wireless transmission. CAP is a multi-level and multi-dimensional modulation format proposed by Bell Labs for short-range communication [26], which is similar to QAM signal, but does not require an RF carrier source. Different from QAM or OFDM intensity modulation, CAP does not use a sinusoidal carrier to generate two orthogonal

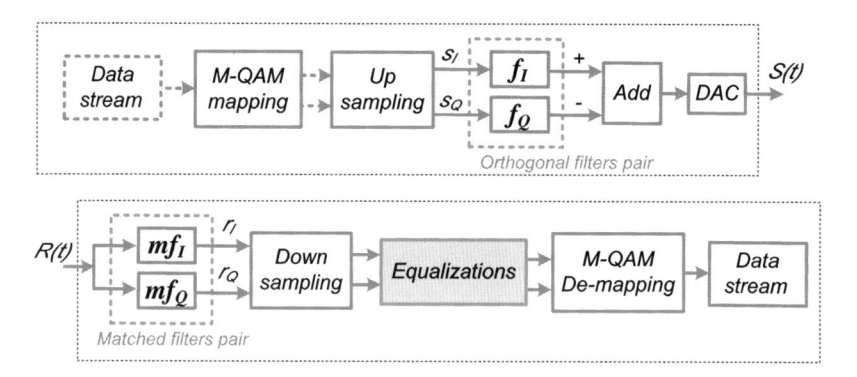

Fig. 3.1. Schematic diagrams of transmitter and receiver based on CAP m-QAM for CAP signal generation and processing.

components I and Q. The two-dimensional CAP can be generated by using two orthogonal filters, f_I and f_Q, as the filter pair (see Fig. 3.1). The original bit sequence is first mapped into complex symbols of m-QAM (m is the order of QAM), and then the mapped symbols are up-sampled to match the sample rate of shaping filters. The sample rate of shaping filters is determined by the data baud rate and DAC sample rate. For CAP generation, the I and Q components of the up-sampled sequence are separated and sent into the digital shaping filters, respectively. The outputs of the filters are subtracted to be combined together as $S(t)$ after DAC to drive the upper MZM$_1$. At the receiver side, the received signal after down-conversion is fed into two matched filters to separate the I and Q components.

The orthogonal and matched filter pairs $f_I(t), f_Q(t), mf_I(t)$ and $mf_Q(t)$ are the corresponding shaping filters and form the so-called Hilbert pair in the transmitter and receiver. The two orthogonal filters are constructed by multiplying a square root raised cosine pulse with a sine and cosine function, respectively, as described in [21–26]. Assuming $s_I(t)$ and $s_Q(t)$ are the I and Q data after QAM mapping and up-sampling, then the combined output signal $S(t)$ of CAP signal can be expressed as

$$S(t) = [s_I(t) \otimes f_I(t) - s_Q(t) \otimes f_Q(t)] \quad (3.1)$$

At the receiver, generally, we have the matched filters with relations as $mf_I^n(t) = f_I^n(-t)$, and $mf_Q^n(t) = f_Q^n(-t)$. In this way, for the CAP receiver, the I and Q data after matched filter pair can be expressed as follows:

$$r_I(t) = R(t) \otimes mf_I(t), \quad r_Q(t) = R(t) \otimes mf_Q(t) \quad (3.2)$$

where $R(t)$ is the CAP signal after down-conversion at the receiver side and $r_I(t)$ and $r_Q(t)$ are the outputs of the matched filter pair. After down-sampling, a linear equalizer is employed for the complex signal and a decoder is utilized to obtain the original bit sequence.

3.2.2. CAP for data-center interconnection and DSP

Since the appropriate sampling time is hard to decide, the sampling time offsets will lead to subsequent signals seriously affected by ISI and the crosstalk between the in-phase and quadrature components. One phase rotation is also produced by the crosstalk. Therefore, after down-sampling, a linear equalizer is employed for the complex signal and a decoder is utilized to obtain the original bit sequence. In the system, the orthogonal filters and matched filters are realized by digital finite impulse response

(FIR) filters with a tap length of transmitter orthogonal filter length (T-OFL) and the receiver-matched filter length (R-MFL), respectively. As analyzed in [20], the tap length of the FIR filters determines the filter time-domain pulse shape and frequency response. The impact of tap length of these FIR filters on the system performance is studied in the experiment.

In [20], they use two-stage equalizations, ISI equalization and phase recovery (PR) to equalize the CAP signal. The ISI equalization is performed using a CMA for pre-convergence followed by the CMMA. However, it is difficult for high-order CAP-QAM signal equalization using CMMA, since the ring spacing in QAM is generally smaller than the minimum symbol spacing. It has been proved that DD-LMS can achieve better signal-to-noise ratio (SNR) performance than CMMA for high-order QAMs [27]. On the other hand, since CMMA is based on the radius of the circles of the symbol, it is a phase-independent algorithm. In this way, additional phase recovery is required following the CMMA to equalize the crosstalk.

Here, we introduce a one-stage equalizer based on DD-LMS after CMA pre-convergence for CAP-QAM signal ISI and crosstalk equalization. Figures 3.2(a) and 3.2(b) show the structure and principle of DD-LMS algorithm. Different from the DD-LMS used in coherent optical system based on four complex-valued FIR filters, a four-real-valued $T/2$-spaced butterfly configured adaptive digital FIR filter structure is used for CAP signal equalization as shown in Fig. 3.2(a). $Z_I(n)$ and $Z_Q(n)$ are the nth outputs of the filters as the equalized in-phase and quadrature signals. $D_I(n)$ and $D_Q(n)$ are the decision results of the in-phase and quadrature signals. Although the in-phase and quadrature signals are input independently, the outputs are dependent on both inputs. The error function of DD-LMS can be expressed as

$$e_{I,Q}(n) = D_{I,Q}(n) - Z_{I,Q}(n) \tag{3.3}$$

where $e_I(n)$ and $e_Q(n)$ are the error functions for in-phase and quadrature signals. The four real-valued FIR filters h_{ii}, h_{iq}, h_{qi} and h_{qq} are updated by the error function after decision as

$$h_{ii}(n) = h_{ii}(n-1) + \mu e_I(n) r_I(n) \tag{3.4}$$

$$h_{qi}(n) = h_{qi}(n-1) + \mu e_I(n) r_Q(n) \tag{3.5}$$

$$h_{iq}(n) = h_{iq}(n-1) + \mu e_Q(n) r_I(n) \tag{3.6}$$

$$h_{qq}(n) = h_{qq}(n-1) + \mu e_Q(n) r_Q(n) \tag{3.7}$$

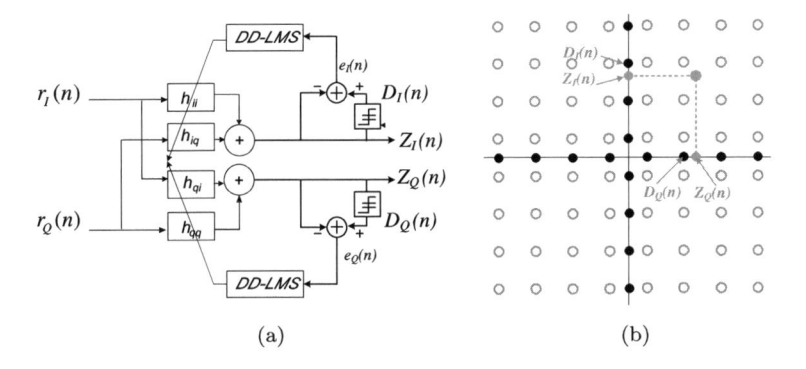

Fig. 3.2. (a) The structure of DD-LMS for CAP-mQAM signal and (b) the principle.

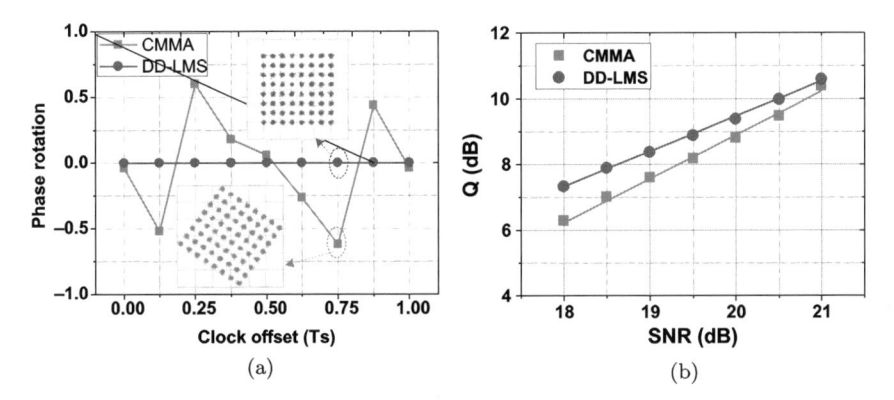

Fig. 3.3. Simulation results of (a) the phase rotation after different equalization schemes versus the clock offset and (b) the Q performance versus different SNR for CAP-64QAM under different equalization schemes.

In this way, both the ISI and crosstalk between in-phase and quadrature signals can be removed.

The two benefits of using DD-LMS for CAP-mQAM signal equalization are investigated as shown in Fig. 3.3 with simulation results of CAP-64QAM signal. Figure 3.3(a) shows the simulation results of the phase rotation results after CMMA and DD-LMS versus the clock offset. The up-sampling rate here is 8 Sa/symbol. We can see that the phase rotation caused by clock offset cannot be handled by CMMA, which requires the additional phase recovery equalization after CMMA. However, the phase is recovered by DD-LMS after equalization without any phase rotation since the DD-LMS is phase-sensitive. Figure 3.3(b) shows the Q performance versus different

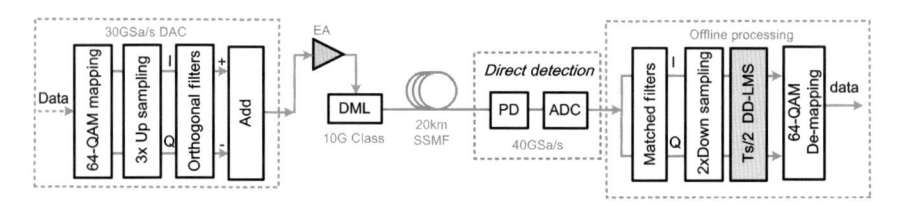

58 *Digital Signal Processing for High-Speed Optical Communication*

Fig. 3.4. Experimental setup of 60-Gb/s CAP-64QAM transmission based on 10G class DML with direct detection and DSP.

SNR for CAP-64QAM under different equalization schemes. We can see that DD-LMS shows better Q performance for CAP-64QAM compared with CMMA, especially under low SNR conditions. It is because the ring spacing in QAM is generally smaller than the minimum symbol spacing. For CAP-64QAM, the error function for DD-LMS is based on symbol spacing, while that for the CMMA is based on ring spacing.

The experimental results from [22] demonstrate the benefit of above advanced equalizations. Figure 3.4 shows the experimental setup of 60-Gb/s CAP-64QAM generation, transmission and detection based on 10 G Class DML with direct detection and DSP. The DML is operated at 1295.14 nm as the carrier source and driven by the CAP signal. The 10-Gbaud CAP-64QAM signal is generated by the programmable 30-GSa/s DAC. The data sequence is first mapped to the 8-level 64-QAM with 16×2 [11] symbols. Then, the 8-level in-phase and quadrature data are up-sampled to 3 Sa/symbol and filtered by the orthogonal Hilbert filter pairs. The filters are FIR filters with length of 10 symbols each. The roll-off coefficient is 0.2 and the excess bandwidth is set to 15%. In this way, the 10-Gbaud baseband CAP-64QAM signal is generated by the DAC. The output power of DML is 8 dBm. After modulated by the DML, the optical signal is transmitted over 20-km SMF.

At the receiver side, the signal is directly detected by a PD and then sampled by a digital scope at 40 GSa/s for offline processing. The sampled signal is first demodulated by the two orthogonal matched filters. After the matched filters, the I and Q signals are then down-sampled to 2-Sa/symbol before the linear equalization. Here, a four-real-valued, 31-tap, $T/2$-spaced butterfly configured adaptive digital FIR filter structure, based on modified DD-LMS after CMA pre-convergence, is used for signal equalization and recovery. The BER is measured after signal de-mapping.

Figure 3.5(a) shows the BTB BER performance versus the DD-LMS filter tap length under different transmitter-orthogonal filter length (T-OFL)

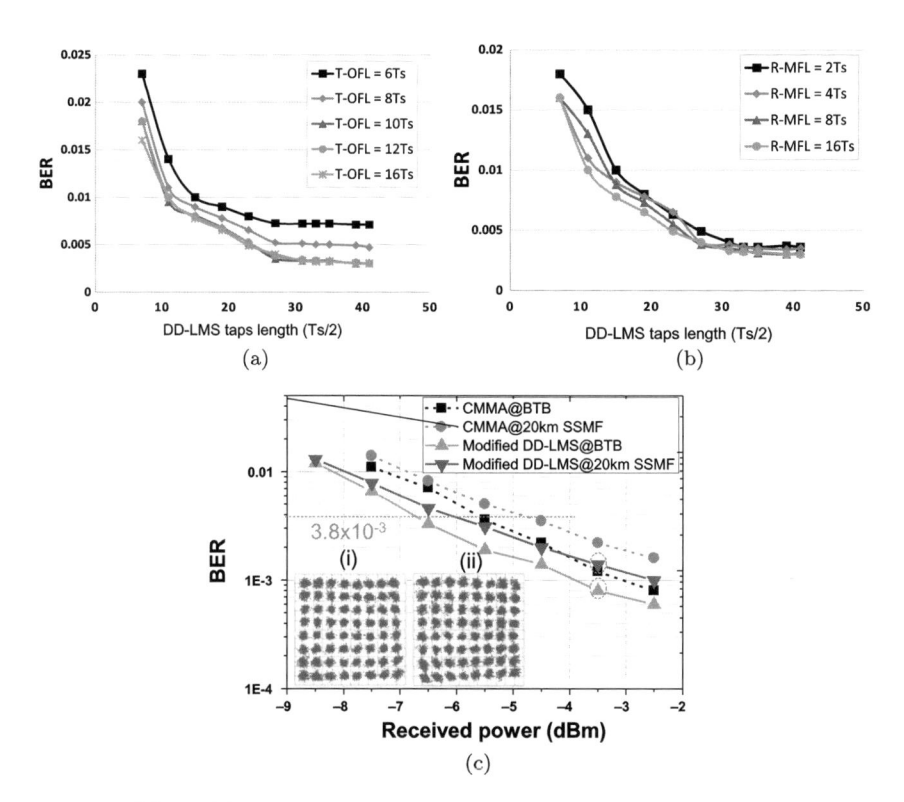

Fig. 3.5. (a) and (b) The BTB BER performances versus the DD-LMS filter tap length under different T-OFL and R-MFL. (c) The BER curves of 60-Gb/s CAM-64QAM signal versus the received optical power before and after 20-km SMF.

for CAP-64QAM signal generation. We can see that larger T-OFL has better BER performance. The optimal filter length for CAP-64QAM generation is about 10Ts. Figure 3.5(b) shows the BER curve with the DD-LMS filter tap length under different matched filter length at the receiver (R-MFL). We can see that the R-MFL has less impact on the receiving BER performance. Both Figs. 3.5(a) and 3.5(b) show that the BER is improved with the increasing of DD-LMS tap length within the range of 1–41. The optimal tap length for DD-LMS is about 31. Here, the received optical power is kept at −6.5 dBm. Figure 3.5(c) shows the BER performance of 60-Gb/s CAM-64QAM signal versus the received optical power before and after 20-km SMF using different equalization methods. We can see that for both back-to-back (BTB) and 20-km SMF transmission cases, the DD-LMS equalization shows better performances compared to the CMMA + PR method. About 1.5-dB and 2.5-dB received power sensitivity improvement can be obtained for BTB and 20-km SMF transmission case at the BER of 3.8×10^{-3}, respectively. The constellations of the processed CAP-64QAM signal at -3.5-dBm received power before and after 20-km SMF transmission using DD-LMS are shown in the insets (a) and (b) of Fig. 3.5(c), respectively. These results clearly demonstrate the feasibility of the proposed CAP-64QAM system based on DML and direct detection with modified DD-LMS for equalization.

3.3. Multi-Band CAP for Access Networks

3.3.1. *Principles of multi-band CAP for access network*

Figure 3.6 shows the principle of downstream for WDM-CAP-PON based on MM-CAP. In the central office, each optical line terminal (OLT) transmitter (Tx) at the ith wavelength λi carries the MM-CAP signals for multiple users. For multi-band CAP, more than one filter pairs are used, which are located on different frequencies.

At the optical network unit (ONU) side, each user can recover the data in each sub-band using a pair of matched filters corresponding to the pulse shaping filters in the transmitter. Only one sub-band signal can be recovered by one matched filter pair. In this way, for each channel, N sub-bands can be assigned to N users without any interference. For WDM-CAP-PON, K wavelengths with N sub-bands can totally be assigned to $K \times N$ users. On the other hand, when considering long distance transmission, OSSB can be used against the spectrum fading effect caused by CD and direct detection, and enables linear equalizations as in [23]. Here, only intensity modulation and direct detection are needed.

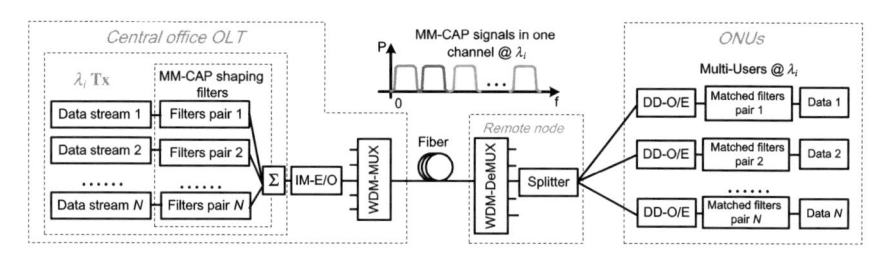

Fig. 3.6. Principle of WDM-CAP-PON based on MM-CAP signal generation and transmission for multi-user access network (IM: intensity modulation; DD: direct detection, λi is the ith wavelength in the WDM-PON).

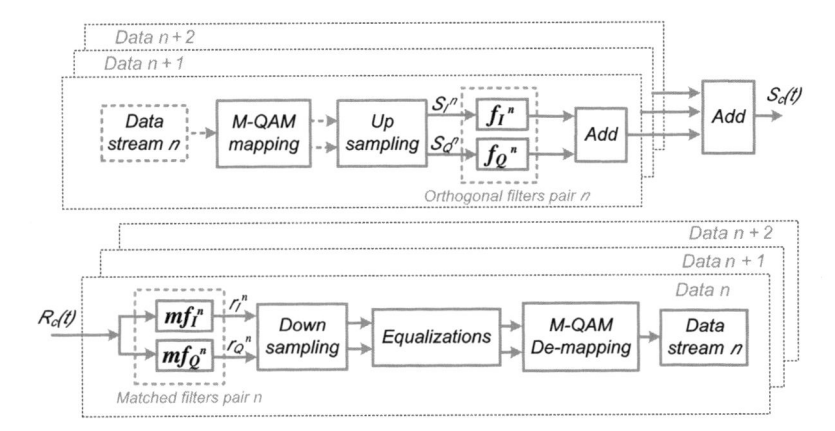

Fig. 3.7. Schematic diagram of transmitter and receiver based on multi-band CAP for one data stream (data n is the data transmitted in the nth sub-band).

This scheme with WDM-CAP-PON for downstream is compatible with the other PON structures for the upstream. There are mainly two schemes for the upstream. Since the bandwidth requirement for the upstream is much less than the downstream, OOK can be directly used with lower data rate.

Figure 3.7 shows the schematic diagram of transmitter and receiver based on multi-band CAP for one data stream. The original bit sequence of a data stream n is first mapped into complex symbols of m-QAM (m is the order of QAM), and then the mapped symbols are up-sampled to match the sample rate of shaping filters. The sample rate of shaping filters is determined by the data baud rate and DAC sample rate. For CAP generation, the in-phase and quadrature components of the up-sampled sequence are separated and sent into the digital shaping filters, respectively. The outputs of the filters are subtracted. For multiple users, the transmitter data for each data steam after the orthogonal filter pairs can be added together before the digital-to-analog converter (DAC). At the ONU side, the received signal after analog-to-digital converter (ADC) is fed into two different matched filters to separate the in-phase and quadrature components. After down-sampling, an equalizer is employed for the complex signal and a decoder is utilized to obtain the original bit sequence. Different and unique matched filter pairs are used for multiple users.

The orthogonal and matched filter pairs $f_I^n(t), f_Q^n(t), mf_I^n(t)$ and $mf_Q^n(t)$ are the corresponding shaping filters and form the so-called Hilbert pair in the transmitter and receiver, as described in [23]. Different from the single band CAP in [20], a multi-band CAP uses multi-filter pairs which are located in different frequency sub-bands [23]. For each sub-band, the two orthogonal filters are constructed by multiplying a square root raised cosine pulse with a sine and cosine function, respectively, as shown in Eqs. (3.1) and (3.2). For MM-CAP with N sub-bands, the orthogonal filter pair of the nth$(1-N)$ sub-band in time domain can be expressed as follows:

$$f_I^n(t) = \frac{\sin[\pi(1-\beta)\frac{t}{T_s}] + 4\beta\frac{t}{T_s}\cos[\pi\frac{t}{T_s}(1+\beta)]}{\pi\frac{t}{T_s}[1-(4\beta\frac{t}{T_s})^2]} \sin\left[\pi(2n-1)(1+\beta)\frac{t}{T_s}\right] \quad (3.8)$$

$$f_Q^n(t) = \frac{\sin[\pi(1-\beta)\frac{t}{T_s}] + 4\beta\frac{t}{T_s}\cos[\pi\frac{t}{T_s}(1+\beta)]}{\pi\frac{t}{T_s}[1-(4\beta\frac{t}{T_s})^2]} \cos\left[\pi(2n-1)(1+\beta)\frac{t}{T_s}\right] \quad (3.9)$$

where T_s is the symbol duration and also the reciprocal of the symbol-rate, β is the roll-off factor generally between 0 and 1.

Taking four sub-band CAP filters as an example, Fig. 3.8 shows the time impulse response and frequency response of these filter pairs of $f_I^n(t)$, $f_Q^n(t)$, for different sub-bands. Here, the up-sample rate is 10, and the roll-off factor is 0.2. We can see that the filter pair for different sub-band has a different waveform in time domain. In the frequency domain, these filter pairs are located in different sub-bands, which can be assigned to different users.

3.3.2. *Experimental demonstration of WDM-CAP-PON*

As a proof of concept, the experimental setup of $11 \times 5 \times 10$-Gb/s WDM-CAP-PON downstream with 55 users over 40-km SMF is shown in Fig. 3.9. Eleven carriers as the source of 11 channels are generated by an optical comb generator [23]. Odd and even channels are separated by a two-port 28GHz-gird WSS and then intensity-modulated by two MZMs. The 5-band MM-CAP-16 signals are generated by the high-speed DAC, which works at 30GSa/s. Five different data sequences are generated and first mapped to the 4-level 16-QAM I and Q signals with 10×2^{11} symbols. Then, ten sets of 4-level sequences are up-sampled to 12 Sa/symbol and filtered by five pairs of shaping MM-CAP filters to generate the MM-CAP signal with five sub-bands. The filters are FIR filters with length of eight symbols each. The roll-off coefficient is 0.2 and the excess bandwidth is set to 15%. The 50-Gb/s MM-CAP-16 signal is generated by simply adding the outputs of

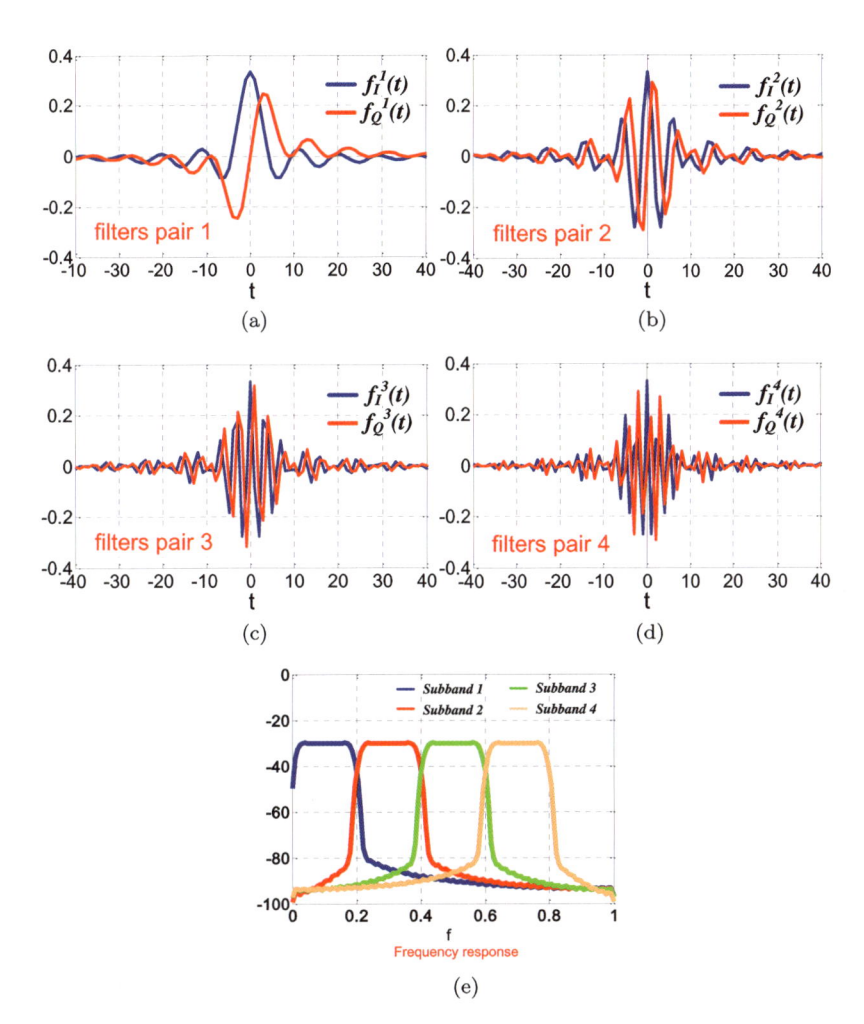

Fig. 3.8. (a)–(d) The time-domain impulse response of different filter pairs 1–4. (e) The frequency response of the four filter pairs located in different sub-bands.

the five filter pairs. It is worth noting that the frequency response of output of the DAC is non-flat with a 3-dB bandwidth less than 13-GHz. Insets (a) and (b) in Fig. 3.9 show the spectrum of the output of DAC with and without frequency-domain pre-equalization, respectively. We can see that, by simply adding weights of each pairs of filters, we can get a more identical frequency response for each sub-band CAP signal. The odd and even channels after MZMs are then filtered to produce the OSSB signals and combined together by a 28GHz-spaced WSS with 14 GHz frequency offset and

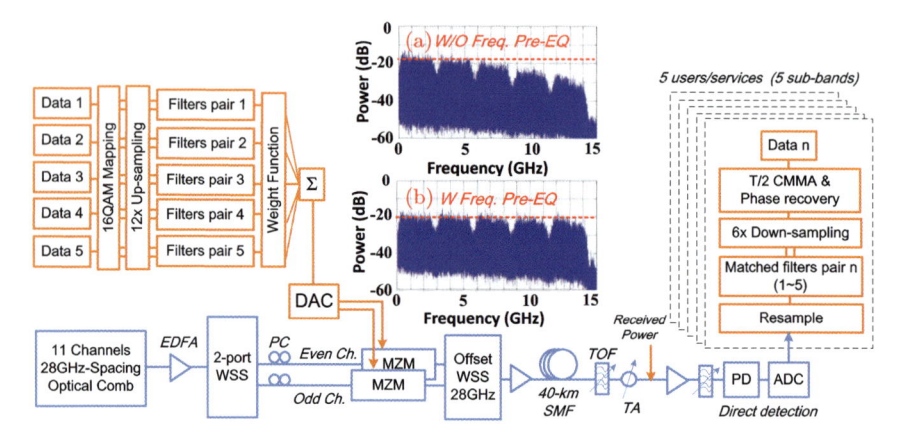

Fig. 3.9. Experimental setup (WSS: wavelength selective switch; MZM: Mach–Zehnder Modulator; TOF: tunable optical filter; TA: tunable attenuator).

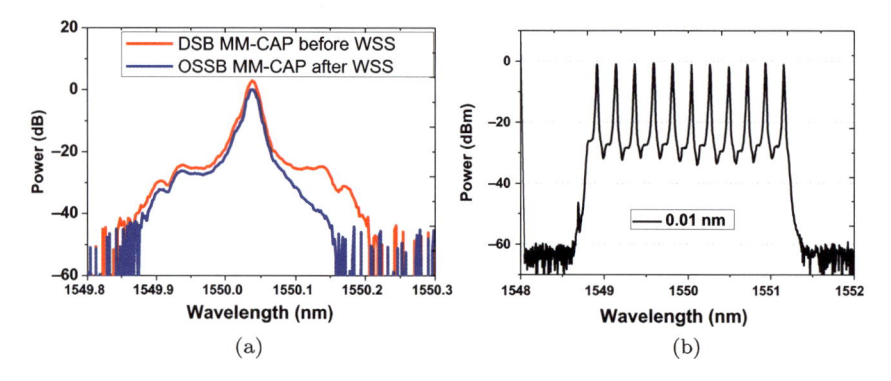

Fig. 3.10. The optical spectrum of (a) single channel and (b) 11-channel WDM OSSB MM-CAP signals.

20 GHz bandwidth. The optical spectrum of single channel and 11-channel WDM OSSB MM-CAP signals is shown as (a) and (b) in Fig. 3.10. The 11 WDM channel signals are transmitted over 40-km SMF with loss of 10 dB.

At the receiver side, the channel is first filtered out by a tunable filter (TOF) with 0.3-nm bandwidth. After direct detection, the signals are sampled by a digital scope at 50-GSa/s for offline processing. A 0.9-nm TOF is used before PD front-end to remove excess ASE noise. The signals are first resampled to 30GSa/s and then demodulated by the assigned matched-filter pairs. Each sub-band CAP is processed by different matched-filters for data recovery. After that, the signals are down-sampled to 2Sa/Symbol before

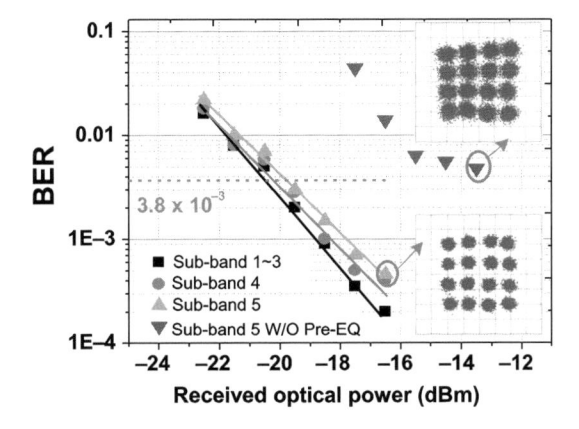

Fig. 3.11. The back-to-back BER results versus receiver optical power for each sub-band.

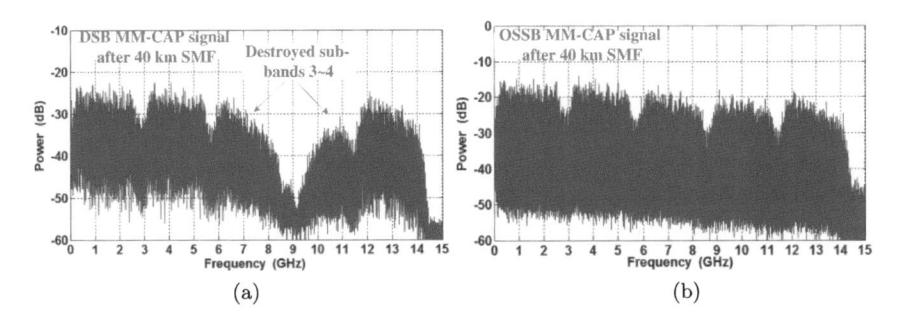

Fig. 3.12. Spectrum of MM-CAP after 40-km SMF for DSB and OSSB signals.

CMMA and phase recovery. Each sub-band is with 2.5 GHz bandwidth and 10-Gb/s data rate, thus the total downstream data rate is 550-Gb/s for 55 users. Figure 3.11 shows the back-to-back (BTB) bit error ratio (BER) as a function of the received optical power for each sub-band of channel 4. We can see that, with frequency-domain pre-equalization by adding weights on each sub-band, identical BER performance for sub-band 1–3 and negligible power penalty for sub-band 4 and 5 at BER of 3.8×10^{-3} are obtained. We also measure the BER of sub-band 5 without pre-equalization, which shows very poor performance.

Figures 3.12(a) and 3.12(b) show the spectrum of the received MM-CAP signal after 40-km SMF for double sideband (DSB) and OSSB signals, respectively. We can see that, without OSSB, sub-bands 3 and 4 are

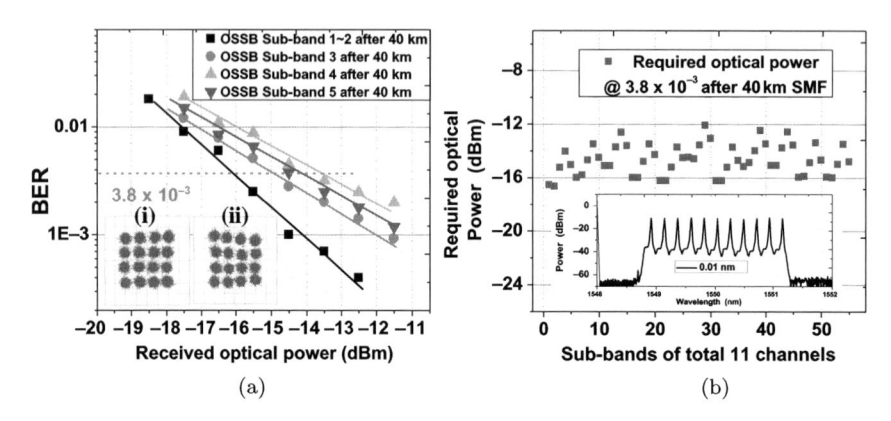

Fig. 3.13. (a) The BER of each sub-band in channel 4 versus the received optical power after 40-km transmission and (b) The required optical power for total 55 sub-bands in 11 channels at the BER of 3.8×10^3 after 40 km SMF transmission.

destroyed due to the power fading effect caused by CD and direct detection. Almost 40% data transmission is cut-off. However, the power fading can be avoided by OSSB, as shown in Fig. 3.12(b).

Figure 3.13(a) shows the measured BER of each sub-band in channel 4 versus the received optical power after 40-km SMF transmission. Insets (i) and (ii) show the constellation of sub-bands 1 and 5 at the received power of -15dBm. Sub-bands 1 and 2 have identical performance with about 4.5-dB power penalty compared with the BTB case. However, sub-bands 3–5 have larger power penalty of about 5.5-dB, 6.5-dB and 6-dB, respectively. It is due to the residual power fading effect under the imperfect OSSB filtering. Figure 3.13(b) shows the required optical power for the total 55 sub-bands in 11 channels at the BER of 3.8×10^{-3} after 40-km SMF transmission. The downstream data rate is 50-Gb/s/ch, and totally $11 \times 5 \times 10$Gb/s for 55 users. Inset shows the optical spectrum of the received 11 channels.

3.4. PAM-4 for High-speed Short-Haul Transmission

In this section, we introduce the PAM-4 for high-speed short-haul transmission. An experiment from [25] is shown and analyzed here. The transmitter and receiver setup of the ETDM 120-GBaud PDM-PAM-4 signal with coherent detection is shown in Fig. 3.14. The 120-Gbaud PAM-4 signal is generated through three stages of ETDM with a combination of two

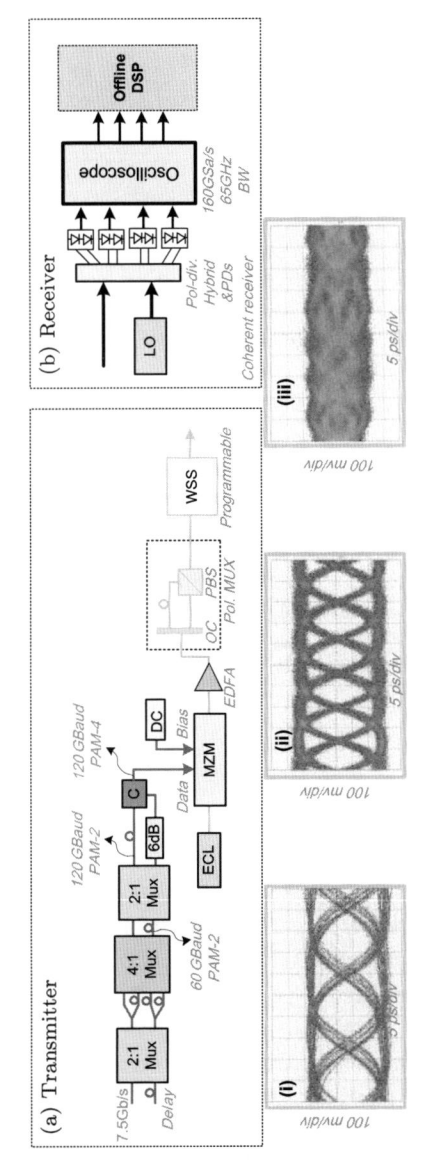

Fig. 3.14. The experimental setup of 120-GBaud PDM-PAM-4 signal generation and coherent detection (ECL: external cavity laser; Mux: time-division multiplexer; MZM: Mach–Zehnder modulator; Pol. MUX: polarization multiplexer; C: coupler; OC: optical coupler; eye-diagrams are measured in 5 ps/div). Insets (i), (ii) and (iii) are the generated ETDM 60-GBaud NRZ signal, 120-GBaud NRZ signal and 120-GBaud PAM-4 signal, respectively.

120-GBaud NRZ signals. The first stage multiplexed signal generates the 15-Gb/s 2-level signal by a 2:1 multiplexer from the 2-delay de-correlated pseudo-random binary sequence (PRBS) signals with a word length of 215-1, and then the 60-GBaud NRZ signal is generated after the second stage ETDM using a 4:1 broadband multiplexer, as shown in inset (i) in Fig. 3.14. After that, another 2:1 multiplexer is used for the final 120-GBaud NRZ signal generation, as shown in inset (ii) in Fig. 3.14. Here, the 4:1 MUX is a 56-Gb/s 4:1 broadband multiplexer module and 2:1 MUX is a 120-Gb/s 2:1 broadband multiplexer module. In our case, the 4:1 MUX works well with 60 Gb/s output and the 2:1 MUX also works well with 120 Gb/s output. The output of the 4:1 MUX has a 500 mv Vpp and the output of 2:1 MUX has a 400 mv Vpp. The 4-level PAM-4 signal is then generated by an electrical combiner from two de-correlated 120-Gb/s NRZ signals and one of NRZ signal is first reduced to half of the amplitude by a 6-dB attenuator. The generated 120-GBaud PAM-4 signal after the 2:1 MUX is shown in inset (iii) in Fig. 3.14. For data decorrelation, the delay between the two 60-Gb/s data sequences is about 60 bits for multiplexing and the delay between the two 120-Gb/s data sequences is about 40 bits for modulation.

For optical PAM-4 signal modulation, one external cavity laser (ECL) is used as the light source with the wavelength at 1549.44 nm, the linewidth less than 100 kHz and the output power of 13.5 dBm. The Mach–Zehnder modulator (MZM) with a 3-dB bandwidth of 34.1 GHz is driven by the generated PAM-4 signal without using an electrical amplifier. Since coherent detection is used at the receiver side, the MZM is biased at null point. After modulation, one erbium-doped fiber amplifier (EDFA) is used to boost the optical power before the polarization multiplexing. The polarization multiplexing is realized via the polarization-multiplexer, which consists of a polarization-maintaining optical coupler (PM-OC), an optical delay line to provide over 100 symbols delay, and a polarization beam combiner (PBC) to recombine the signals. Due to the cascaded bandwidth narrowing effect caused by the bandwidth limitation of optic-electronic devices, the system performance is seriously degraded by ISI, noise and inter-channel crosstalk enhancement.

To overcome the filtering effect caused by bandwidth limitation, we use a programmable wavelength-selective switch (WSS) to perform the transmitter-side pre-equalization by enhancing the power of high-frequency components while attenuating the low-frequency components [28–31]. Here, we use a scheme similar to the one we proposed for digital pre-equalization based on receiver-side channel estimation [31]. Namely, we can obtain the channel inverse as the function of taps number, which is also the frequency value. Although the optical pre-equalization is done offline in a manual manner in this experiment, the whole process can be adaptively optimized at the startup calibration since the optical filter is fully programmable.

At the receiver side, an ECL with a linewidth less than 100 kHz and output power of 13.5 dBm is utilized as the local oscillator (LO). A polarization-diverse 90° hybrid is used for polarization- and phase-diversity coherent detection. The bandwidth of the balanced detector is 50 GHz. The sampling and digitization (A/D) is realized by the real-time digital oscilloscopes with 160-GSa/s sample rate and 65-GHz electrical bandwidth. After the ADC with an oversampling rate of about 1.33 for the 120 GBaud PDM-PAM-4 signals, the offline DSP is then applied to four-channel 160GSa/s sampled data sequence. The data is first resampled to 240GSa/s and then processed by adaptive equalizers for channel estimation, as shown in Fig. 3.15(a). Since the constellation of the PAM-4 signals shows two moduli, the $T/2$-tap spaced CMMA [32] is used for pre-convergence with 21 taps, and then we use the $T/2$-tap spaced DD-LMS algorithm also with 21 taps to further improve the PAM-4 results. It is worth noting that the phase rotation caused by laser linewidth and frequency offset are still introduced to the received signals after coherent detection. And in the DD-LMS loop, phase rotation caused by laser frequency offset and phase noise is removed by a

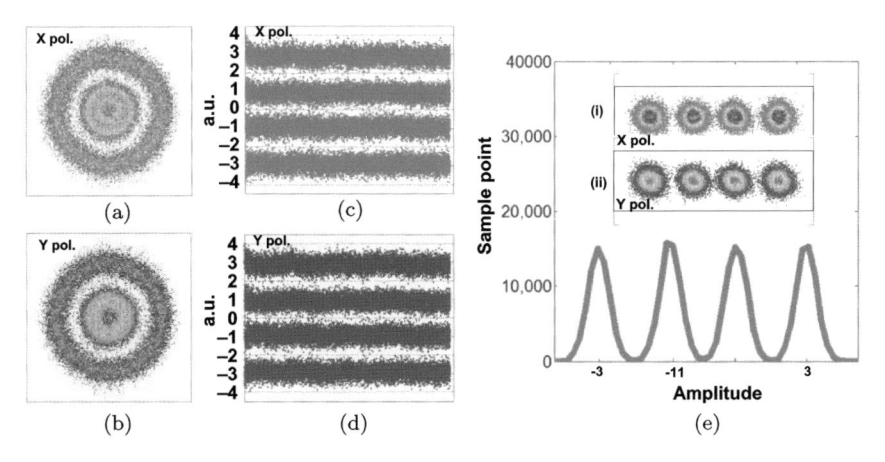

Fig. 3.15. The receiver-side signal processing results: (a) and (b) are the constellations of recovered X- and Y-pol. signals after CMMA; (c) and (d) are the recovered X- and Y-pol. signals after DD-LMS and (e) the histogram recovered and the constellations of X and Y pol. signals.

feed-forward Mth power carrier recovery algorithm [32]. The symbol decision for DD-LMS taps updating is performed after the carrier recovery. The channel estimation is redone considering the fiber length using digital chromatic compensation before CMMA [33]. Since the symbol rate is more than 100 GBaud, the signal is very sensitive to the fiber length due to the dispersion.

The receiver-side DSP results are shown in Fig. 3.15. The 120-GBaud PDM-PAM-4 measured signal with OSNR larger than 30-dB is used here for processing. Figures 3.15(a) and 3.15(b) show the constellations of X- and Y-polarization signals after the first stage CMMA processing. We can see that the signals with different amplitudes are separated by CMMA. After that, a second stage DD-LMS is used to further improve the performances. Figures 3.15(c) and 3.15(d) show the PAM-4 signals recovered after DD-LMS with clear amplitude distribution. We also plot the histogram of the recovered signals as shown in Fig. 3.15(e) and the symbol statistics is checked following a Gaussian distribution. The probability density of the four levels in the generated PAM signal is the same. Insets (i) and (ii) show the constellation of the 120-GBaud PAM-4 signals after DSP for X- and Y-pol.

In order to overcome filtering effect and avoid noise enhancement by receiver-side adaptive equalizations, they use the transmitter-side optical pre-equalization, as mentioned above. The back-to-back BER performance of WSS-based pre-equalization scheme with the pre-emphasis strength is shown in Fig. 3.16(a). Here, they use the alpha factor to adjust the pre-emphases strength; no pre-equalization when alpha is 0, and full pre-equalization when alpha is 1. The frequency response of full pre-equalization is obtained based on the receiver-side channel estimation. Similar setup and analysis can be found in the previous work [33]. The optimal alpha factor for the polarization multiplexed PAM-4 signal is ∼0.8. The BER performance gets worse when alpha is less than 0.8 due to the ISI penalty and it slightly degrades due to the over compensation. Figures 3.16(b) and 3.16(c) are the optical spectra of the 120-GBaud PAM-4 signal without (alpha is 0) and with optical-pre-equalization (alpha is 0.8), respectively. There is an obvious enhancement of the high frequency components after the receiver-side optical pre-equalization.

Finally, Fig. 3.17 shows the BER results versus the OSNR. A large error-floor exists when there is no optical pre-equalization, and larger than 5-dB OSNR improvements are observed for the 120-GBaud PAM-4 signals with pre-equalization. Here, the pre-emphasis strength is kept at 0.8.

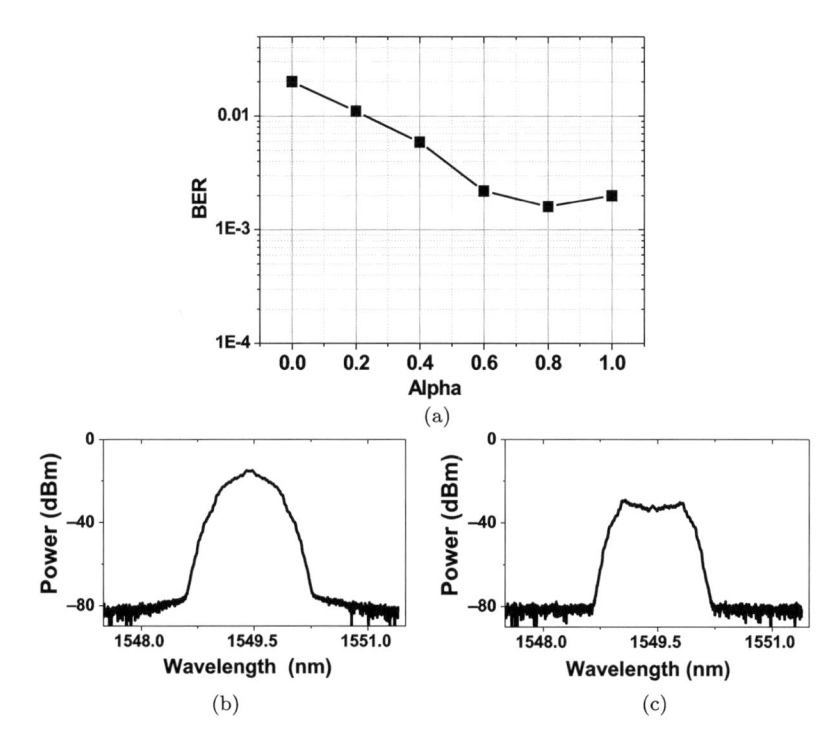

Fig. 3.16. (a) The back-to-back BER at 29.5-dB OSNR with different pre-equalization strengths; (b) and (c) are the optical spectra of the 120-GBaud PAM-4 signal without and with transmitter-side optical-pre-equalization, respectively. The resolution of the optical spectra is 0.1 nm.

Assuming 20% SD-FEC, the required OSNR is about 24.8-dB at the BER of 2×10^{-2}, and the 120-GBaud PDM-PAM-4 signals can provide a net information rate of 400 Gb/s. They also measured the BER performances of the 120-GBaud PDM-PAM-4 signals after 20-km SMF transmission versus the OSNR. Negligible OSNR penalty is observed compared with the BTB case. Considering the high OSNR achieved at the transmitter-side, there is still large margin for both short range and metro/regional networks. In this experiment, the insertion loss of WSS for optical pre-equalization is more than 17-dB. Therefore, optical amplifier is required. If large bandwidth optical or electrical devices are used without pre-equalization, then it can be realized without the amplifier. Since the optical amplifier is used, it shows the results with OSNR for coherent detection.

The theoretical BER curves of 120-GBaud PDM-QPSK, 120-GBaud PDM-PAM-4, and 60-GBaud PDM-16QAM signals are also plotted in

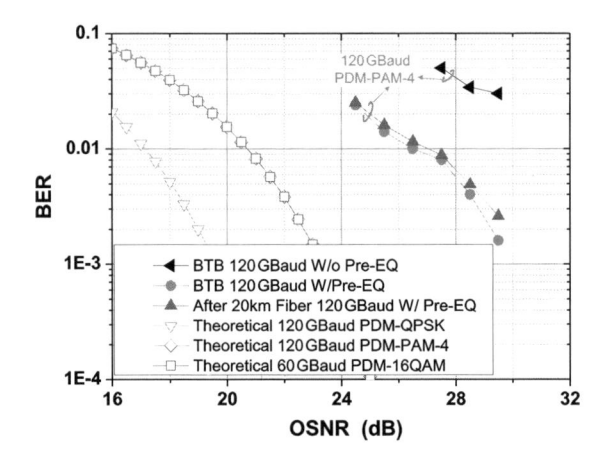

Fig. 3.17. BER results as a function of OSNR for different signals.

Fig. 3.17. Compared with the theory curve, there is about 5.5-dB implementation penalty in this experiment. There is still a large space for further improvement by utilizing large bandwidth optical and electronic components. Compared with the 120 GBaud PDM-QPSK, there is 3.5-dB OSNR penalty for the 120-GBaud PDM-PAM-4 and 60-GBaud PDM-16QAM signals at the same BER level of 1×10^{-2}. It is believed that the PDM-QPSK is much more suitable for long-haul transmission and PDM-16QAM, and PAM-4 is more suitable for short-distance system. However, since only amplitude modulation is used based on MZM for PAM-4, the optical part is simpler. Furthermore, since PAM signals are modulated on the amplitude, the frequency-offset and phase noise of laser sources have less impact on the signal. Therefore, the tolerance of PAM-4 to laser phase noise is higher than that of QPSK and 16QAM signals, which makes it a competitive scheme for high speed, short range optical communications.

3.5. Summary

In this chapter, the advanced DSP at the transmitter and receiver side for signal pre-equalization and post-equalization is introduced for short haul and access optical communication network. In Section 3.2, we show the idea of advanced DSP for single-band CAP for short-haul data center interconnections by using high-order QAMs. Multi-band CAP is introduced in Section 3.3 for optical access networks. The novel DSP-based digital and optical pre-equalization schemes are discussed and introduced in Section 3.4 with PAM-4 signals.

References

[1] M. Iglesias Olmedo, Z. Tianjian, J. Bevensee Jensen, Z. Qiwen, X. Xu and I. T. Monroy, Towards 400GBASE 4-lane solution using direct detection of MultiCAP signal in 14 GHz bandwidth per lane, presented at the Opt. Fiber Commun., Nat. Fiber Opt. Eng. Conf., Anaheim, CA, 2013, paper PDP5C.10.

[2] J. Estaran, M. Iglesias, D. Zibar, X. Xu and I. Tafur, First experimental demonstration of coherent CAP for 300-Gb/s metropolitan optical networks, in *Optical Fiber Communication Conference*, OSA Technical Digest (online) (Optical Society of America, 2014), paper Th3K.3.

[3] L. Suhr, J. J. Vegas Olmos, B. Mao, X. Xu, G. N. Liu and I. Tafur Monroy, Direct modulation of 56 Gbps duobinary-4-PAM, in *Optical Fiber Communication Conference*, OSA Technical Digest (online) (Optical Society of America, 2015), paper Th1E.7.

[4] K. Zhong, X. Zhou, T. Gui, L.Tao, Y. Gao, W. Chen, J. Man, L. Zeng, A. P. Tao Lau and C. Lu, Experimental study of PAM-4, CAP-16, and DMT for 100 Gb/s short reach optical transmission systems, *Opt. Express* **23** (2015) 1176–1189.

[5] X. Xu, E. Zhou, G. N. Liu, T. Zuo, Q. Zhong, L. Zhang, Y. Bao, X. Zhang, J. Li and Z. Li, Advanced modulation formats for 400-Gbps short-reach optical inter-connection, *Opt. Express* **23** (2015) 492–500.

[6] M. Morsy-Osman, M. Chagnon, M. Poulin, S. Lessard and D.V. Plant, 1λ × 224 Gb/s 10 km transmission of polarization division multiplexed PAM-4 signals using 1.3 3m SiP intensity modulator and a direct-detection MIMO-based Receiver, in *Proc. ECOC 2014*, PD.4.4

[7] C. Xie, S. Spiga, P. Dong, P. J. Winzer, A. Gnauck, C. Gréus, C. Neumeyr, M. Ortsiefer, M. Müller and M. Amann, Generation and transmission of 100-Gb/s PDM 4-PAM using directly modulated VCSELs and coherent detection, in *Optical Fiber Communication Conference*, OSA Technical Digest (online) (Optical Society of America, 2014), paper Th3K.2.

[8] C. Xie, S. Spiga, P. Dong, P. Winzer, M. Bergmann, Be. Kögel, C. Neumeyr and M.-C. Amann, 400-Gb/s PDM-4PAM WDM system using a monolithic 2 × 4 VCSEL array and coherent detection, *J. Lightwave Technol.* **33** (2015) 670–677.

[9] C. Xie, S. Spiga, P. Dong, P. Winzer, A. Andrejew, B. Kögel, C. Neumeyr and M.-C. Amann, All-VCSEL based 100-Gb/s PDM-4PAM coherent system for applications in metro networks, in *Proc. ECOC 2014*, paper P.4.3.

[10] K. Szczerba *et al.*, 37 Gbps transmission over 200 m of MMF using single cycle subcarrier modulation and a VCSEL with 20 GHz modulation bandwidth, in *Proc. 36th Eur. Conf. Opt. Commun.* (2010), paper We.7.B.2.

[11] A. S. Karar and J. C. Cartledge, Generation and detection of a 56 Gb/s signal using a DML and half-cycle 16-QAM Nyquist-SCM, *IEEE Photon. Technol. Lett.* **25**(8) (2013) 757–760.

[12] R. Rodes *et al.*, High-speed 1550 nm VCSEL data transmission link employing 25 GBd 4-PAM modulation and hard decision forward error correction, *J. Lightwave Technol.* **31** (2013) 689–695.

[13] T. Tanaka, M. Nishihara, T. Takahara, L. Li, Z. Tao and J. C. Rasmussen, 50 Gbps class transmission in single mode fiber using discrete multi-tone modulation with 10G directly modulated laser, in *Proc. Conf. Opt. Fiber Commun.* (March 2012), pp. 1–3, paper Oth4G.

[14] R. P. Giddings, X. Q. Jin, E. Hugues-Salas, E. Giacoumidis, J. L. Wei and J. M. Tang, Experimental demonstration of a record high 11.25 Gb/s real-time optical OFDM transceiver supporting 25 km SMF end-to-end transmission in simple IMDD systems, *Opt. Express* **18**(6) (2010) 5541–5555.

[15] J. L. Wei, J. D. Ingham, D. G. Cunningham, R. V. Penty and I. H. White, Performance and power dissipation comparisons between 28 Gb/s NRZ, PAM, CAP and optical OFDM systems for data communication applications, *J. Lightwave Technol.* **30** (2012) 3273–3280.

[16] J. D. Ingham, R. Penty, I. White and D. Cunningham, 40 Gb/s carrierless amplitude and phase modulation for low-cost optical data communication links, in *Proc. OFC 2011*, paper OThZ3.

[17] R. Rodes, M. Wieckowski, T. T. Pham, J. B. Jensen, J. Turkiewicz, J. Siuzdak and I. T. Monroy, Carrierless amplitude phase modulation of VCSEL with 4 bit/s/Hz spectral efficiency for use in WDM-PON, *Opt. Express* **19**(27) (2011) 26551–26556.

[18] M. B. Othman, X. Zhang, L. Deng, M. Wieckowski, J. Jensen and I. T. Monroy, Experimental investigations of 3D/4D-CAP modulation with DM-VCSELs, *IEEE Photon. Technol. Lett.* **24**(22) (2012) 2009–2012.

[19] J. Wei, L. Geng, D. G. Cunningham, R. V. Penty and I. White, 100 Gigabit Ethernet transmission enabled by carrierless amplitude and phase modulation using QAM receivers, in *Proc. OFC 2013*, paper OW4A.5.

[20] L. Tao, Y. Wang, Y. Gao, A. P. Tao Lau, N. Chi and C. Lu, Experimental demonstration of 10 Gb/s multi-level carrier-less amplitude and phase modulation for short range optical communication systems, *Opt. Express* **21**(5) (2013) 6459–6465.

[21] J. Zhang, X. Li, J. Xiao, G.-K. Chang and F. Li, Demonstration of 24-Gb/s Carrier-less amplitude and phase modulation (CAP) 64QAM radio-over-fiber system over 40-GHz Mm-wave fiber-wireless transmission, in *Proc. OFC 2014*, paper M2D.5.

[22] J. Zhang, X. Li, Y. Xia, Y. Chen, X. Chen, J. Yu and J. Xiao, 60-Gb/s CAP-64QAM transmission using DML with direct detection and digital equalization, in *Proc. OFC 2014*, paper W1F.3.

[23] J. Zhang, J. Yu, F. Li, N. Chi, Z. Dong and X. Li, $11 \times 5 \times 9.3$ Gb/s WDM-CAP-PON based on optical single-side band multi-level multi-band carrier-less amplitude and phase modulation with direct detection, *Opt. Express* **21** (2013) 18842–18848.

[24] G. Stepniak and J. Siuzdak, Transmission beyond 2 Gbit/s in a 100 m SI POF with multilevel CAP modulation and digital equalization, in *Proc. OFC 2013*, paper NTu3J.5.
[25] J. Zhang, J. Yu, F. Li, X. Li and Y. Wang, Demonstration of single-carrier ETDM 400GE PAM-4 signals generation and detection, *IEEE Photon. Technol. Lett.* **27**(24) (2015) 2543–2546.
[26] G. H. Im, D. D. Harman, G. Huang, A. V. Mandzik, M. H. Nguyen and J. J. Werner, 51.84 Mb/s 16-CAP ATM LAN standard, *IEEE J. Sel. Areas Comm.* **13**(4) (1995) 620–632.
[27] X. Zhou, J. Yu, M.-F. Huang, Y. Shao, T. Wang, L. Nelson, Peter Magill, Martin Birk, Peter I. Borel, David W. Peckham, R. Lingle and B. Zhu, 64-Tb/s, 8 b/s/Hz, PDM-36QAM transmission over 320 km using both pre- and post-transmission digital signal processing, *J. Lightwave Technol.* **29** (2011) 571–577.
[28] G. Raybon, A. Adamiecki, P. Winzer, C. Xie, A. Konczykowska, F. Jorge, J. Dupuy, L. L. Buhl, C. Sethumadhavan, S. Draving, M. Grove and K. Rush, Single-carrier 400G interface and 10-channel WDM transmission over 4800 km using all-ETDM 107-Gbaud PDM-QPSK, in *Proc. OFC 2013*, PDP5A.5.
[29] J. Zhang, Z. Dong, H. Chien, Z. Jia, Y. Xia and Y. Chen, Transmission of 20 × 440-Gb/s super-Nyquist-filtered signals over 3600 km based on single-carrier 110-GBaud PDM QPSK with 100-GHz grid, in *Proc. OFC 2014*, paper Th5B.3.
[30] G. Raybon, A. Adamiecki, P. J. Winzer, M. Montoliu, S. Randel, A. Umbach, M. Margraf, J. Stephan, S. Draving, M. Grove and K. Rush, All ETDM 107-Gbaud PDM-16QAM (856-Gb/s) transmitter and coherent receiver, in *Proc. ECOC 2013*, paper PD 2.D.3.
[31] J. Zhang, J. Yu, N. Chi, and H.-C. Chien, Time-domain digital pre-equalization for band-limited signals based on receiver-side adaptive equalizers, *Opt. Express* **22** (2014) 20515–20529.
[32] X. Zhou, J. Yu, M.-F. Huang, Y. Shao, T. Wang, P. Magill, M. Cvijetic, L. Nelson, M. Birk, G. Zhang, S. Ten, H. B. Matthew and S. K. Mishra, Transmission of 32-Tb/s capacity over 580 km using RZ-shaped PDM-8QAM modulation format and cascaded multimodulus blind equalization algorithm, *J. Lightwave Technol.* **28** (2010) 456–465.
[33] J. Zhang, J. Yu, B. Zhu, F. Li, H.-C. Chien, Z. Jia, Y. Cai, X. Li, X. Xiao, Y. Fang and Y. Wang, Transmission of single-carrier 400G signals (515.2-Gb/s) based on 128.8-GBaud PDM QPSK over 10,130- and 6,078 km terrestrial fiber links, *Opt. Express* **23** (2015) 16540–16545.

Chapter 4

DSP for Direct-Detection OFDM System

4.1. Introduction

There are two modulation formats in optical communication system, which are direct modulation [1, 2] and external modulation [3–9], as shown in Fig. 4.1. Direct modulation refers to the modulation of light intensity achieved by the injection current of the semiconductor laser, and the structure is simple, economical and easy to realize. However, because of the limited bandwidth, low extinction ratio and large line width of the direct modulation semiconductor laser, direct modulation is often used in short distance direct-detection optical orthogonal frequency division multiplexing (DDO-OFDM) optical fiber communication system [1–9]. In the DDO-OFDM system with transmission distance of more than 100 km and the CO-OFDM optical fiber communication system with large capacity and long distance [10–47], external modulation is generally used. External modulation is realized by using an external modulator independent of the laser source, which has the advantages of high bandwidth, high extinction ratio and narrow linewidth. The current commonly used external modulator in optical fiber communication system mainly includes the Mach–Zehnder Modulator (MZM) based on the electro-optic effect and the electric absorption modulator (EAM) based on the electric absorption effect.

MZM is employed to convert electrical signal to optical signal based on the electro-optic effect of the material. Figure 4.2 shows the structure of the MZM. Optical modulation changes the refractive index of the material

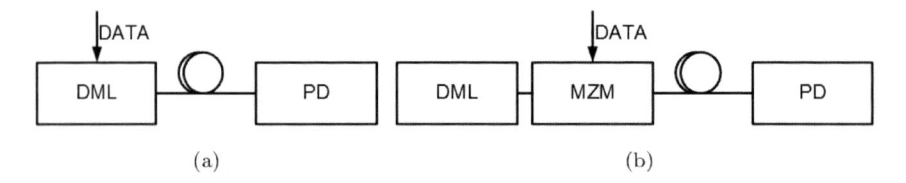

78 Digital Signal Processing for High-Speed Optical Communication

Fig. 4.1. Two different modulation modes of DDO-OFDM system. (a) Direct modulation and (b) external modulation. DML: direct-modulated laser, PD: photodiode, MZM: Mach–Zehnder modulator.

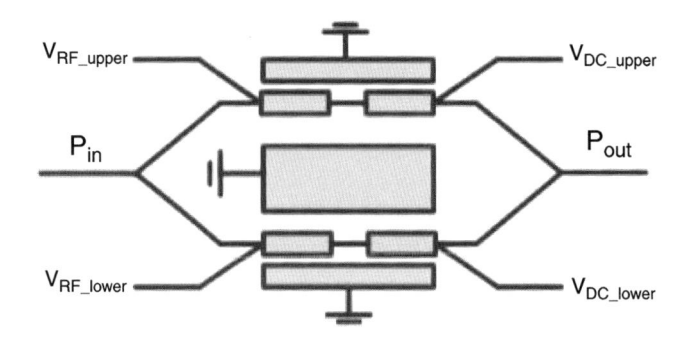

Fig. 4.2. MZM structure.

by adjusting the applied voltage of the photoelectric material loaded on the two electrodes, so as to achieve the purpose of controlling the light intensity of the output signal.

The output light signal of MZM can be expressed as follows:

$$E_{\text{out}}(t) = \frac{1}{2} E_{\text{in}}(t) \{ e^{j\pi \frac{V_1(t)}{V_\pi}} + \gamma e^{j\pi \frac{V_2(t)}{V_\pi}} \} \tag{4.1}$$

where $E_{\text{in}}(t)$ is the input optical signal, $E_{\text{out}}(t)$ is the output optical signal, $V_1(t) = V_{\text{RF\_upper}}(t) + V_{\text{DC\_upper}}$ is the sum of the input signal of the upper arm and the DC offset of the upper arm, $V_2(t) = V_{\text{RF\_lower}}(t) + V_{\text{DC\_lower}}$ is the sum of the input signal of the lower arm and the DC offset of the lower arm, $\gamma = (\sqrt{\delta} - 1)/(\sqrt{\delta} + 1)$ is the symmetry factor of the upper and lower branches, δ is the extinction ratio of modulator, V_π is the half-wave voltage of modulator. In the ideal case, δ is greater than 20 dB, and γ is approximately equal to 1.

When two different input signals are loaded on the two arms of the MZM, we can get

$$E_{out}(t) = \frac{1}{2}E_{in}(t)\left\{\begin{array}{l}\cos\left(\pi\dfrac{V_1(t)}{V_\pi}\right) + j\sin\left(\pi\dfrac{V_1(t)}{V_\pi}\right) \\ + \cos\left(\pi\dfrac{V_2(t)}{V_\pi}\right) + j\sin\left(\pi\dfrac{V_2(t)}{V_\pi}\right)\end{array}\right\}$$

$$= E_{in}(t)\cos\left\{\pi\frac{V_1(t) - V_2(t)}{2V_\pi}\right\}\exp\left\{j\pi\frac{V_1(t) + V_2(t)}{2V_\pi}\right\} \quad (4.2)$$

In Eq. (4.2), $\cos[\pi(V_1(t) - V_2(t))/2V_\pi]$ is the amplitude modulation component, and $\exp\{j\pi(V_1(t) + V_2(t))/2V_\pi\}$ is the phase modulation component. When MZM works in the push-pull mode, which means $V_1(t) + V_2(t) = 0$, the modulator chirp is 0. Making $V(t) = V_1(t) - V_2(t)$ in the direct detection system, the transfer function is the ratio of the output power to the input power, which can be expressed as

$$\frac{P_{out}}{P_{in}} = \frac{|E_{out}(t)|^2}{|E_{in}(t)|^2} = \cos^2\left\{\pi\frac{V_1(t) - V_2(t)}{2V_\pi}\right\} = \cos^2\left\{\pi\frac{V(t)}{2V_\pi}\right\} \quad (4.3)$$

In the coherent detection system, the transfer function is the ratio of the output electric field to the input electric field, which can be expressed as

$$\frac{E_{out}(t)}{E_{in}(t)} = \cos\left\{\pi\frac{V_1(t) - V_2(t)}{2V_\pi}\right\} = \cos\left\{\pi\frac{V(t)}{2V_\pi}\right\} \quad (4.4)$$

Transfer function curves of direct detection and coherent detection represented by Eqs. (4.3) and (4.4) are shown in Fig. 4.3. As we can see, the MZM transmission curve is nonlinear. In order to make the efficiency and performance of the MZM high, the modulation signal should be made, as far as possible, in the region of high linearity of the MZM transfer function, which can be controlled by adjusting the peak value of the modulation signal and the bias voltage of the MZM. When the amplitude of the signal is located in the region of high nonlinearity of the MZM transfer function, the signal will be distorted seriously. Thus, the BER performance of the system is worse. For the DDO-OFDM system, the optimal bias point should be chosen as the positive intersection (Quadrature point).

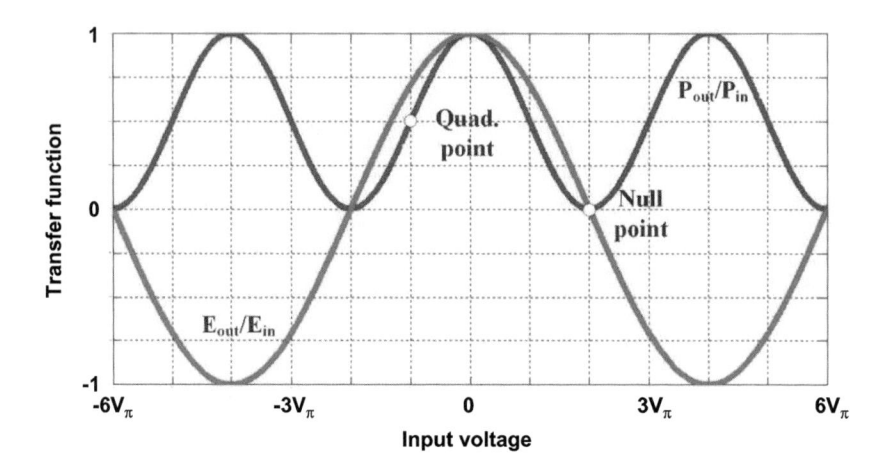

Fig. 4.3. Transfer functions curve of different detection methods.

For the CO-OFDM system, the optimal bias point should be chosen as zero (Null point).

According to the different detection methods, the OFDM optical communication system can be divided into two categories: DDO-OFDM and CO-OFDM.

Compared with the CO-OFDM system, the DDO-OFDM system does not need to provide high cost devices such as optical hybrid and balance detector at the receivers [11, 14–17], so the cost will be well controlled. Due to the sub-carrier beat noise and the dispersion in fiber transmission which will lead to frequency selective fading, the fiber transmission distance of DDO-OFDM system is very limited. Therefore, it is mainly used in the transmission of short distance wired access network and peer-to-peer data center.

The main advantage of the CO-OFDM system is that it can realize the modulation of intensity and phase at the same time, so as to realize the high-speed signal transmission and the signal-to-noise ratio (SNR) of the received signal can be adjusted by adjusting the power of the local oscillation signal so that the system can realize ultra-long distance transmission and ensure high receiving sensitivity of the receiver. CO-OFDM is mainly used in the ultra-high speed and long distance backbone transmission network and the metro transmission with medium distance (<1000 km).

DDO-OFDM system mainly includes five parts: OFDM transmitter, OFDM signal electro-optic modulation, optical fiber transmission of the

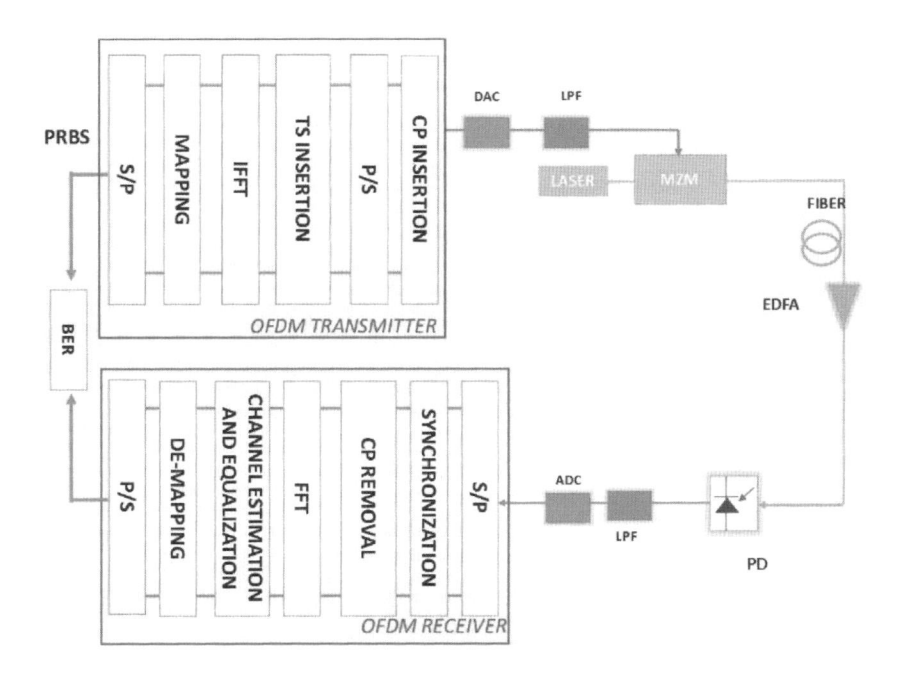

Fig. 4.4. DDO-OFDM system schematic.

OFDM signal, OFDM signal photoelectric conversion and OFDM receiver. Figure 4.4 shows DDO-OFDM system schematic.

The OFDM transmitter mainly includes the following processes: (1) pseudo-random sequence parallel conversion; (2) fast inverse Fourier transform, which transforms signal from frequency domain to time domain; (3) inserting a training sequence at the beginning of the signal, which is mainly used for symbol synchronization and channel estimation; (4) parallel serial conversion of the obtained signal; (5) inserting cyclic prefix (CP) into OFDM signals in the time domain for resistance of inter-symbol interference (ISI) and inter-carrier interference (ICI).

The generated OFDM signals are transformed into analog signals by digital-to-analog conversion (DAC) transform, and the analog signals passing through low pass filter (LPF) and the amplifier are then inserted into an external modulator to realize electro-optic conversion. A scheme of photoelectric conversion is realized by direct modulation. After the optical OFDM signals are transmitted through the optical fiber, the optical signals are amplified by Erbium-doped fiber amplifier (EDFA) before entering the photodiode (Photodiode, PD) to achieve photoelectric conversion.

After the photoelectric conversion, the electrical OFDM signals firstly pass through an LPF to filter out of band noise, and then an analog-to-digital conversion (ADC) cascading behind the LPF converts analog OFDM signals into digital OFDM signals. The OFDM receiver digital signals need to carry out the following digital signal processing to achieve demodulation: (1) series parallel conversion, that is, converting the digital signal after ADC into a parallel signal; (2) symbol synchronization which is used to determine the length of the OFDM signal; (3) remove the CP of the OFDM signal; (4) fast Fourier transform, which transforms the signal from time domain into frequency domain; (5) extracting the training sequence for channel estimation and channel equalization; (6) achieving inverse mapping according to the mapping rules of the transmitter; (7) parallel serial conversion of the signal and calculating the BER compared with the original bit sequence.

A very popular DDO-OFDM application is currently used to implement the point-to-point transmission between short distance access network and data center with a single optical wavelength, single polarization and bit rate exceeding 100 Gbit/s [3]. To achieve this kind of program, the cost is a key factor, so that the system of Polarization Division Multiplexing (PDM) is not highly regarded because of the high cost. Thus this scheme usually uses direct modulation and direct detection of the structure.

DDO-OFDM is mainly used in the access network [1–5] and ultra-high-speed point-to-point optical fiber communication service because of its simple structure, low cost and low complexity [1–9]. DDO-OFDM system has some problems of its own, mainly due to subcarrier–subcarrier mixing interference (SSMI) [7–9] and optical fiber dispersion and device imperfect frequency response caused by high-frequency attenuation. In short-range DDO-OFDM systems, fiber dispersion is not so obvious that the high-frequency attenuation caused by dispersion can be ignored. The application scenarios of DDO-OFDM discussed and analyzed in this chapter are very short transmission distances, and imperfect frequency response will be the most important factor leading to high frequency attenuation. With the promotion and popularization of the broadband communication service in the daily life mentioned in Chapter 1, it will drive the rapid increase of the transmission capacity of the DDO-OFDM system which realizes the access network and the high-speed point-to-point optical fiber communication service. It is a feasible scheme to eliminate the subcarrier's noise by using the Digital Signal Processing (DSP) method to improve the transmission performance of this large-capacity DDO-OFDM system. In addition, the high frequency attenuation caused by the imperfect frequency response of the device in a large capacity DDO-OFDM system will be very noticeable. In general, increasing the capacity of the system increases the order of the modulation format to improve the spectral efficiency, while the higher order modulation format will be more sensitive to the ISI caused by the high frequency attenuation. In order to ensure the use of high-end modulation format broadband DDO-OFDM system transmission performance, we must study how to overcome this high-frequency attenuation.

In Section 4.2, we propose a method to eliminate the SSMI without extra overhead. In a typical DDO-OFDM system, OFDM signal will be interfered by SSMI after square-law detection via the PD at the receiver [7–9]. As reported in [7], the frequency guard-band was proposed to prevent SSMI from OFDM signal. Another scheme, which was called interleaved OFDM, was applied in the DDO-OFDM to eliminate the impact of SSMI by inserting data only in even subcarriers [7–9]. Both these schemes can effectively mitigate the distortions introduced by SSMI, while the system SE will be decreased to half. In order to maintain high SE, the bit interleaver and turbo code techniques are proposed to combat the SSMI [7]. These techniques can mitigate the SSMI effectively, but the SE will still be degraded due to the utilization of turbo code. Moreover, the complexity due to forward-error-correction (FEC) decoding will limit its applications. There is also a proposed method based on the subcarrier of the SNR of the different modulation of different modulation formats of the signal, so that the subcarrier at low frequencies can carry lower signal modulation format if SSMI is more serious. This scheme requires a complex calibration process at the beginning of the system, in order to find out the distribution of subcarrier noise. Therefore, it is not suitable for DDO-OFDM system with simple structure, low cost and low complexity. The half-cycled DDO-OFDM scheme is proposed in Section 4.2 in order to achieve overcoming SSMI without reducing spectral efficiency and increasing system complexity. The scheme was successfully demonstrated to resist SSMI without SE reduction. The receiver sensitivity was improved by 2 and 1.5 dB in QPSK and 16-Quadrature Amplitude Modulation (16QAM) OFDM with 40-km standard single-mode fiber-28 (SSMF-28) transmission, respectively.

In Section 4.3, we analyze the difficulties of low-cost high-order modulation format DDO-OFDM system and put forward the corresponding solution. Direct-detection optical OFDM (DDO-OFDM) systems tend to be applied in access networks with short reaches for its simple and cost-effective configurations [1–9]. Moreover, unlike long haul and metro networks where expense can be shared by a large number of users, access applications desire low-capital expenditures (CapEx) and low-operational expenditures (OpEx) expense. Consequently, DML is considered as a competitive candidate in access applications due to its advantages of lower cost, compact size, low power consumption, and high output power when compared with other transmitter sources using external modulators (EMs). With the emergence of bandwidth-hungry services such as high-definition (HD) television streaming, cloud computing, interactive HD online gaming, the bit rate requirement of the next generation short reach optical fiber system is up to tens of Gbit/s or even 100 Gbit/s. In order to achieve such high bit rate under limited bandwidth of optoelectronic devices, increasing the spectral efficiency with high-level modulation formats is preferred. As far as we concerned, 128-ary quadrature amplitude modulation (128QAM) is the highest modulation format level ever reported in the DDO-OFDM systems [1]. In the coherent optical OFDM (CO-OFDM) transmission system, 1024QAM and 2048QAM are realized with large FFT size. In the CO-OFDM systems, large FFT size can be used to improve the robustness to the ISI, while the impact of ICI becomes much more obvious. In the CO-OFDM systems with 1024QAM and 2048QAM, fairly complicated frequency offset and phase noise compensation scheme are proposed to mitigate the impact of ICI. Fortunately, the ICI induced by the phase noise and frequency offset in the DDO-OFDM system can be ignored as the beating subcarriers and optical carrier are highly correlated when the fiber transmission distance is very short. The performance of the DDO-OFDM system with high level QAM modulation format can be improved with large size FFT due to enhanced robustness to ISI. The peak-to-average power ratio (PAPR) of OFDM signal is proportional to the FFT size, the large FFT size will lead to high PAPR [6]. OFDM with high PAPR is vulnerable to the nonlinear distortion, and it will lead to dramatic degradation of high-level QAM-OFDM signal performance. In order to reduce the PAPR of high-level QAM-OFDM signal, Discrete Fourier Transform-spread (DFT-spread) is applied before the OFDM modulation with IFFT in the transmitter. Apart from these, intra-symbol frequency-domain averaging (ISFA) is used to suppress the noise during the optical to electrical (O/E) conversion with PD [15].

Section 4.4 is based on DFT-spread to achieve the research of large-capacity DDO-OFDM signal short-distance transmission. In the

DDO-OFDM system with short distance transmission, the direct modulation method can greatly reduce the cost of the system. Therefore, the direct modulation laser is currently used in the short distance access network based on direct detection. At present, the advanced modulation technology which can be used in the direct detection system to improve the system spectral efficiency include OFDM [1–6], half cycle 16QAM [7], carrierless amplitude phase (CAP) modulation [48] and pulse amplitude modulation (PAM) [49]. Since the OFDM signal exhibits good resistance to dispersion and polarization-mode dispersion and is transparent to the modulation format of the signal based on the frequency domain equalization, it is the most competitive candidate for the short reach optical transmission systems. However, the PAPR problem undoubtedly degrades the transmission performances of OFDM signal. In order to reduce the PAPR, discrete-flourier-transform-spread (DFT-spread) without any distortion is introduced in the IM/DD systems utilizing DML. This scheme does not cause signal distortion and computational complexity is not high. In this chapter, we have completed two aspects of the work: optimization of training sequence for the DFT-spread DDO-OFDM signal; pre-emphasis in the DDO-OFDM system with large bandwidth DFT-spread and performance analysis of the DFT-spread technology in resisting high frequency attenuation.

In the DFT-spread OFDM transmission system, an additional pair of DFT/IDFT is added in the offline transceiver compared to the conventional OFDM transmission system. The digital signal mapped from pseudorandom binary sequence is converted to the analog signal after that additional DFT in the transmitter. This does not affect the OFDM signal symbols generation and transmission, while it is essential for the training sequence (TS) based FDE. With that additional DFT in the TS symbol generation, digital TS symbol will be transformed to analog TS symbol. This can help to control the PAPR of TS symbol, but it may lead to the channel estimation performance degradation as digital TS symbol is obviously robust to the interference of noise during channel estimation. Without that additional pair of DFT/IDFT for TS symbol, the PAPR of OFDM signal may increase a little, while the overall performance will be very good with the strong digital TS symbol. 79.86 Gbit/s DFT-spread 32QAM-OFDM signal demonstrated the best performance with binary-phase-shift-keying/quadrature-phase-shift-keying (BPSK/QPSK) digital TS symbol and the strongest robustness against optical noise in FDE. With the optimized TS symbol, no penalty is observed and the bit error ratio (BER) is well below the requirement of hard decision Forward Error Correction (HD-FEC) threshold (3.8×10^{-3}) after 20-km SSMF.

4.2. Research on SSMI Cancellation in Direct-Detection Optical OFDM with Novel Half-Cycled OFDM

Figure 4.5(a) shows the concept of three different kinds of OFDM signal in DDO-OFDM. The first one is traditional OFDM signal; after PD the SSMI will spread in the whole OFDM signal band, and the BER performance of signal will be seriously degraded. The second one is guard-band enabled OFDM signal proposed in [8, 9], for which the SSMI will locate only in the guard-band and the BER performance can be improved with half reduced SE. The third one is interleaved OFDM signal [7], and the data were only modulated onto the even subcarriers. After the PD, the SSMI will distribute only in the odd subcarriers, and data only modulated on the even subcarriers are immune to the SSMI. In the SSMI cancellation scheme with the interleaved OFDM signal, the OFDM symbol will exhibit symmetry in the time domain when data only modulated onto the even/odd subcarriers, and based on this theory we proposed the half-cycled DDO-OFDM.

The IFFT size during the OFDM modulation is N and the time length of one OFDM symbol is T; after IFFT, the OFDM signal can be expressed as

$$s(t) = \sum_{k=0}^{N-1} c_k \exp(j2\pi f_k t) \quad (1 \le t \le T) \tag{4.5}$$

where k represents the index of subcarriers, f_k is the frequency of kth subcarrier and can be expressed as

$$f_k = k\Delta f = \frac{k}{T} \tag{4.6}$$

In the time domain, the first half and the second half of one OFDM symbol can be expressed as

$$s(t_1) = \sum_{k=0}^{N-1} c_k \exp(j2\pi f_k t_1) \quad \left(1 \le t_1 \le \frac{T}{2}\right) \tag{4.7}$$

$$s(t_2) = \sum_{k=0}^{N-1} c_k \exp(j2\pi f_k t_2) \quad \left(\frac{T}{2}+1 \le t_2 \le T\right) \tag{4.8}$$

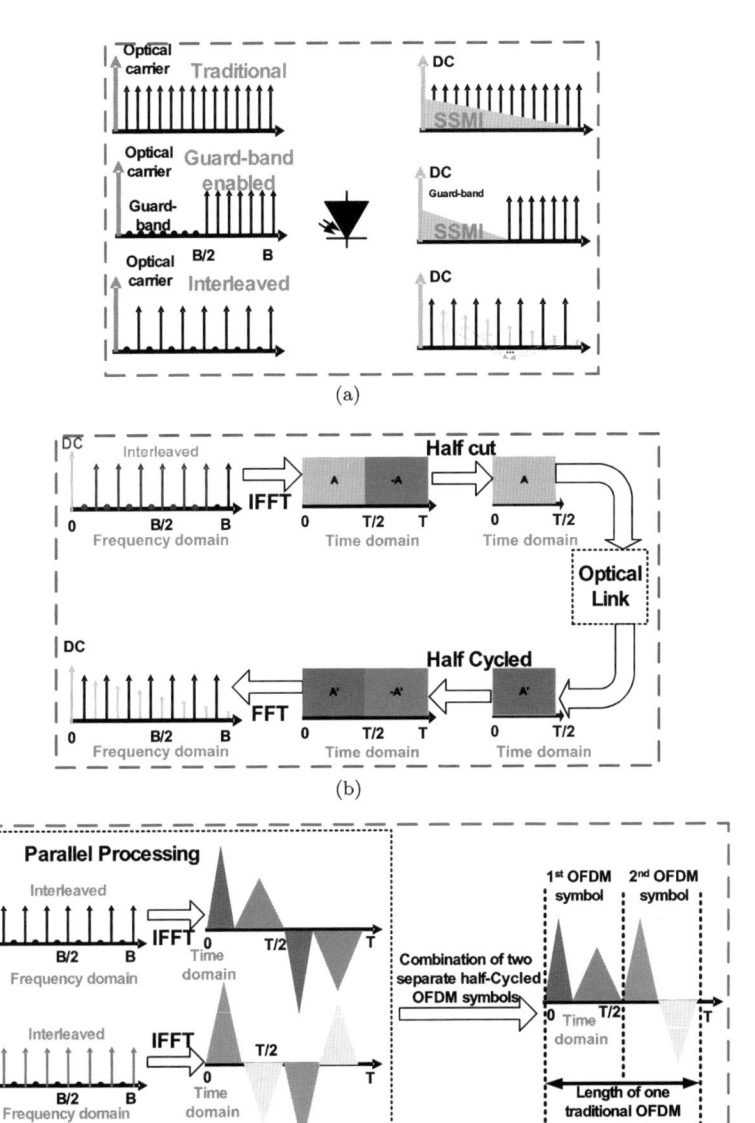

Fig. 4.5. (a) Concept of three different types of DDO-OFDM, (b) structure of the proposed half-cycled DDO-OFDM, and (c) combination of two parallel half-cycled OFDM symbols.

As $t_2 = t_1 + T/2$, the second half can be expressed as

$$s\left(t_1 + \frac{T}{2}\right) = \sum_{k=0}^{N-1} c_k \exp\left(j2\pi f_k \left(t_1 + \frac{T}{2}\right)\right)$$

$$= \sum_{k=0}^{N-1} c_k \exp(j2\pi f_k t_1 + jk\pi)$$

$$= \sum_{k=0}^{N-1} c_k (\cos k\pi + j \sin k\pi) \exp(j2\pi f_k t_1)$$

$$= \sum_{k=0}^{N-1} c_k \cos k\pi \exp(j2\pi f_k t_1) \quad \left(1 \le t_1 \le \frac{T}{2}\right) \quad (4.9)$$

In order to further simplify the formula, the index for subcarriers is from 0 to $N - 1$, the index of even and odd subcarriers can be expressed as m and n, respectively. The first half and the second half of one OFDM symbol can be expanded as

$$s(t_1) = \sum_{n=1}^{N-1} c_n \exp(j2\pi f_n t_1) + \sum_{m=0}^{N-2} c_m \exp(j2\pi f_m t_1) \quad \left(1 \le t_1 \le \frac{T}{2}\right)$$
(4.10)

$$s(t_2) = s\left(t_1 + \frac{T}{2}\right)$$

$$= \sum_{k=0}^{N-1} c_k \cos k\pi \exp(j2\pi f_k t_1)$$

$$= \sum_{n=1}^{N-1} c_n \cos n\pi \exp(j2\pi f_n t_1) + \sum_{m=0}^{N-2} c_m \cos m\pi \exp(j2\pi f_m t_1)$$

$$= \sum_{n=1}^{N-1} -c_n \exp(j2\pi f_n t_1) + \sum_{m=0}^{N} c_m \exp(j2\pi f_m t_1) \quad \left(1 \le t_1 \le \frac{T}{2}\right)$$
(4.11)

In the interleaved OFDM scheme, subcarriers with even index are reserved without data mapping to resist SSMI, which means $c_m = 0$, the first half and the second half of one OFDM symbol can be simplified as

$$s(t_1) = \sum_{n=1}^{N-1} c_n \exp(j2\pi f_n t_1) \quad \left(1 \leq t_1 \leq \frac{T}{2}\right) \tag{4.12}$$

$$s(t_2) = \sum_{n=1}^{N-1} -c_n \exp(j2\pi f_n t_1) = -s(t_1) \quad \left(t_2 = \frac{T}{2} + t_1\right) \tag{4.13}$$

From Eqs. (4.12) and (4.13), we can find that the first half and the second half in one OFDM symbols demonstrate the inverted amplitude in the time-domain, and the second half was cut during the transmission. By this way, the time length used to convey data transmission was shortened to $T/2$, the second half of the symbol can be replaced with an independent "new first half" carrying additional data. Thus, the data rate is doubled and the SE will maintain the same as traditional OFDM signal after this process.

Figure 4.5(b) shows the structure of the proposed half-cycled DDO-OFDM. First, data were modulated only onto even subcarriers just the same as the interleaved OFDM signal in the frequency domain, and then IFFT, was applied to realize OFDM modulation. After IFFT, the period of one OFDM symbol is T while the first half and the second half demonstrate the inverted amplitude in the time domain. The second half was cut during the transmission and thus the time length used to convey data transmission was shortened to $T/2$, while the SE will maintain the same as traditional OFDM signal after this process. After electrical-to-optical (E/O) and optical-to-electrical (O/E) conversions, as the fiber link can be regarded as an approximate time invariant system, the second half of the OFDM symbol can be recovered by the half cycle of the first half of the OFDM symbol. Then FFT was applied to realize OFDM demodulation, and after demodulation the SSMI will distribute in the odd subcarriers. In the proposed half-cycled DDO-OFDM, the cancellation of SSMI can be implemented without SE decrease. Figure 4.5(c) shows the combination of two half-cycled OFDM symbols with parallel processing; after IFFT, two different half-cycled OFDM symbols dropped the second half of the OFDM symbols in the time domain and two parallel remaining parts of symbols are combined to be a new symbol within a period T which is the length of the traditional OFDM symbol. After this processing, two different half-cycled OFDM symbols are transmitted within a period T, if the length of CP after OFDM modulation is also cut half during the half cut in time domain, and the total capacity of the system still maintain the same as the traditional OFDM system. The main drawback of the half-cycled scheme compared to the traditional OFDM symbol is that the resistance of ISI induced by the chromatic dispersion (CD) of fiber is limited as the length of CP is cut half. While in the access networks with direct direction, the span is usually not very long, this drawback is not very critical.

In [7] the half-cycled technique was successfully demonstrated to overcome SSMI without SE reduction. In this scheme, two half-cycled OFDM symbols with parallel processing are combined after the second half of the OFDM symbols in the time domain are dropped in the transmitter. After this processing, two different half-cycled OFDM symbols are transmitted within the same period as the traditional OFDM symbol and the total capacity of system still maintain the same as the traditional OFDM system. The receiver sensitivity was improved by 2 and 1.5 dB in 9 Gb/s QPSK and 18 Gb/s 16QAM OFDM with 40-km SSMF-28 transmission, respectively [7].

4.3. Research on Transmission of Directly Detected High-Order QAM-OFDM Signals

Figure 4.6 shows the experimental setup for a high-level QAM-OFDM transmission utilizing DML in intensity modulation and direct-detection (IM-DD) system [1]. At the transmitter side, the optical carrier at 1537.92 nm generated from a commercial DML is directly modulated by the high-level QAM-OFDM signal. In this section, 256QAM, 512QAM, 1024QAM and 2048QAM are respectively demonstrated in the system. The OFDM signal is generated offline in MATLAB and then uploaded into an arbitrary waveform generator (AWG). Here, the FFT size for OFDM generation is 8192. Among the 8192 subcarriers, 2048 subcarriers are used to transmit effective data in the positive frequency bins. Other corresponding 2048 subcarriers in the negative frequency bins are filled with Hermitian symmetric data to generate the real value OFDM signal. The first subcarrier is set to zero for DC-bias and the rest 4095 null subcarriers at the edge are reserved for oversampling. A 14-sample CP is added to the 8192 samples, giving 8206 samples per optical OFDM (OOFDM) symbol. One TS is inserted before every 25 OFDM data symbols to realize time synchronization and obtain the channel response. The bandwidth of the generated electrical OFDM signal is 3 GHz (2048/8192 ×

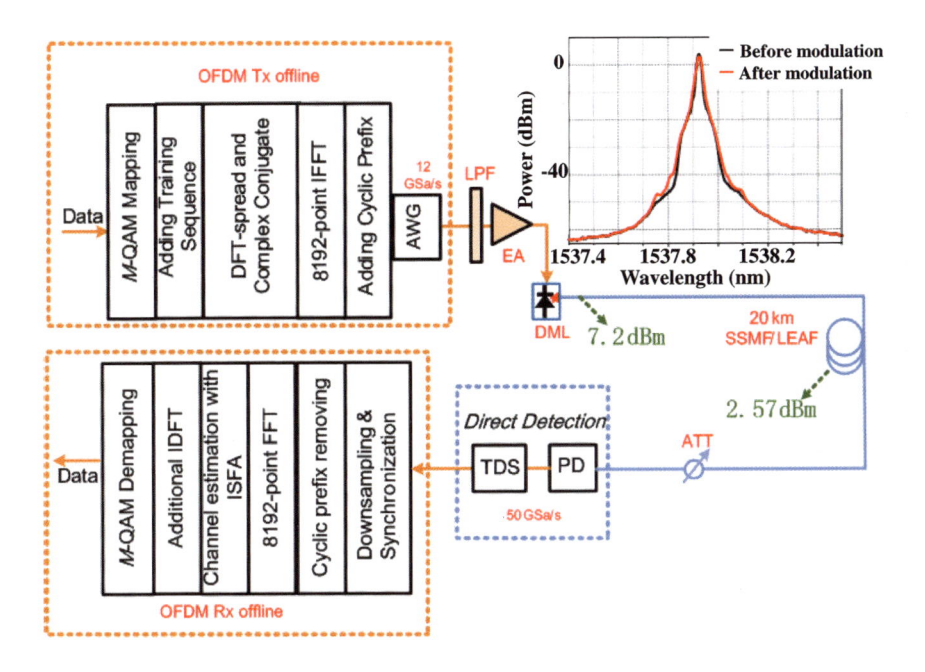

Fig. 4.6. Experimental setup for direct detection of high-order QAM-OFDM signal.

12 GHz = 3 GHz). The original bit rate of 2048QAM OFDM signal can be calculated as

$$\text{Raw data rate}$$
$$= \frac{N_{\text{OFDM symbol}}}{N_{\text{Training}} + N_{\text{OFDM symbol}}} \times \frac{N_{\text{data\_subcarrier}}}{N_{\text{total\_subcarrier}} + N_{\text{CP}}}$$
$$\times \text{Sampling rate} \times \frac{\text{bits}}{\text{Sample}}$$
$$= 25/(25+1) \times 2048/(8192+14) \times 12 \times 11 \, \text{Gbit/s}$$
$$\approx 31.7 \, \text{Gbit/s} \qquad (4.14)$$

where $N(\,\cdot\,)$ represents the number of the specified subcarriers, symbols and samples. The bits/sample for 2048QAM-OFDM is 11. One LPF with 3-GHz bandwidth is used to filter vestigial sideband after DAC and then the peak-to-peak voltage of OFDM signal is amplified to 2.4 V by one electrical amplifier before being injected into distributed feedback (DFB)-based DML. For OOFDM modulation, the DML with 10-GHz bandwidth is biased

Table 4.1. Fiber parameters.

Type of fiber	D [ps/km/nm]	S [ps/nm²/km]	α [dB/km]
SSMF	17	0.06	0.2
LEAF	4	0.106	0.21

at 88 mA to produce 7.2-dBm average output power. The optical spectra (0.01-nm resolution) of optical carrier before and after modulation are shown in Fig. 4.6 as inset with black and red lines, respectively. The generated signal is then injected into the 20-km LEAF/SSMF. As the fiber span is really short, EDFA is not necessary in the link. Table 4.1 summarizes the main parameters of the used fiber types at 1550 nm. The optical power is 2.57 dBm after fiber transmission. An optical ATT is applied to adjust the received optical power for sensitivity measurement and then the O/E conversion is implemented via an optical receiver with 3-dB bandwidth of 10 GHz. The converted signal was captured by a real-time oscilloscope (Tektronix DPO72004B) with 50 GSa/s sampling rate. The captured signal is then further processed with offline DSP. The offline DSP procedure contains CP removal, FFT, channel estimation with ISFA, one-tap equalization, high-level QAM de-mapping and BER calculation. In this section, one 256/512/1024/2048 OFDM symbol contains 16384/18432/20480/22528 bits (2048 × 8/9/10/11 bits = 16384/18432/20480/22528 bits) and the BER was obtained by error counting with 25 OFDM symbols (25 × 16384/18432/20480/22528 bits = 409600/460800/512000/563200 bits).

Channel estimation is an important procedure in OFDM transmission systems. With accurate channel response estimation, the physical impairments of the fiber transmission link such as CD and PMD can be obtained, and subsequent channel equalization can be performed to restore the signal. In this short reach DDO-OFDM system, the receiver noise from PD is the dominant noise component after O/E conversion. The ISFA is applied to improve the accuracy of channel estimation by the receiver noise suppression. The noise power is reduced along with the increase of ISFA taps as the average sample should be large enough to suppress the power level of noise. In order to reduce the impact of noise on the channel estimation, the ISFA taps should be large. While the ISFA is implemented based on the assumption that the samples used for averaging are highly correlated, the average samples cannot be very large as the correlation will decrease with large

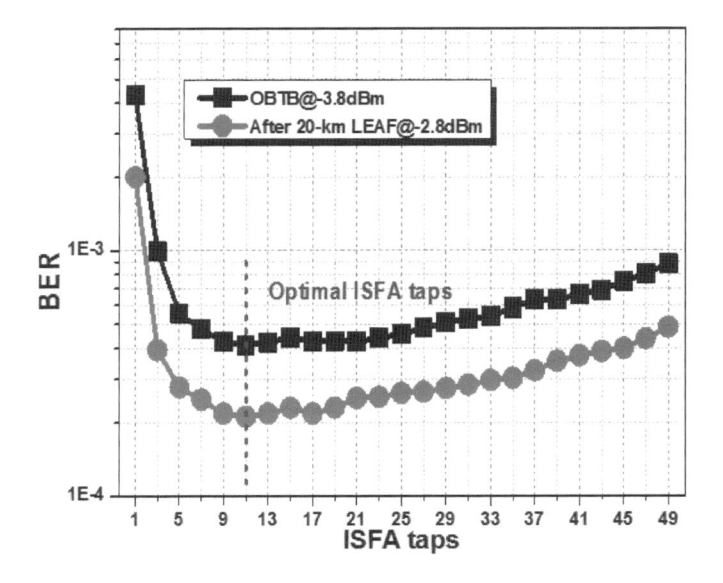

Fig. 4.7. Measured BER versus ISFA taps.

ISFA taps. There is a tradeoff between the noise suppression and the channel estimation accuracy when the ISFA taps are selected. In order to find out the optimal ISFA taps, the BER of 256QAM OFDM signal in the system is measured in both OBTB and after 20-km LEAF transmission, as shown in Fig. 4.7. It can be seen that the optimal ISFA tap number is 11, which will be applied in the rest of the section as the TS-based channel estimation is transparent to the modulation format.

The number of subcarriers used is a critical parameter for system performance due to the high-level QAM used. In the OFDM modulation, the spectrum becomes more rectangular with shaper roll-off when the number of subcarriers which is the same as FFT size is increased [1]. As we know, when a signal is transmitted in a bandwidth limitation link, the ISI appears. The electronic-optical-electronic link of DDO-OFDM system should be regarded as a bandwidth limitation link as the bandwidth of DAC and LPF is insufficient to the OFDM signal in this section. In the time domain, the spread of an individual symbol ISI caused by the DAC and LPF is only decided by their frequency response. When the system is established, the frequency response is unchangeable and the shape of the spectrum of signal is the vital factor for the ISI evaluation. Therefore, an OFDM symbol with a large FFT size has more tolerance to this kind of

ISI. Apart from this, the DDO-OFDM link which can be regarded as a time invariant system is really stable and the phase noise induced ICI is negligible in the ultra-short span system. In order to achieve high spectral efficiency in the ultra-short span direct detection system with a high-level QAM-OFDM, the FFT size should be very large. In order to verify the theoretical analysis, the BER of 512QAM OFDM signal versus different FFT sizes in the system is measured in the OBTB case and the measurement is shown in Fig. 4.8. It can be seen that the BER performance is improved when the FFT size in the OFDM modulation is increased. The electrical spectra of received OFDM signal with FFT size of 256, 2048 and

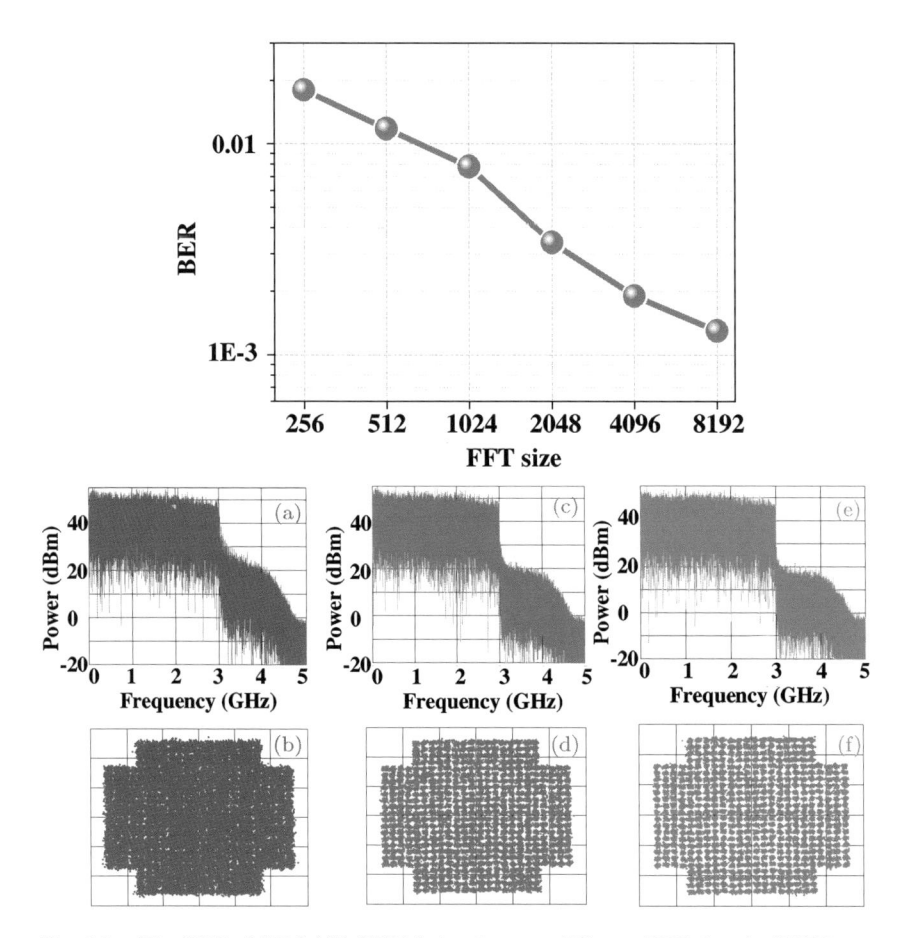

Fig. 4.8. The BER of 512QAM OFDM signal versus different FFT sizes in OBTB case and measured spectra and constellations in different FFT sizes.

8192 are shown in Figs. 4.8(a), 4.8(c) and 4.8(e), respectively. The corresponding constellations are inserted in Figs. 4.8(b), 4.8(d) and 4.8(f), respectively. Figure 4.8(f) shows that the spectrum of OFDM signal with 8192 FFT size rolls off much more quickly than the other two and the constellation of OFDM signal shown in Fig. 4.8(f) demonstrates the best performance.

Figure 4.9 shows the BER versus received optical power of 256QAM-OFDM signal. The BERs of 256QAM-OFDM signal in OBTB case, after 20-km LEAF and SSMF, are measured to evaluate the performance of this DML-based IM-DD system, respectively. It can be seen that there is no penalty between OBTB and after 20-km LEAF, while the penalty between OBTB and after 20-km SSMF is more than 4 dB at the BER of 3.8×10^{-3}. After the 20-km SSMF transmission, an error floor is observed. We believe the large penalty and error floor are caused by the CD-induced serious ICI. As the dispersion coefficient (D in Table 4.1) of LEAF is significantly less than that of SSMF, the impact of CD-induced ICI after 20-km LEAF is significantly less than that after 20-km SSMF, and the penalty after 20-km LEAF can be neglected. The constellations of 256-QAM OFDM signal after 20-km LEAF and SSMF transmission at 0.2-dBm received optical power are shown in Figs. 4.10(a) and 4.10(b), respectively.

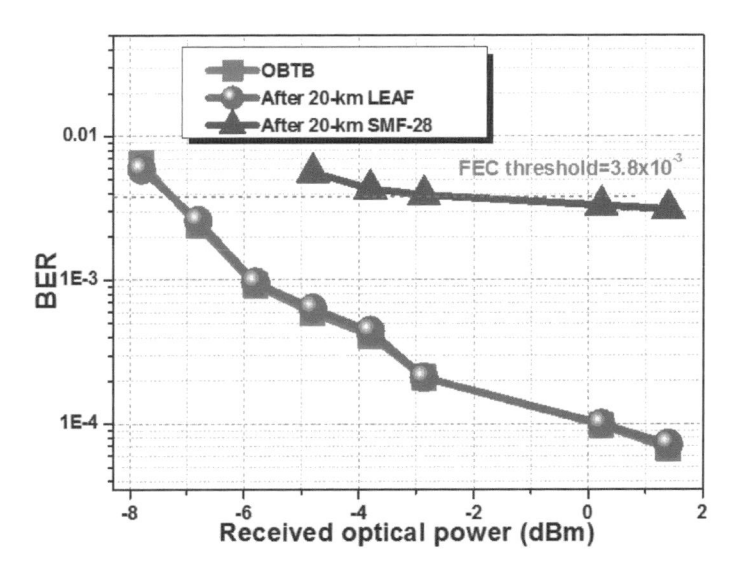

Fig. 4.9. BER versus received optical power of 256QAM-OFDM signal.

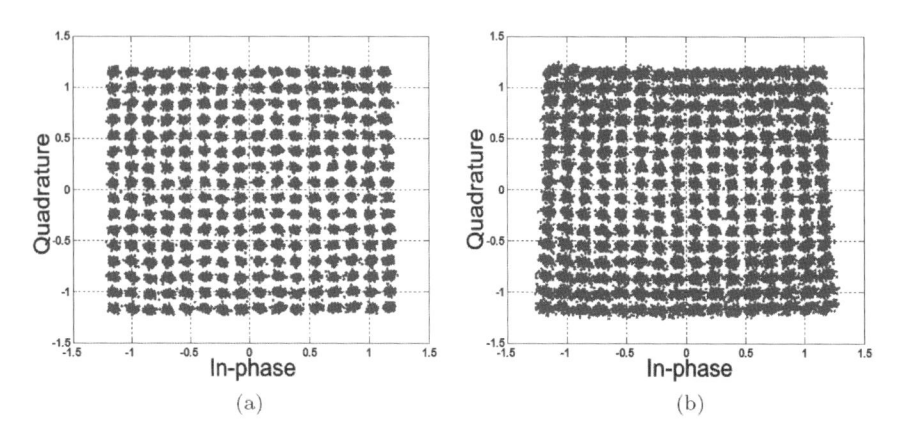

Fig. 4.10. Constellations of 256QAM-OFDM after 20-km (a) LEAF and (b) SSMF.

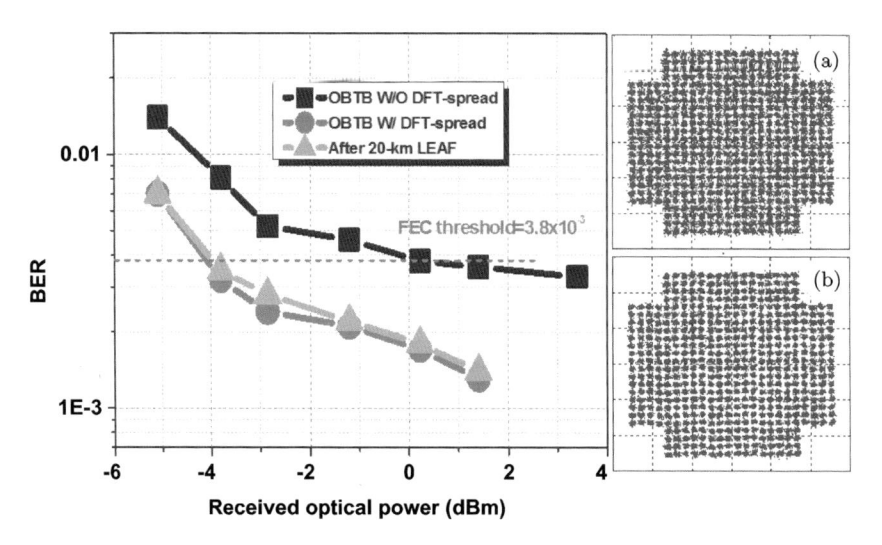

Fig. 4.11. BER versus received optical power of 512QAM-OFDM signal and constellations.

It can be seen that the CD-induced ICI after 20-km SSMF distorts the constellation.

The BER versus received optical power of 512QAM-OFDM signal and constellations are shown in Fig. 4.11. In order to demonstrate that the PAPR reduction can be used to improve the BER performance of the system, the OFDM signals with and without the DFT-spread technique are transmitted in the OBTB case. The BER of signal with DFT-spread after the 20-km LEAF is also measured and given in Fig. 4.11. We evaluate the PAPR performance by complementary cumulative distribution function (CCDF) that denotes the probability distribution of which the PAPR of current OFDM symbol is over a certain threshold. The PAPRs of DFT-spread OFDM signal and traditional OFDM signal at the probability of 1×10^{-3} are 11.45 dB and 14.16 dB, respectively. The measurement shows that there is 4-dB receiver sensitivity improvement at the FEC threshold of 3.8×10^{-3} when the DFT-spread technique is applied to reduce the PAPR and there is still no penalty observed after the 20-km LEAF transmission. The constellations of OFDM signal without and with DFT-spread technique are inserted as insets (a) and (b) in Fig. 4.11, respectively. The constellation becomes more convergent with the DFT-spread technique.

The BER versus the received optical power of 1024/2048QAM-OFDM signal and constellations are shown in Fig. 4.12. For both kinds of OFDM signal, there are no penalties between OBTB and after 20-km LEAF transmission. The required received optical power is −5.5 dBm and 0.23 dBm at BER of 2.4×10^{-2}. After removing the 20% FEC overhead, the net data rate is 26.4 Gbit/s (Net data rate = Raw data rate $\times 1/(1+0.2) \approx 26.4$ Gbit/s). The constellations of 1024QAM-OFDM signal at 1.6-dBm received optical power and 2048QAM-OFDM signal at 2.6-dBm received optical power after 20-km LEAF transmission are shown in Figs. 4.12(a) and 4.12(b), respectively.

Summary: In this section, we have demonstrated the largest QAM constellation reported to date for IM-DD system utilizing low cost DML, and 31.7-Gbit/s 2048QAM OFDM signal was successfully transmitted over 20-km LEAF under 20% SD-FEC threshold. This experiment was enabled by ISFA-based channel estimation, where large FFT size was used to enhance the resistance of ISI and DFT-spread technique to reduce PAPR.

4.4. DDO-OFDM Signal Based on DFT-Spread

The importance of channel estimation in OFDM based on frequency domain equalization has been discussed in the previous section of this chapter. The PAPR of the OFDM signal can be reduced by using the DFT-spread technique in a large-capacity DDO-OFDM system. Figures 4.13(a)

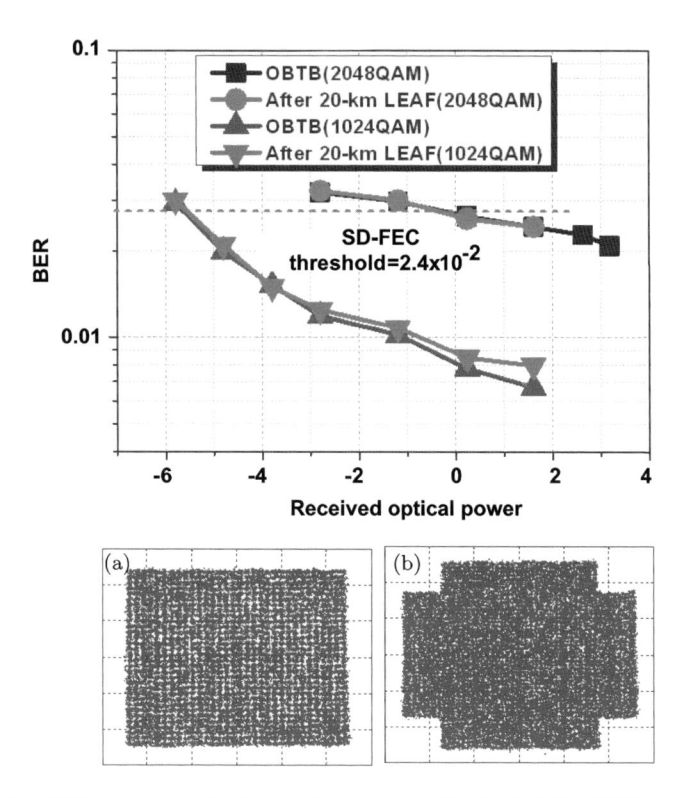

Fig. 4.12. BER versus received optical power of 1024/2048QAM-OFDM signal and constellations.

and 4.13(b) show the block diagram of conventional and DFT-spread DDO-OFDM systems, respectively. Here, the FFT size for OFDM signal generation and demodulation is N. L subcarriers in the positive frequency bins are used to convey 32-QAM data, L subcarriers in the negative frequency are filled with Hermitian symmetric data to generate the OFDM signal and the remaining subcarriers are reserved null for dc-bias and oversampling [3, 6]. Relative to conventional OFDM, the DFT-spread OFDM transceiver has an extra pair of L-point DFT/IDFT in the OFDM transmitter and receiver, respectively.

In the TS-based FDE for OFDM, it is ensured that the DSP operations for TS and OFDM signal symbols are exactly consistent. In the conventional OFDM transceiver, the DSP operations for TS and OFDM symbols are completely identical and the modulation format for TS is chosen to be QPSK. While in the DFT-spread transceiver shown in Fig. 4.13(b), the DSP

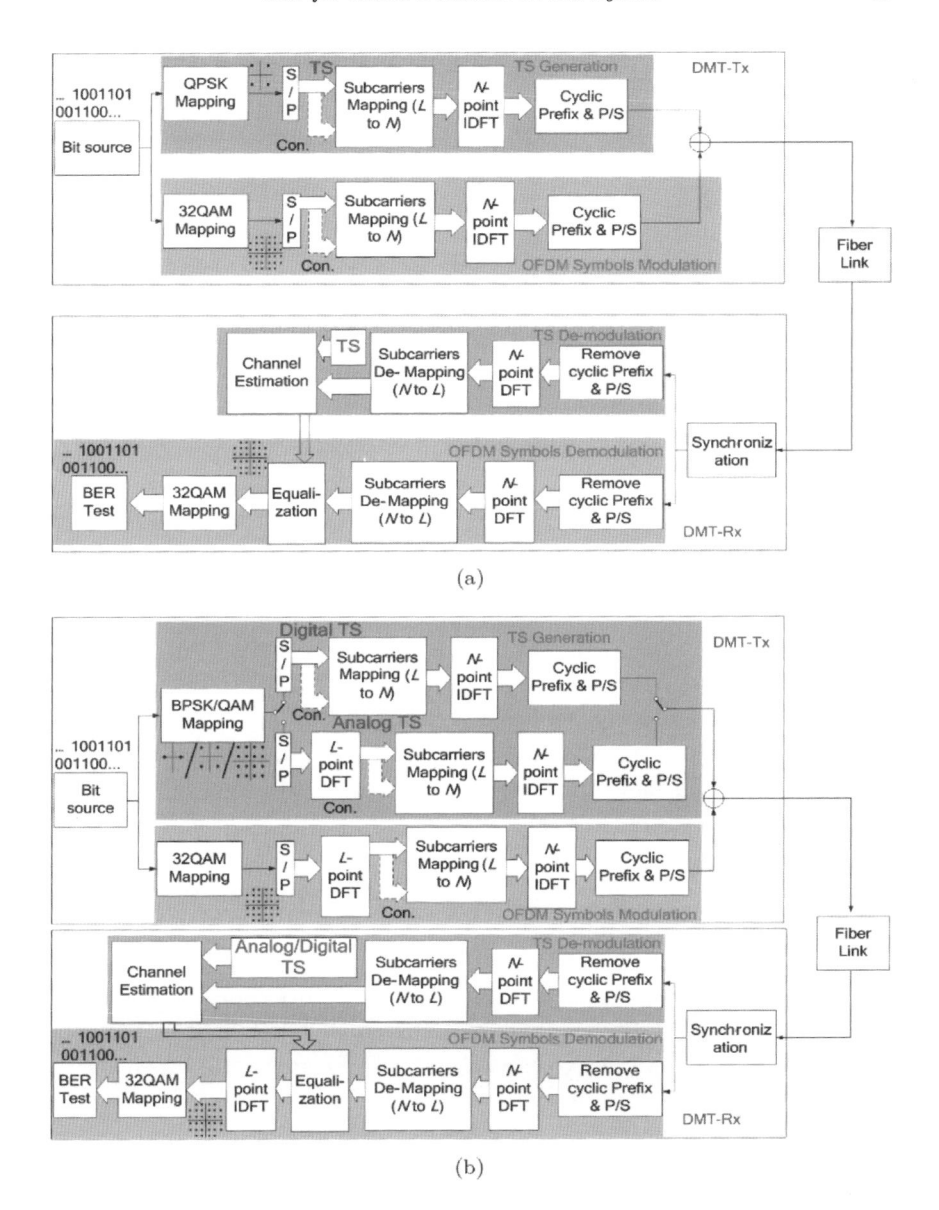

Fig. 4.13. Block diagram of DDO-OFDM system.

operations for TS are discussed due to the additional pair of DFT/IDFT in the OFDM transceiver. The main concern is related to two aspects: whether to join a group of additional DFT/IDFT in the training sequence of the transmitter and the receiver; Which modulation format of the training sequence should be selected to ensure optimal performance of the system. The digital and analog TSs are classified according to the type of the data before OFDM modulation. In the DFT-Spread OFDM scheme, if L-point DFT/IDFT is added during TS generation and reception process, analog TS will be generated, and if not, digital TS will be generated. If the modulation formats of the data are BPSK/QPSK/16QAM, this type of TS is classified as digital TS. The TS should be regarded as analog TS when DFT-spread is applied in the TS generation. With that extra L-point DFT, the TS becomes an analog signal in the frequency domain and the PAPR of the corresponding time domain signal after OFDM modulation becomes lower. Without that extra L-point DFT, the TS is a digital signal in the frequency domain and is the same as the TS of the conventional OFDM signal. Unfortunately, the PAPR of the corresponding time domain signal is a little higher. We will discuss the performance of digital and analog TSs in channel response estimation. In the digital TS schemes, BPSK/QPSK/16QAM-TSs will be discussed. In analog TS schemes, we only choose the analog TS generated by an extra L-point DFT with QPSK for discussion as the distribution characteristics of any analog TSs are similar. Apart from this, conventional OFDM with QPSK TS is also discussed for comparison.

Figure 4.14 gives the probability distribution of the amplitude of the OFDM signal: DFT-spread OFDM with BPSK/QPSK/16QAM/analog TS symbol and conventional OFDM. For the DFT-spread OFDM, six high discrete peaks appearing after that extra L-point DFT are applied for 32-QAM data. From this point of view, the DFT-spread OFDM can be regarded as the combination of one high probability PAM-6 signal and one low probability analog signal, and it explains why the PAPR of the DFT-spread OFDM is lower than the conventional OFDM. The probability distributions of the DFT-spread OFDM are quite similar except for slight differences in their TSs. In Fig. 4.14(d), another two small discrete peaks like PAM-2 can be seen as extra L-point DFT is applied for QPSK data in TS. The probability distribution of conventional OFDM in Fig. 4.14(e) likes Gaussian distribution. CCDF denotes that a probability distribution of the PAPR of the current OFDM symbol is over a certain threshold. During the

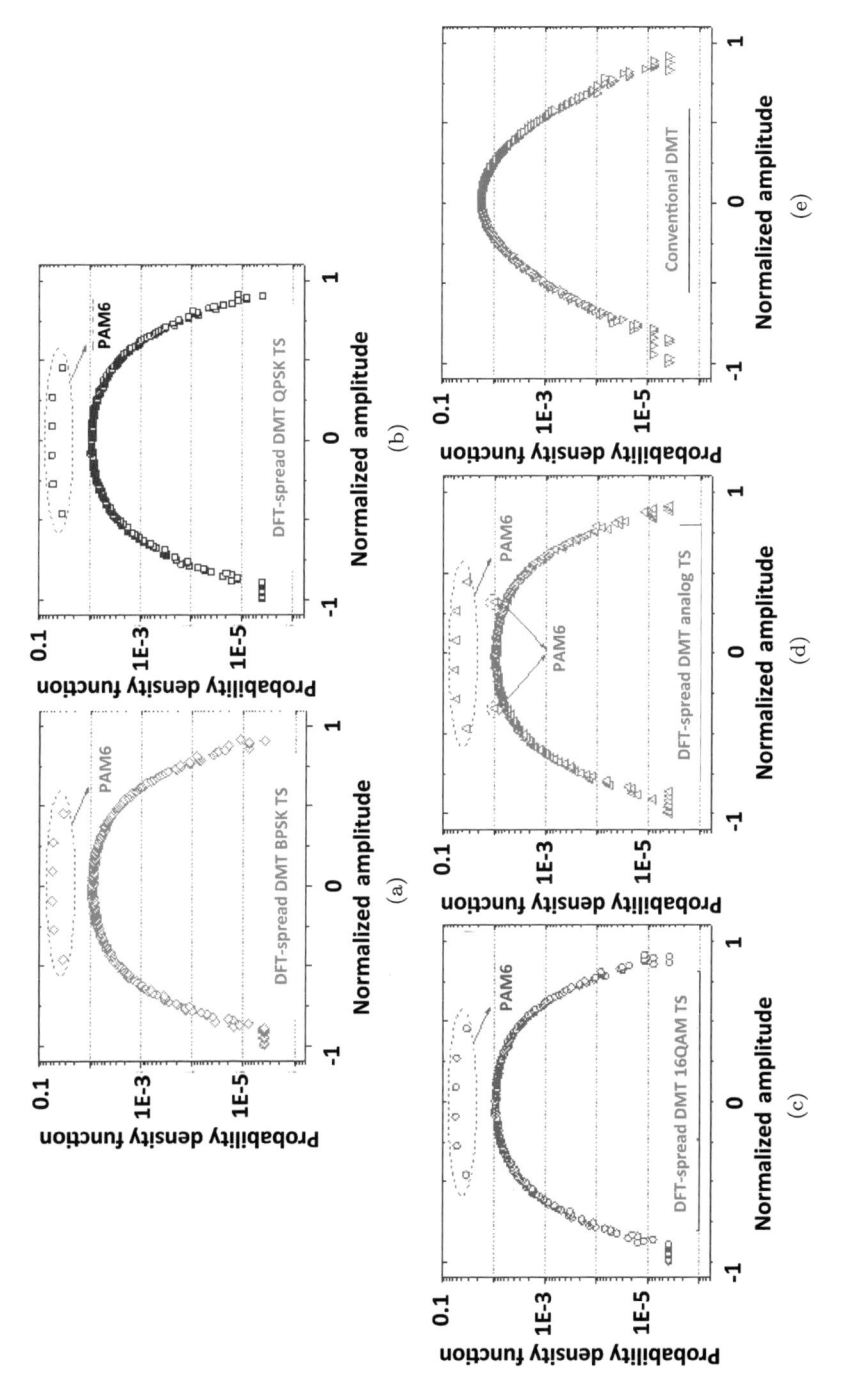

Fig. 4.14. Probability density function of OFDM signal.

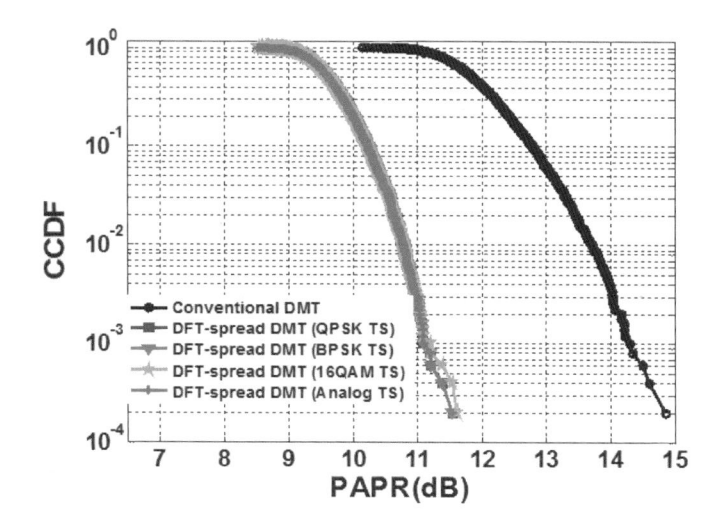

Fig. 4.15. CCDF curve of the OFDM signals with different types of TS.

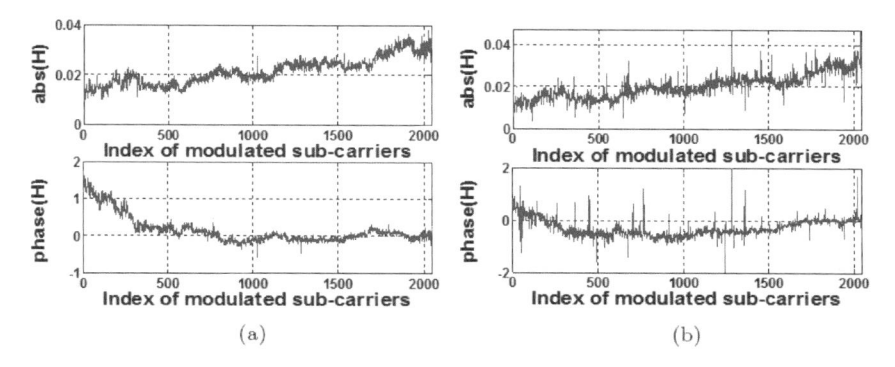

Fig. 4.16. Estimated channel response of DFT-spread OFDM with (a) QPSK TS and (b) analog TS.

process of the CCDF calculation, N and L are 8192 and 2048, respectively. Figure 4.15 gives the calculated CCDF curves for the conventional OFDM and DFT-spread OFDM with different types of TS. The PAPRs of DFT-spread OFDM with different types of TS are quite similar except for some slight differences in the TS and outperform that of the conventional OFDM. This means the additional pair DFT/IDFT for the TS symbol will not affect the PAPR of the DFT-spread OFDM. 3.4-dB PAPR improvement is attained at the probability of 2×10^{-4}.

In [3], the performance of channel estimation with the digital TS and analog TS is compared. The estimated channel response of the DFT-spread 32QAM-OFDM signal with digital QPSK TS and analog TS is shown in Figs. 4.16 (a) and 4.16(b), respectively. Compared to the channel response obtained with digital QPSK TS, both the amplitude and phase of the channel response acquired with analog TS exhibit high-frequency fluctuation as the analog TS is much more vulnerable than the digital TS to the noise in the optical fiber link. Thus, the performance of channel estimation with the digital TS is much better than that with the analog TS.

In [3], the measured BER shows that the BER performance of the DFT-spread OFDM with digital TSs is better than that of the DFT-spread OFDM with analog TSs and conventional OFDM. DFT-spread OFDM with BPSK/QPSK TS symbol demonstrated the best performance.

References

[1] F. Li, X. Li, L. Chen, Y. Xia, C. Ge and Y. Chen, High-level QAM OFDM system using DML for low-cost short reach optical communications, *IEEE Photon. Technol. Lett.* **26**(9) (2014) 941–944.

[2] F. Li, X. Li, J. Yu and L. Chen, Optimization of training sequence for DFT-spread DMT signal in optical access network with direct detection utilizing DML, *Opt. Express* **22**(19) (2014) 22962–22967.

[3] F. Li, X. Li, J. Zhang and J. Yu, Transmission of 100-Gb/s VSB DFT-spread DMT signal in short-reach optical communication systems, *IEEE Photon. J.* **7**(5) (2015) 1–7.

[4] L. Xu, D. Qian, J. Hu, W. Wei and T. Wang, OFDMA-based passive optical networks (PON), in *Proc. Technol. Digest. IEEE/LEOS Summer Topical Meetings*, 2008, pp. 159–160.

[5] D. Qian, J. Hu, P. Ji, T. Wang and M. Cvijetic, 10-Gb/s OFDMA PON for delivery of heterogeneous services, in *Proc. OFC*, 2008, paper OWH4.

[6] F. Li, J. Yu, Z. Cao, J. Zhang, M. Chen and X. Li, Experimental demonstration of four-channel WDM 560 Gbit/s 128QAM-DMT using IM/DD for 2-km optical interconnect, *J. Lightwave Technol.* **35**(4) (2017) 941–948.

[7] F. Li, Z. Cao, J. Yu, X. Li and L. Chen, SSMI cancellation in direct-detection optical OFDM with novel half-cycled OFDM, *Opt. Express* **21**(23) (2013) 28543–28549.

[8] W. R. Peng, X. Wu, V. R. Arbab, K. M. Feng, B. Shamee, L. C. Christen and J. Y. Yang, Theoretical and experimental investigations of direct-detected RF-tone-assisted optical OFDM systems, *J. Lightwave Technol.* **27**(10) (2009) 1332–1339.

[9] W. R. Peng, Analysis of laser phase noise effect in direct-detection optical OFDM transmission, *J. Lightwave Technol.* **28**(10) (2010) 2526–2536.

[10] T. Okoshi and K. Kikuchi, *Coherent Optical Fiber Communications* (KTK, 1988).
[11] X. Zhou and J. Yu, Digital signal processing for coherent optical communication, in *IEEE WOCC*, 2009.
[12] S. Haykin, *Adaptive Filter Theory* (Prentice-Hall, Englewood Cliffs, NJ, 2001).
[13] M. Oerder and H. Meyr, Digital filter and square timing recovery, *IEEE Trans. Commun.* **36**(5) (1988) 605–612.
[14] S. L. Jansen, I. Morita, T. C. Schenk and H. Tanaka, Long-haul transmission of 16 52.5 Gbits/s polarization-division-multiplexed OFDM enabled by MIMO processing (Invited), *J. Opt. Netw.* **7** (2008) 173–182.
[15] X. Liu and F. Buchali, Intra-symbol frequency-domain averaging based channel estimation for coherent optical OFDM, *Opt. Express* **16** (2008) 21944–21957.
[16] A. J. Lowery, Fiber nonlinearity pre- and post-compensation for longhaul optical links using OFDM, *Opt. Express* **15** (2007) 12965.
[17] A. Sano, E. Yamada, H. Masuda, E. Yamazaki, T. Kobayashi, E. Yoshida, Y. Miyamoto, R. Kudo, K. Ishihara and Y. Takatori, No-guard-interval coherent optical OFDM for 100-Gb/s long-haul WDM transmission, *J. Lightwave Technol.* **27**(16) (2009) 3705–3713.
[18] I. B. Djordjevic and B. Vasic, Orthogonal frequency division multiplexing for high-speed optical transmission, *Opt. Express* **14** (2006) 3767–3775.
[19] S. L. Jansen, I. Morita, N. Takeda and H. Tanaka, 20-Gb/s OFDM transmission over 4 160-km SSMF enabled by RF-pilot tone phase noise compensation, in *Proc. Opt. Fiber Commun. Conf.*, Anaheim, CA (2007), paper PDP15.
[20] X. Liu, S. Chandrasekhar, B. Zhu, P. Winzer, A. Gnauck and D. Peckham, 448-Gb/s reduced-guard-interval CO-OFDM transmission over 2000 km of ultra-large-area fiber and five 80-GHz-Grid ROADMs, *J. Lightwave Technol.* **29**(4) (2010) 483–490.
[21] W. Shieh, H. Bao and Y. Tang, Coherent optical OFDM: Theory and design, *Opt. Express* **16**(2) (2008) 841–859.
[22] W. Shieh and I. Djordjevic, *OFDM for Optical Communications* (Academic Press, 2009).
[23] W. Shieh and C. Athaudage, Coherent optical orthogonal frequency division multiplexing, *Electron. Lett.* **42**(10) (2006) 587–589.
[24] F. Gardner, A BPSK/QPSK timing-error detector for sampled receivers, *IEEE Trans. Commun.* **34**(5) (1986) 423–429.
[25] D. Godard, Passband timing recovery in an all-digital modem receiver, *IEEE Trans. Commun.* **26**(5) (1978) 517–523.
[26] G. Proakis, *Digital Communications*, 4th edn., Chapter 6, McGraw-Hill Education.
[27] D. N. Godard, Self-recovering equalization and carrier tracking in two-dimentional data communication systems, *IEEE Trans. Commun.* **28**(11) (1980) 1867–1875.

[28] C. R. Johnson, P. Schniter, T. J. Endres, J. D. Behm, D. R. Brown and R. A. Casas, Blind equalization using the constant modulus criterion: A review, *Proc. IEEE* **86** (1998) 1927–1950.

[29] X. Zhou, J. Yu and P. D. Magill, Cascaded two-modulus algorithm for blind polarization de-multiplexing of 114-Gb/s PDM-8-QAM optical signals, in *Proc. OFC 2009*, paper OWG3.

[30] I. Fatadin and S. J. Savory, Compensation of frequency offset for 16-QAM optical coherent systems using QPSK partitioning, IEEE *Photon. Technol. Lett.* **23**(17) (2011) 1246–1248.

[31] T. Nakagawa, Frequency-domain signal processing for chromatic dispersion equalization and carrier frequency offset estimation in optical coherent receivers, in *Advanced Photonics Congress*, OSA Technical Digest (online) (Optical Society of America, 2012), paper SpTh1B.4.

[32] T. Pfau, S. Hoffmann and R. Noe, Hardware-efficient coherent digital receiver concept with feedforward carrier recovery for M-QAM constellations, *J. Lightwave Technol.* **27**(8) (2009) 989–999.

[33] K.-P. Ho and J. M. Kahn, Electronic compensation technique to mitigate nonlinear phase noise, *J. Lightwave Technol.* **22**(3) (2004) 779–783.

[34] E. Ip and J. M. Kahn, Compensation of dispersion and nonlinear impairments using digital backpropagation, *J. Lightwave Technol.* **26** (2008) 3416–3425.

[35] E. Ip and J. M. Kahn, Fiber impairment compensation using coherent detection and digital signal processing, *J. Lightwave Technol.* **28**(4) (2010) 502–519.

[36] X. Li, X. Chen, G. Goldfarb, E. F. Mateo, I. Kim, F. Yaman and G. Li, Electronic post-compensation of WDM transmission impairments using coherent detection and digital signal processing, *Opt. Express* **16**(2) (2008) 880–888.

[37] S. Zhang, M. Huang, F. Yaman, E. Mateo, D. Qian, Y. Zhang, L. Xu, Y. Shao, I. Djordjevic, T. Wang, Y. Inada, T. Inoue, T. Ogata and Y. Aoki, 40 × 117.6 Gb/s PDM-16QAM OFDM transmission over 10,181 km with soft-decision LDPC coding and nonlinearity Compensation, in *Proc. OFC 2012*, paper PDP5C.4.

[38] G. Bosco, V. Curri, A. Carena, P. Poggiolini and F. Forghieri, On the performance of Nyquist-WDM terabit superchannels based on PM-BPSK, PM-QPSK, PM-8QAM or PM-16QAM subcarriers, *J. Lightwave Technol.* **29** (2011) 53–61.

[39] Y. Gao, J. C. Cartledge, A. S. Karar, S. S.-H. Yam, M. O'Sullivan, C. Laperle, A. Borowiec and K. Roberts, Reducing the complexity of perturbation based nonlinearity pre-compensation using symmetric EDC and pulse shaping, *Opt. Express* **22** (2014) 1209–1219.

[40] J. Wang, C. Xie and Z. Pan, Generation of spectrally efficient Nyquist-WDM QPSK signals using digital FIR or FDE filters at transmitters, *J. Lightwave Technol.* **30** (2012) 3679–3686.

[41] Z. Dong, X. Li, J. Yu and N. Chi, 6 × 144-Gb/s Nyquist-WDM PDM-64QAM generation and transmission on a 12-GHz WDM Grid equipped with Nyquist-band pre-equalization, *J. Lightwave Technol.* **30** (2012) 3687–3692.
[42] Z. Dong, X. Li, J. Yu and N. Chi, 6×128-Gb/s Nyquist-WDM PDM-16QAM generation and transmission over 1200-km SMF-28 with SE of 7.47 b/s/Hz, *J. Lightwave Technol.* **30** (2012) 4000–4005.
[43] X. Zhou, J. Yu, M.-F. Huang, Y. Shao, T. Wang, L. Nelson, P. Magill, M. Birk, P. I. Borel, D. W. Peckham, R. Lingle and B. Zhu, 64-Tb/s, 8 b/s/Hz, PDM-36QAM transmission over 320 km using both pre- and post-transmission digital signal processing, *J. Lightwave Technol.* **29** (2011) 571–577.
[44] X. Zhou, L. E. Nelson, P. Magill, B. Zhu and D. W. Peckham, 8 × 450-Gb/s, 50-GHz-spaced, PDM-32QAM transmission over 400 km and one 50 GHz-grid ROADM, in *Proc. OFC 2011*, paper PDPB3.
[45] D. McGhan, C. Laperle, A. Savchenko, C. Li, G. Mak and M. O'Sullivan, 5120 km RZ-DPSK transmission over G652 fiber at 10 Gb/s with no optical dispersion compensation, in *Proc. OFC 2005*, PDP27.
[46] J. Zhang and H. Chien, A novel adaptive digital pre-equalization scheme for bandwidth limited optical coherent system with DAC for signal generation, in *Proc. OFC 2014*, paper W3K.4.
[47] J. Zhang, J. Yu, N. Chi and H.-C. Chien, Time-domain digital pre-equalization for band-limited signals based on receiver-side adaptive equalizers, *Opt. Express* **22** (2014) 20515–20529.
[48] J. Zhang, J. Yu, F. Li, N. Chi, Z. Dong and X. Li, 11 × 5 × 9.3 Gb/s WDM-CAP-PON based on optical single-side band multi-level multi-band carrier-less amplitude and phase modulation with direct detection, *Opt. Express* **21**(16) (2013) 18842–18848.
[49] J. Zhang and J. Yu, EML-based IM/DD 400G (4 × 112.5-Gbit/s) PAM-4 over 80 km SSMF based on linear pre-equalization and nonlinear LUT pre-distortion for inter-DCI applications, in *OFC* (2017), W4I.4.

Chapter 5

DFT-Spread for OFDM Based on Coherent Detection

5.1. Introduction

The implementation of 400+ Gb/s or 1+ Tb/s bit rate per wavelength is expected in near future to satisfy enormous capacity demand from diversified bandwidth-hungry services in next generation optical networks. Along with the advent and promotion of high-speed digital-to-analog conversion (DAC), coherent detection of OFDM systems (CO-OFDM) with high spectrum efficiency in more than 100Gb/s transmission system has attracted the interest of many researchers [1-21]. But in order to increase system capacity to 400Gb/s, we need to increase the spectral efficiency (SE) [8-10]. It is well known that SE in coherent-detection systems can be improved with the use of high-order modulation formats [8-10, 20, 21]. Using high-order modulation formats requires higher OSNR on the receiver and increases the complexity of the transmission system. Consequently, the transmission distance will be reduced by using high-order modulation formats. Considering all these factors, 16QAM is a very feasible alternative modulation format to realize over 100 Gb/s or even 400 Gb/s signal transmission in the future backbone network [11-14].

High-frequency power attenuation is an inevitable problem when transmitting broadband orthogonal frequency division multiplexing (OFDM) in the CO-OFDM system. It is mainly caused because the bandwidths of DAC, electrical amplifier, modulator and analog-to-digital conversion (ADC) in the CO-OFDM system are insufficient [12]. High-frequency subcarrier of broadband OFDM signal will appear in different degrees of power attenuation because of insufficient bandwidth. Reduction of signal-to-noise ratio (SNR) will lead to the emergence of bit error rate (BER) [15] The BER performance of the overall system will be seriously deteriorated if high-frequency power attenuation is very serious. Pre-equalization can be used to compensate high-frequency power attenuation in optical fiber communication systems with coherent detection [12, 13]. In order to achieve pre-equalization, we need to obtain the accurate transfer function by complicated channel estimation during optical back-to-back (OBTB). Another problem will appear when peak-to-average power ratio (PAPR) of OFDM signal is too high [14, 16, 17]. In this case, it will cause severe nonlinear noise in optical fiber link. We must reduce PAPR of OFDM signal in order to increase the distance of optical transmission. In this chapter, we find that PAPR of the signal after pre-equalization is a little bit higher than that of the original signal when we compensate high-frequency power attenuation by pre-equalization. So the transmission distance of broadband OFDM signal in optical fiber cannot increase even if we have compensated high-frequency power attenuation because PAPR of the OFDM signal is not decreased. DFT-spread (DFT-S) technique is proposed to simultaneously overcome high-frequency power attenuation and reduce PAPR of the OFDM signals in the CO-OFDM systems when transmitting wideband OFDM signals [1, 7]. DFT-spread technique is used to reduce the PAPR of OFDM signals in wireless communication [17]. It was introduced in optical communication because it was able to effectively reduce the OFDM signal PAPR [17]. In the DFT-S OFDM scheme, all subcarriers carrying the signal can be divided into 1 band or multiple bands, and then the DFT operation is performed in the band units, respectively. It should be noted that when the DFT-S 16QAM-OFDM transmits, it has a very good robustness to high-frequency power attenuation because each 16QAM data signal after the DFT-S operation is distributed to all subcarriers of the sub-band [17]. So we conclude that transmission performance of the 16QAM-OFDM signal using the DFT-S technique is better than that of 16QAM-OFDM signal using the pre-equalization technique.

In this chapter, we experimentally compare the BER performance during OBTB and nonlinear noise tolerance after fiber transmission by using PDM-16QAM-OFDM signals with one or more band DFT-S PDM-16QAM-OFDM signal. Experimental results showed that DFT-S OFDM 1 band has the best BER performance because it can simultaneously overcome high-frequency power attenuation and reduce PAPR of the OFDM signals. In addition, it has been experimentally demonstrated that DFT-S OFDM with 1 band has the best resistance to narrowband optical filtering.

We demonstrated wavelength division multiplexing (WDM) transmission with 244.2 Gb/s 16QAM-OFDM per channel and transmitted 400+ Gb/s 16QAM-OFDM when we considered two WDM channels as a whole. During the measurement, the BER of 8 × 244.2 Gb/s WDM pre-equalization 16QAM-OFDM signal transmitted over 2 × 420 km SMF is less than 2.4×10^{-2}. The transmission distance over optical fiber can be increased to 3×420 km. This indicates that system performance of 16QAM-OFDM using the DFT-spread technique is better than that of the 16QAM-OFDM signal using the pre-equalization technique in the WDM CO-OFDM system.

5.2. DFT-Spread Technique and Pre-equalization Technique

5.2.1. *DFT-spread technique*

In this chapter, the bandwidth of all subcarriers in the OFDM signal is 32 GHz, and the sampling rate of the DAC generating OFDM signal is set to 64 GSa/s. If the size of the IFFT in the OFDM signal generation process is assumed to be N, the number of subcarriers carrying signals in the system is $N/2$. At the transmitter of DFT-S OFDM, all subcarriers carrying signals are divided into K sub-bands, which have $N/2K$ samples. Figure 5.1 shows the subcarrier structure of K ($K < 8$) bands DFT-spread OFDM signal. K groups of $N/2K$ samples after FFT are inserted into subcarriers carrying signals. Subcarriers at high frequencies are used for oversampling by inserting zeros, subcarriers at zero frequency and nearby are reserved

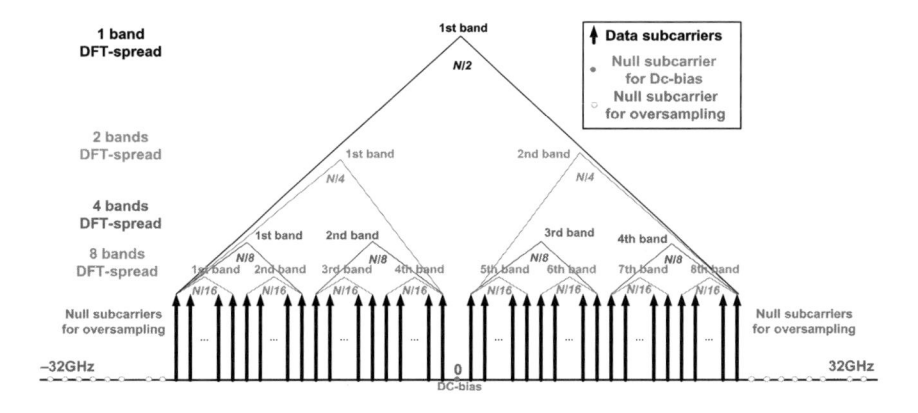

Fig. 5.1. Subcarrier structure of K ($K < 8$) bands DFT-spread OFDM signal.

Table 5.1. Increased multiplication percentage at corresponding systems with different K.

K	1	2	4	8
η	$\left(\frac{100\times\log_2\frac{N}{2}}{2\times\log_2 N}\right)\%$	$\left(\frac{100\times\log_2\frac{N}{4}}{2\times\log_2 N}\right)\%$	$\left(\frac{100\times\log_2\frac{N}{8}}{2\times\log_2 N}\right)\%$	$\left(\frac{100\times\log_2\frac{N}{16}}{2\times\log_2 N}\right)\%$

for RF pilot for frequency offset estimation and phase noise estimation, and then an N-point IFFT is used to generate the OFDM signal. The PAPR using the DFT-spread technique is lower than that of the original OFDM signal [7, 17]. DFT-spread OFDM employed identical digital signal processing (DSP) with original OFDM signal except that it added K pairs of $N/2K$-points FFT/IFFT at transmitter and receiver. We compared the additional computational complexity caused by DFT-spread with different K. N-point FFT needs to calculate complex multiplication $N/2 \log_2 N$ times. The amount of multipliers of FFT/IFFT is used to estimate its computational complexity. When subcarriers carrying signals are divided into K sub-bands, the proportion of multiple operations added versus that in FFT/IFFT of the original OFDM signal can be expressed as

$$\eta = \left(\frac{100 \times \log_2 \frac{N}{2K}}{2 \times \log_2 N}\right)\% \qquad (5.1)$$

When subcarriers carrying signals are divided into 1, 2, 4, and 8 subbands, the proportions of increased multiplication at corresponding systems are given in Table 5.1. Figure 5.2 showed the curves of extra multiplication operation with different sub-bands when N of FFT changes. We can see that one band scheme brought the most complicated extra computational complexity. But the complexity gap between different bands reduced as N increased. We need to select a large size of FFT in order to ensure that the resolution during channel estimation and equalization at frequency domain of the system is high enough because the OFDM signal has pretty large bandwidth. However, the size of FFT cannot be too large because PAPR of the OFDM signal will be very high and the complexity of the system will increase. In our experiment, we choose N to be 1024 so that the amount of subcarriers carrying signals is 512.

5.2.2. Pre-equalization technique

Pre-equalization technique is proposed to compensate high frequency at the CO-OFDM system [12, 13, 17]. Bandwidth-limitation devices mainly

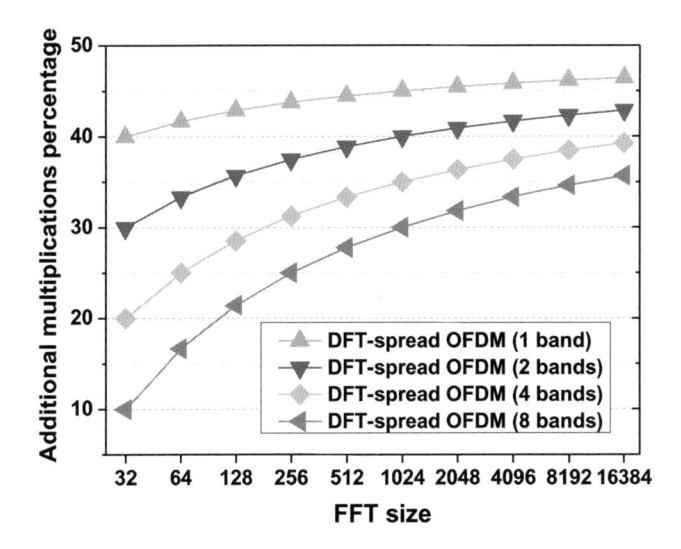

Fig. 5.2. FFT size N versus additional multiplication percentage.

include DAC, electric amplifier, modulator and ADC. These devices have attenuation and there is no polarization crosstalk when single polarization signal is transmitted. So the channel response estimated using the single polarization scheme should be more accurate. We only compensate the intensity with phase unconcerned when pre-equalizing the OFDM signal, since the amplitude response of the system is relatively stable and the phase-frequency response is varied. We first need to accurately estimate the channel response in order to compensate for high-frequency power attenuation. The time-domain average method is applied to the channel estimation to improve the accuracy of the channel estimation. The time-domain average method is repeated to send M OFDM symbols in common in time domain, then we carry out time synchronization after coherent detection without frequency offset and phase noise in the receiver. When the signal synchronized, we average the same M OFDM symbols in time domain and estimate the channel after the average samples have been OFDM demodulated. It should be noted that the samples averaged will no longer be accurate when the value of M is too large because optical fiber is a slowly varying channel. In this experiment, we set the number M of OFDM symbols repeated in time domain to be 124 and choose QPSK which has the strongest robustness against noise to be the modulation format. In order to meet the condition without frequency deviation, we set signal laser and local oscillator to be one external cavity laser (ECL). But phase noise cannot be completely

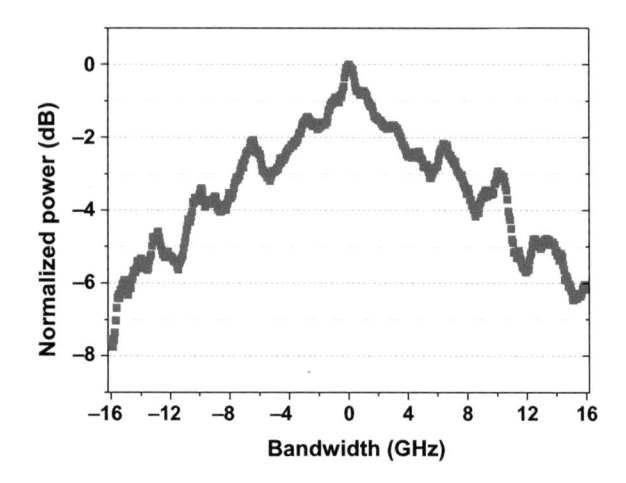

Fig. 5.3. Amplitude-frequency response curve of system channel.

avoided because the bandwidth of ECL in practice cannot be zero. Thus, an ECL with 400 Hz linewidth is chosen to be optical laser to overcome the effect of phase noise on channel estimation accuracy as far as possible. We can first estimate the channel response curve at the receiver based on above description exactly. Figure 5.3 shows the amplitude response curve of system channel, and the 3-dB bandwidth is about 8 GHz as shown in Fig. 5.3. Thus, the high-frequency power attenuation will be obvious when 32 GHz OFDM signal is transmitted in the system with 3-dB bandwidth of 8 GHz. According to the estimated channel amplitude-frequency response, we can calculate the amplitude-frequency characteristic of pre-equalization function as shown in Fig. 5.4, in which only the subcarriers carrying the amplitude of the signal need to be compensated.

5.3. Experimental Comparison of Two Optical Carrier 400 Gb/s OFDM Signal Transmission Performance Based on DFT-Spread and Pre-equalization Techniques

5.3.1. *Experimental setup and results*

Figure 5.5 shows the experimental setup of eight WDM PDM-16QAM-OFDM with 37.5-GHz channel spacing transmitted in the CO-OFDM system. At the transmitter, eight ECLs with 100 kHz linewidth and <14.5 dBm

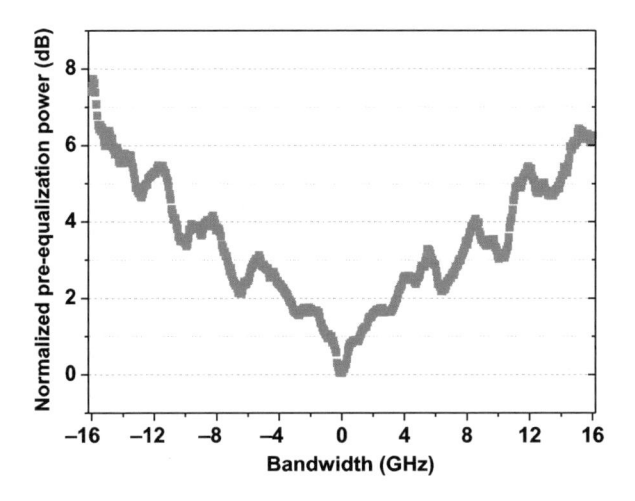

Fig. 5.4. Amplitude-frequency characteristic of pre-equalization function.

output power are used to be optical laser source. The odd group of ECLs includes ECL1, ECL3, ECL5, ECL7 from 1553.875 nm to 1555.677 nm, and the even group of ECLs includes ECL2, ECL4, ECL6, ECL8 from 1554.175 nm to 1555.977 nm. The channel spacing between two ECLs of each group is 75 GHz. The odd or even groups of ECLs are firstly combined by a polarization maintaining optical coupler (PM-OC) and then modulated by I/Q modulator, the I and Q components of 32-GHz 16QAM-OFDM signal generated by each of the two 64-GSa/s DACs are firstly amplified and then used to drive an I/Q modulator. As shown in Fig. 5.5, the signal is generated by offline DSP in the transmitter. In the OFDM modulation, the IFFT size is 1024. In all 1024 subcarriers, 512 subcarriers are filled with 16QAM data. Seven subcarriers around the zero frequency are reserved for RF-pilot, and other high-frequency subcarriers are set to zero for oversampling. A Cyclic Prefix (CP) of 32 samples is added before every OFDM symbol to resist CD and PMD after converting signal to time domain by IFFT, giving an OFDM symbol size of 1056. The 3-dB bandwidth of DAC generating OFDM signal is 11.5 GHz, and the bandwidth of I/Q modulator is 27 GHz. Thus, DAC causes the most serious high-frequency power attenuation. In the E/O modulation of the OFDM signal, the bias points of two parallel Mach–Zehnder modulators in I/Q modulator are controlled to slightly deviate from the zero point to generate RF-pilot, and the third modulator keeps bias at the orthogonal point ensuring that the phase difference between upper and lower

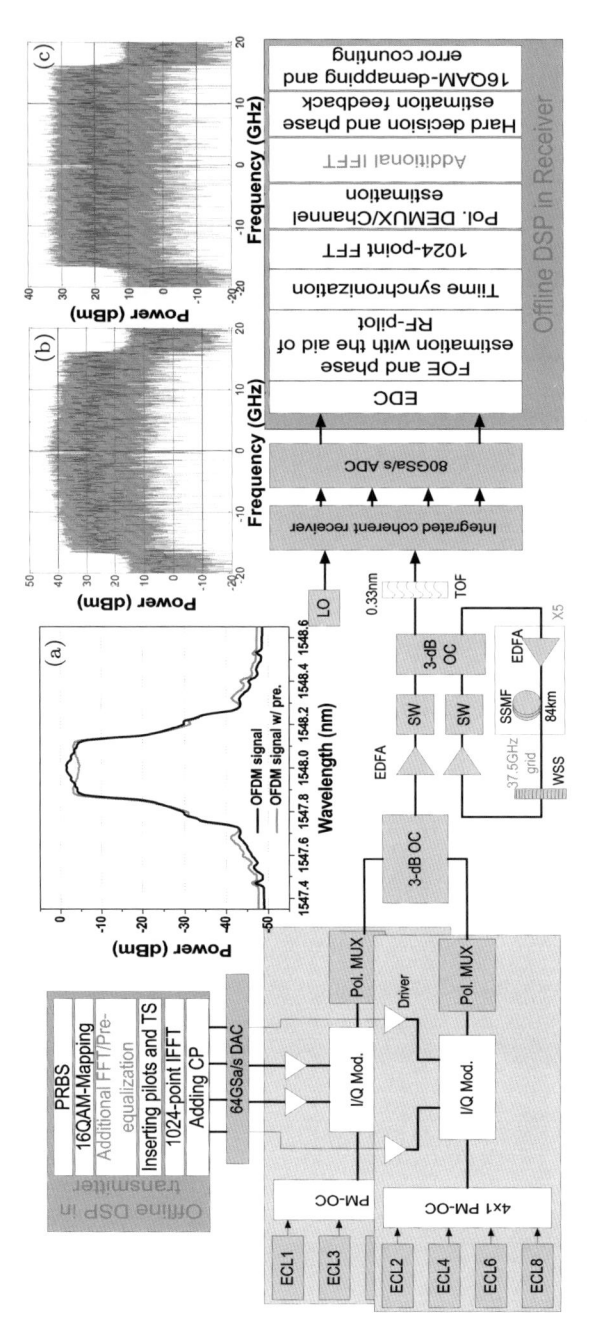

Fig. 5.5. Experimental setup of eight WDM PDM-16QAM-OFDM with 37.5-GHz channel spacing transmitted in the CO-OFDM system.

branches is controlled in $\pi/2$. The polarization multiplexing is realized by a polarization multiplexer. The simulated polarization multiplexer includes PM-OC that can equal the signal into two branches, an optical delay line (DL) that can provide a delay between two polarizations identical to the OFDM symbol length $(1/80 \times (1024 + 16))$ns = 13 ns) to eliminate the correlation between X-polarization and Y-polarization and constructing a time-domain orthogonal training sequence (TS), and a polarization beam combiner (PBC). One pair of TSs is inserted before every set of 122 OFDM symbols. The line data rate of PDM-16QAM-OFDM signal is 64 Gb/s $\times$ 512/1056 $\times$ 122/124 $\times$ 4 $\times$ 2 = 244.2 Gb/s after excluding overheads. The generated PDM-16QAM-OFDM optical signals in odd and even channels are combined to 8 $\times$ 244.2 Gb/s WDM PDM-16QAM-OFDM signal and then amplified by an EDFA before launching into a fiber loop which have five spans of 84-km SMF and five EDFA (NF = 5 dB). The signal output from fiber loop is then input to the integrated coherent receiver for the photoelectric detection. The total attenuation of each SMF span is about 18 dB and dispersion coefficient is 17 ps/km/nm at 1550 nm. A wavelength selective switch (WSS) is used to filtering ASE noise of the signal after five SMF span transmission. Each WSS is followed by an EDFA to compensate for the fiber loss. So optical signal will transmit through five SMF spans and a WSS in the loop every time.

In the receiver, a tunable optical filter (TOF) with 3-dB bandwidth of 0.33 nm is used to select the desired channel. An ECL is used as LO with <100-kHz linewidth for the selected channel. O/E detection of the signal is implemented with an integrated coherent receiver. The ADC is realized in the real-time oscilloscope with 80-GSa/s sampling rate and 30-GHz bandwidth. The captured data is then processed with offline DSP. The offline DSP can be divided into eight stages: (1) Electronic Dispersion Compensation (EDC) [18]; (2) Frequency Offset Equalization (FOE) and phase estimation with the aid of RF-pilot [19, 21]; (3) time synchronization; (4) 1024-point FFT; (5) polarization demultiplexing (Pol. DEMUX); (6) additional IFFT corresponding to additional FFT in the transmitter; (7) hard decision and phase estimation feedback; (8) 16QAM-demapping and error counting. In our experiment, BER is counted over 10 $\times$ 499712 bits (10 data frames, and each frame contains 440832 bits).[19,20–22]

We need to perform an additional FFT/pre-equalization in the OFDM transmitter in order to generate DFT-spread OFDM or pre-compensated OFDM signals. A conventional OFDM signal is also generated and transmitted to compare with other signals at the same time. DFT-spread OFDM

(1 band and 2, 4, and 8 bands) is generated in this experiment and transmitted in the system. Respectively, the 512 subcarriers carrying the signal are divided into groups 1, 2, 4 and 8. Pre-equalization technique has been introduced in the previous section. The number of TSs used for channel estimation in the pre-equalization is 124, with the linewidth of ECL used in the channel estimation as 400 Hz. Figure 5.5(a) shows the spectrum of a single-channel conventional optical OFDM and pre-equalized optical OFDM signal (0.02 nm resolution). As shown in Fig. 5.5, the pre-equalized OFDM signal is slightly overequalized at high frequencies, while this overequalization is exactly offset from the high-frequency attenuation in the ADC. The spectrum of the received conventional OFDM signal and the pre-equalized OFDM signal are given in Figs. 5.5(b) and 5.5(c), respectively. After pre-equalization, all subcarriers have uniformly distributed power.

The complementary cumulative distribution function (CCDF) represents the probability that PAPR of current OFDM symbol exceeds a certain threshold. Figure 5.6 shows the CCDF curves for different OFDM signals. The PAPR of pre-equalized OFDM signals is a bit higher than the PAPR of original OFDM signal, while the PAPR of DFT-spread OFDM signal is smaller than the PAPR of original OFDM signal. This means that DFT-spread OFDM signal has better nonlinear tolerance when transmitted

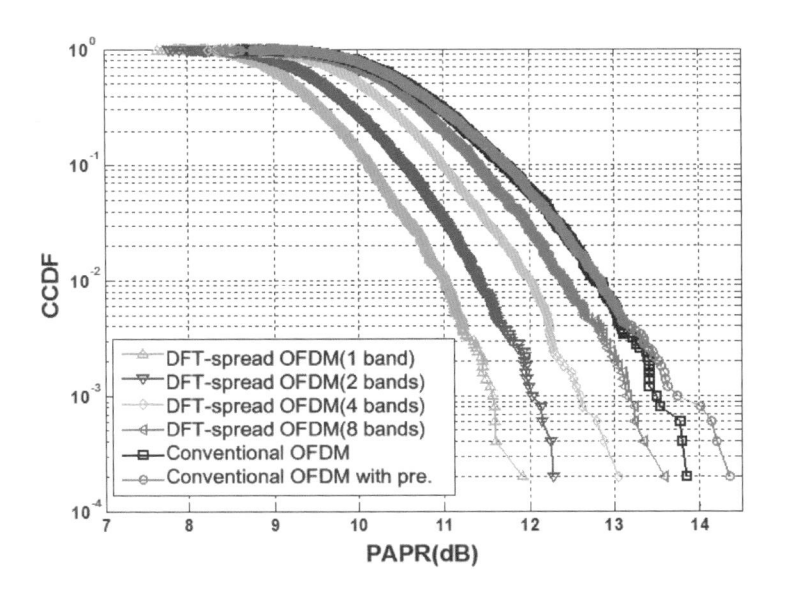

Fig. 5.6. CCDF curves of different OFDM signals.

in the fiber. In the DFT-spread OFDM scheme, the lower the number of bands divided by the signal subcarriers, the lower the PAPR of the OFDM signal. The PAPR of DFT-spread OFDM (1 band) signal is improved by 2 dB and 2.5 dB with respect to the conventional OFDM signal and the OFDM signal after the pre-equalization, respectively, when the CCDF probability is 2×10^{-4}. The 16QAM data in each sub-band is spread over all subcarriers in the band after DFT-spread, so DFT-spread can be used to resist high-frequency power attenuation. All 16QAM data in DFT-spread OFDM (1 band) signal is distributed to all 512 subcarriers. The DFT-spread OFDM signal of 1 band should have the best performance in resisting high-frequency attenuation compared to multiple bands DFT-spread OFDM signal. DFT-spread OFDM (1 band) signal can simultaneously overcome the high-frequency power attenuation and reduce the PAPR of the OFDM signal as mentioned in the two points above.

Figure 5.7(a) shows the optical spectra of output signal from IQ modulator when DAC output is 0 (0.01 nm resolution). We find two peaks on both sides of the residual optical carrier. The frequency difference between the residual optical carrier and two peaks is 16 GHz, which means there is a 16 GHz sine wave signal. But the power of this sine wave signal is very large, so it can be seen as a narrowband interference in the transmission of broadband OFDM. In order to overcome this interference, we will set ±16 GHz and the left and right sides of the two subcarriers to zero to eliminate the narrowband interference on the signal. When we set these six subcarriers to zero to eliminate the narrowband interference, the practical bandwidth of 32 GHz 16QAM-OFDM signal will be $(512 + 7 + 6) \times 64/1024 = 32.8125$ GHz. The spectrum of 32 GHz 16QAM-OFDM without frequency offset is presented in Fig. 5.7(b) at this time. We can see two peaks at the edge of the signal spectrum. We can find that these two peaks are just at ±16 GHz when partially expanding the spectrum within the range. Figures 5.7(c) and 5.7(d) show the peaks at −16 GHz and +16 GHz respectively. In order to verify the effectiveness of this zero-setting method against narrowband interference, the constellations of 32 GHz 16QAM-OFDM signal when transmitted without eliminating narrowband interference and using zero-setting method against narrowband interference (OSNR = 35dB) are shown in Figs. 5.8(a) and 5.8(b) as well as corresponding error vector magnitude (EVM), respectively. When the signal is set to zero for resisting narrowband interference of ±16 GHz, the signal constellation becomes more concentrated and the EVM value drops from 10.9269% to 9.8617%.

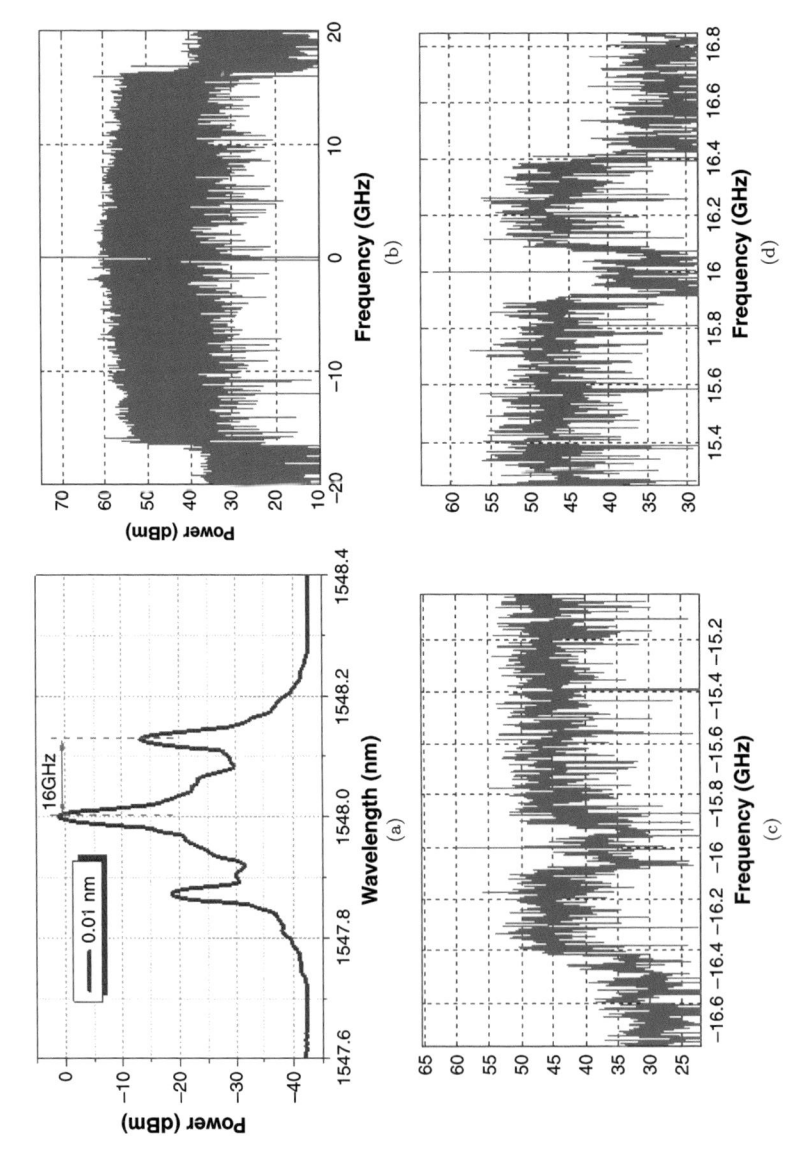

Fig. 5.7. (a) Optical spectra of output signal from IQ modulator when DAC output is 0, (b) spectrum of 32 GHz 16QAM-OFDM without frequency offset, (c) narrowband interference spectrum at −16 GHz, and (d) narrowband interference spectrum at +16 GHz.

EVM=10.9269%   EVM=9.8617%
     (a)   (b)

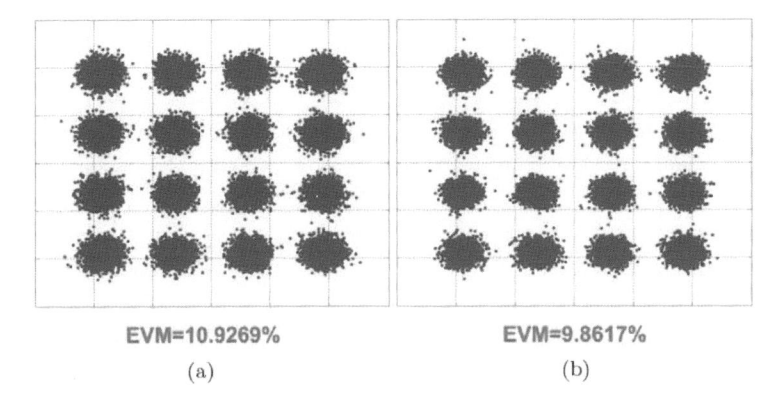

Fig. 5.8. Constellation of 32 GHz 16QAM-OFDM signal (a) without eliminating narrowband interference and (b) using zero-setting method against narrowband interference.

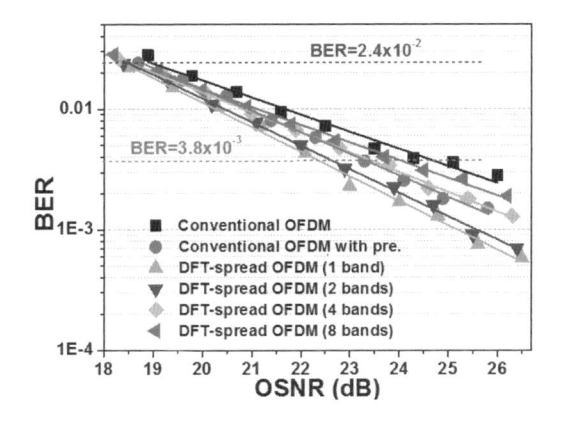

Fig. 5.9. Measured OSNR versus BER of 32 GHz 16QAM-OFDM signal in OBTB.

The measured OSNR versus BER of 32 GHz single channel in 16QAM-OFDM format are shown in Fig. 5.9. DFT-spread OFDM (1 band) signal exhibits the best bit error performance. The less the number of bands of the DFT-spread OFDM signal, the better the BER performance of the signal. Since the high frequency power attenuation is compensated, pre-compensated OFDM signal has better BER performance than conventional OFDM signal. High PAPR of pre-equalized signal limits the degree of system performance improvement. When pre-equalized OFDM signal is transmitted at back-to-back (BTB) position, the nonlinear distortion caused by its high PAPR in the electrical amplifier will limit the improvement of BER performance. Comparing conventional OFDM signal and pre-equalized OFDM signal shows that the receiving sensitivity of the DFT-spread OFDM (1 band) signal is improved by 2.3 dB and 1.2 dB at BER of 3.8×10^{-3}.

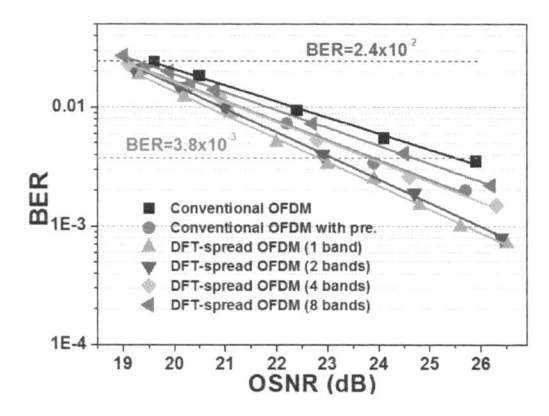

The measured OSNR versus BER of 32 GHz 16QAM-OFDM signal in the third channel of the WDM system with 37.5-GHz channel spacing at OBTB position is shown in Fig. 5.10. Compared with the performance of 32 GHz single channel 16QAM-OFDM signal, eight WDM 32 GHz PDM-16QAM-OFDM with 37.5-GHz channel spacing will not cause any OSNR penalty. In this situation, DFT-spread OFDM (1 band) signal has the best BER performance.

We filter tested the 32 GHz PDM-16QAM-OFDM signal at single channel BTB position, of which we replaced TOF (0.33 nm bandwidth) with a bandwidth-tunable WSS. The 3-dB bandwidth of WSS versus required OSNR and OSNR penalty at BER of 3.8×10^{-3} is shown in Fig. 5.11. We can see that OSNR and OSNR penalty required in the system will increase rapidly when the bandwidth of WSS is less than the bandwidth required for conventional 16QAM-OFDM signal transmission at 32 GHz. In six different types of 32 GHz 16QAM-OFDM signals, the speed of required OSNR and OSNR penalty increasing in DFT-spread OFDM(1 band) signal is the slowest when WSS decreased gradually. We should mention that pre-equalization for signal is achieved before the WSS is added, high-frequency power attenuation appears again if the bandwidth of WSS is insufficient when adjusted to the required bandwidth for 32 GHz 16QAM-OFDM signal

Fig. 5.10. Measured OSNR versus BER of 32 GHz 16QAM-OFDM signal in the third channel of the WDM system with 37.5-GHz channel spacing at OBTB position.

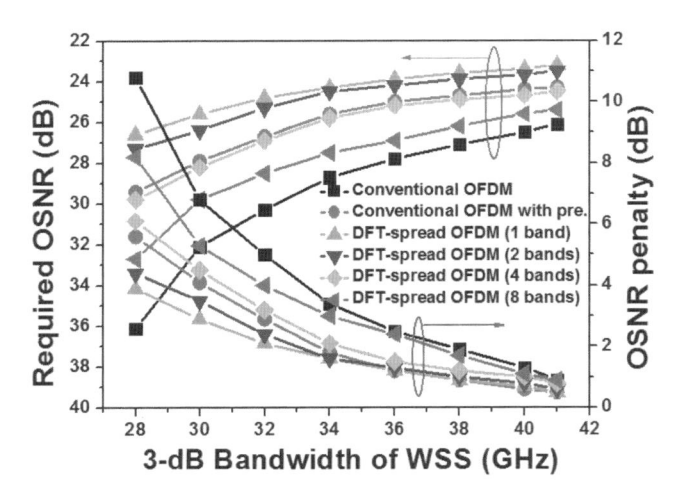

Fig. 5.11. 3-dB bandwidth of WSS versus required OSNR and OSNR penalty at BER of 3.8×10^{-3}.

transmission. When the bandwidth of WSS is adjusted to be 32 GHz, the OSNR penalty of conventional OFDM, pre-equalization OFDM, DFT-spread OFDM (1 band), DFT-spread OFDM (2 bands), DFT-spread OFDM (4 bands) and DFT-spread OFDM (8 bands) are, respectively, 5 dB, 2.9 dB, 2.1 dB, 2.5 dB, 3.2 dB, and 4 dB.

The 3-dB bandwidth of WSS versus required OSNR and OSNR penalty at BER of 2.4×10^{-3} is shown in Fig. 5.12. When the bandwidth of WSS is adjusted to be 32 GHz, the OSNR penalty of DFT-spread OFDM (1 band) is the lowest in the six types of OFDM signals mentioned above. These results indicate that DFT-spread OFDM (1 band) signal has the best resistance to narrowband optical filtering.

The optical spectra of eight WDM 32GHz PDM-16QAM-OFDM signal at OBTB position and over 420-km SMF are shown in Fig. 5.13, and the input optical power/channel of the third channel with 32 GHz 16QAM-OFDM signal transmitted over 420-km SMF versus Q-factor calculating from BER is shown in Fig. 5.14. The optimized input optical power/channel of the conventional OFDM signal and pre-equalized OFDM signal is -2 dBm, while the optimized input optical power/channel of DFT-spread OFDM (1 band), DFT-spread OFDM (2 bands), DFT-spread OFDM (4 bands) and DFT-spread OFDM (8 bands) is -1 dBm. The optimized input optical power/channel increases because the PAPR of the OFDM signal after DFT-spread processing decreases. Compared

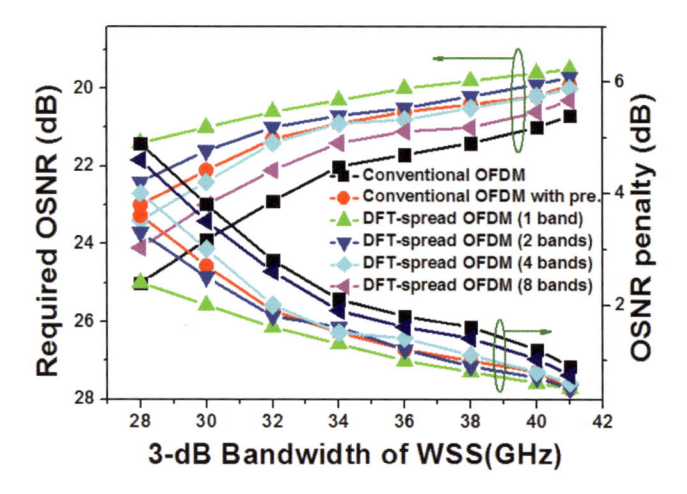

122   *Digital Signal Processing for High-Speed Optical Communication*

Fig. 5.12. 3-dB bandwidth of WSS versus required OSNR and OSNR penalty at BER of 2.4×10^{-3}.

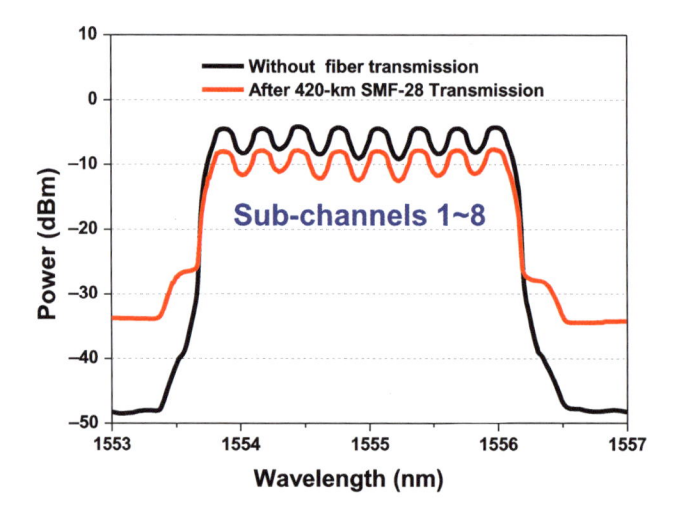

Fig. 5.13. Optical spectra of eight WDM 32GHz PDM-16QAM-OFDM signal.

with conventional OFDM and pre-equalized OFDM signal, the Q-factor of DFT-spread OFDM (1 band) signal over 420-km SMF is improved by 0.9 dB and 0.4 dB, respectively.

The Q-factor of 16QAM-OFDM signal in the third channel versus optical transmission distance is shown in Fig. 5.15. We tested three types of

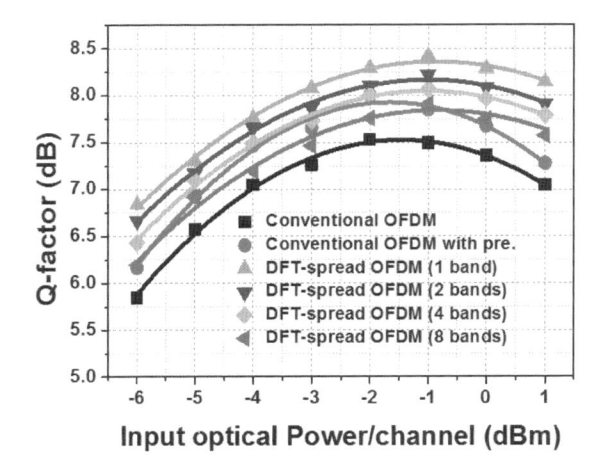

Fig. 5.14. Q-factor versus input optical power/channel of the third channel with 32 GHz 16QAM-OFDM signal transmitted over 420-km SMF.

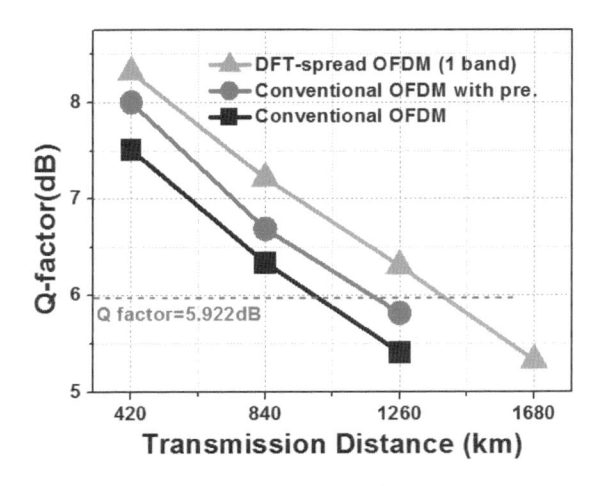

Fig. 5.15. Q-factor of 16QAM-OFDM signal in the third channel versus optical transmission distance.

signals, conventional OFDM, pre-equalization OFDM and 1 band OFDM DFT-spread signal. The transmission performance is improved because high-frequency power attenuation is overcome after pre-equalization. The Q-factor of conventional 32 GHz PDM-16QAM-OFDM signal over 1260 km is 5.406 dB and can be improved to 5.815 dB after pre-equalization. But this value is still lower than the Q-factor threshold of soft decision forward error correction coding (5.922 dB). For the conventional OFDM signal and pre-equalized OFDM signal transmission in the system, the farthest SMF transmission distance is 840 km. and the Q-factor of the signal with DFT-spread OFDM (1 band) signal of the same bandwidth over 1260-km SMF transmission is 6.304 dB, so the transmission distance can be increased to 1260 km.

The measured BERs of all subchannels over 1260-km SMF transmission is lower than 2.4×10^{-2}. Our results indicate that DFT-spread OFDM (1 band) signal can simultaneously overcome high-frequency power attenuation and reduce the PAPR effectively.

5.3.2. Scheme summary

We experimentally compared the performance of one/multiple bands DFT-spread 16QAM-OFDM, conventional 16QAM-OFDM and pre-equalized 16QAM-OFDM on resisting high-frequency power attenuation and non-linear effects in fiber link. The results indicated that DFT-spread OFDM (1 band) signal has the best performance to overcome high-frequency power attenuation and reduce the PAPR, as well as resist narrowband optical filtering. When transmitting conventional 16QAM-OFDM signal and pre-equalized 16QAM-OFDM signal using 8×244.2 Gb/s WDM-PDM system, the farthest signal transmission distance is 840 km at Q-factor of >5.922 dB. Under the same conditions, the transmission distance of 1 band DFT-spread 16QAM-OFDM can be increased to 1260 km.

5.4. 82.29-Tb/s (182 × 560-Gb/s) Transmission of 42 GHz-Spaced WDM PDM-128-QAM OFDM Signals

5.4.1. Experimental setup and results

The experimental setup is shown in Fig. 5.16. At the transmitter, 98 C-band (1530.98-1563.86 nm) and 84 L-band (1576.90-1605.84 nm) Continual Wave (CW) optical carriers with 42-GHz spacing are generated. Two sets of seven consecutive channels are selected to be measured during OBTB and after fiber transmission both in C-band and L-band. Other optical carriers in C-band and L-band are generated with multi-carrier generation techniques as shown in Figs. 5.16(a) and 5.16(b), respectively. 13 C-band and 11 L-band 294 GHz-spaced ECLs with 100-kHz linewidth are injected into

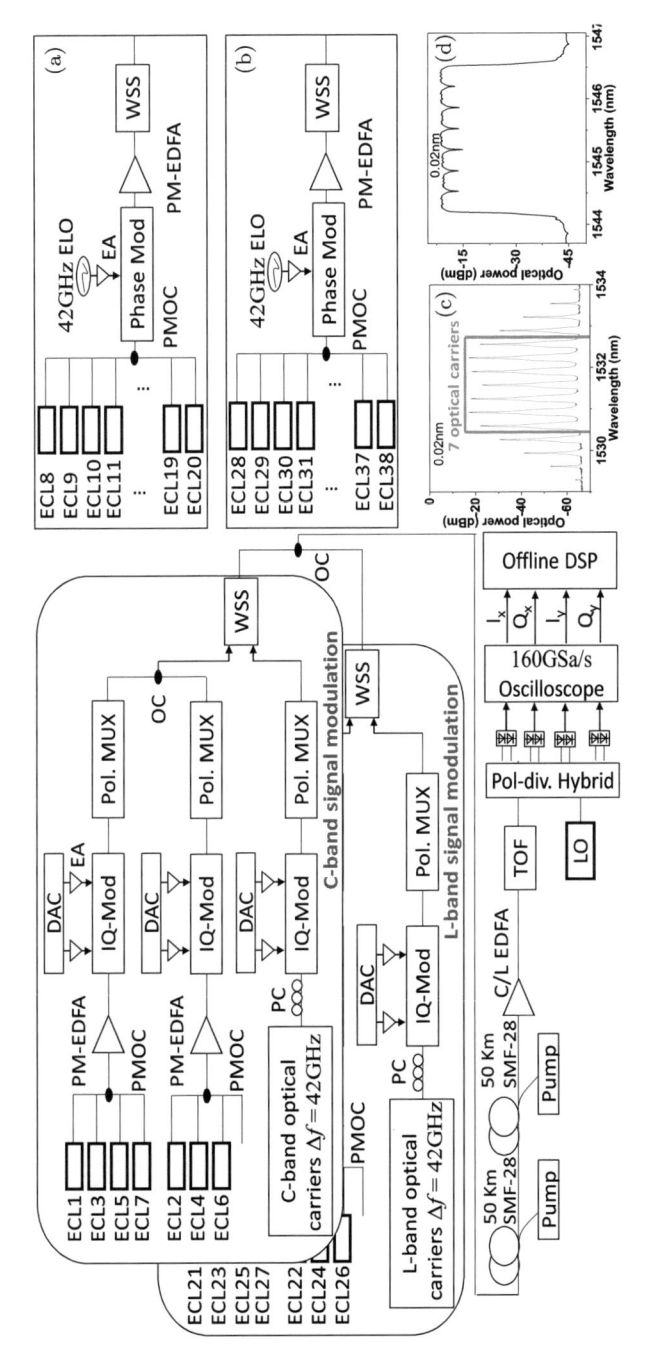

Fig. 5.16. Experimental setup: (a) C-band multi-carrier generation, (b) L-band multi-carrier generation, (c) optical spectra of multiple optical carriers generated with one C-band CW, and (d) spectra of measured seven consecutive channels in C-band.

two independent phase modulators (PMs) to generate C-band and L-band optical carriers. Each PM is driven by an RF clock signal with a fixed frequency of 42 GHz. The amplitude of the RF signal after one booster electrical amplifier is four times greater than the half-wave voltage of the PM to generate multiple carriers with high OSNR. The optical spectra of multiple optical carriers generated with one C-band CW is shown in Fig. 5.16(c), and only the central 7 optical carriers with high OSNR are chosen. Thus, the total number of generated C-band and L-band multiple carriers is 91 and 77, respectively. One PM-EDFA and one WSS are cascaded after phase modulator to produce the optical carriers with high and uniformly distributed OSNR. The measured 7 consecutive channels in C-band and L-band can both be divided into odd and even groups. The odd and even groups of ECLs are firstly combined by two sets of PM-OCs and then modulated by two independent I/Q modulators, respectively. In C-band/L-band signal modulation, the I and Q components of 40-GHz 128QAM-OFDM signal generated by each of the two 80-GSa/s DACs are firstly amplified and then used to drive an I/Q modulator. The number of subcarriers during the OFDM signal generation is 1024, and a CP of 16 samples is added to resist PMD and relax the optimization of EDC, giving an OFDM symbol size of 1040. Around 512 low frequency subcarriers excluding five subcarriers around the zero frequency reserved for RF-pilot for phase noise estimation are filled with 128QAM data, and the total bandwidth of this signal is 40.55 GHz. The polarization multiplexing is realized by a polarization multiplexer, and the delay between two polarizations is identical to the OFDM symbol length $(1/80 \times (1024 + 16)\,\mathrm{ns} = 13\,\mathrm{ns})$. One pair of TSs is inserted before every set of 123 128QAM-OFDM symbols. The line data rate on each optical carrier is $40\,\mathrm{Gb/s} \times 1024/1040 \times 123/125 \times 7 \times 2/(1+20\%) \approx 452.14\,\mathrm{Gb/s}$ after excluding 20% SD-FEC and other overheads and the SE is 10.77 bit/s/Hz. After polarization multiplexing, seven odd and even consecutive channels signal are combined by an OC both in C- and L-band signal modulations.

The optical spectra of seven measured C-band signals are shown in Fig. 5.16. In C-band signal modulation, 91 optical carriers generated with multi-carrier generation technique are firstly modulated by 40-GHz 128QAM-OFDM and then polarization multiplexed as mentioned above. In C-band signal modulation, the measured 7 optical carriers and other 91 optical carriers are coupled with a WSS and the frequency range of measured 7 optical carriers are changed by adjusting both the wavelength of the measured 7 optical carriers and the 13 C-band ECLs in C-band multi-carrier generation. The coupling of 7 measured optical carriers and other 77 optical carriers and adjustment of frequency range of measured 7 optical carriers in L-band signal modulation are similar to those of C-band signal modulation. The generated 98 C-band signal and 84 L-band signal are then combined by a 3-dB OC. Pre-equalization-based pre-equalizer is added in the transmitter during the system calibration stage to compensate for bandwidth limitation-induced high-frequency power attenuation. To avoid IQ imbalance induced by timing misalignment between I and Q components of the OFDM signal, two broadband adjustable time delays are added before two electronic amplifiers and one user-defined SSB signal is used to be loaded into DAC during the calibration stage. The generated 128 × 560-Gb/s, 42 GHz-spaced channel signals are then launched into two 50-km spans of SMF-28 with an average loss of 10.5 dB. Each SMF-28 span is followed by a post-commercial Raman amplifier with 10-dB on–off gain to compensate for the fiber loss. At the receiver, a C-band or L-band EDFA is used to boost the received signal and a TOF with tunable 3-dB bandwidth of 0.4 nm is used to select the desired channel. An ECL is used as LO with 100-kHz linewidth for the selected channel. O/E detection of the signal is implemented with an integrated coherent receiver. The ADC is realized in the real-time oscilloscope with 160-GSa/s sampling rate and 65-GHz bandwidth. The captured data is then processed with offline DSP. After frequency domain equalization, a post-DD-LMS equalizer is used afterward to further compensate channel response and mitigate devices' implementation penalty. BER is counted over 10 × 2 × 440832 bits (10 data frames on two polarizations, and each frame contains 440832 bits).

The measured BER versus OSNR of dual polarization 40-GHz DFT-spread single channel in 128QAM-OFDM format at OBTB position is shown in Fig. 5.17. We should mention that, since the SNR of pre-equalized signal has decreased, an optimized pre-equalization scheme is needed for a full compensation of the high-frequency power attenuation. In our experiment, we used a linear function $f(m) = 1 + m * \text{abs}([-1 : 2/(Ns - 1) : 1])$ to obtain the pre-equalization coefficients, where Ns is the number of data-carrying subcarriers, which is 512 in this case. During the test, we have found that we can achieve the best BER performance with $m = 1.6$, which would correspond to an optimized scenario. The tap size of T-spaced post-DD-LMS equalizer is also investigated and the optimized tap size of 119 is found. We can note the achieved optimization yields from results presented in Fig. 5.17, namely, the BER of conventional

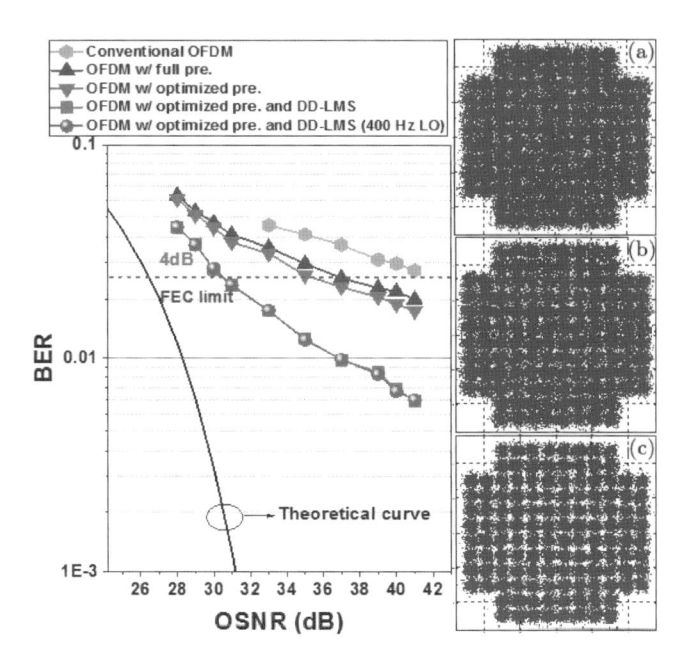

Fig. 5.17. BER versus OSNR of single channel in OBTB and constellations.

OFDM signal is equal to 2.4×10^{-2} at required OSNR of approximately 41 dB, while the required OSNRs for OFDM with full pre-equalization, OFDM with optimized pre-equalization, and OFDM with optimized pre-equalization and DD-LMS post-equalizer (all at BER of 2.4×10^{-2}) are 36.9 dB, 35.3 dB, and 30.5 dB, respectively. The constellations of conventional OFDM, OFDM with optimized pre-equalization, and OFDM with optimized pre-equalization and DD-LMS post-equalizer at 41-dB OSNR are shown in Figs. 5.17(a), 5.17(b) and 5.17(c), respectively. Therefore, the employment of both the pre-equalizer and post-equalizer can significantly improve the receiver sensitivity. There is 4-dB implementation penalty observed for the case of OFDM with optimized pre-equalization and DD-LMS post-equalizer at the BER of 2.4×10^{-2}. We should mention that we also used a 400-kHz linewidth LO laser instead of the original one, and there was no OSNR penalty observed when optimized pre-equalization and DD-LMS post-equalizer are used. This proved that the RF-pilot-based phase noise estimation is a robust scheme with respect to the impact of receiver LO linewidth. The signal spectra for both C- and L-bands observed before transmission are shown in Fig. 5.18. We also measured the required OSNRs

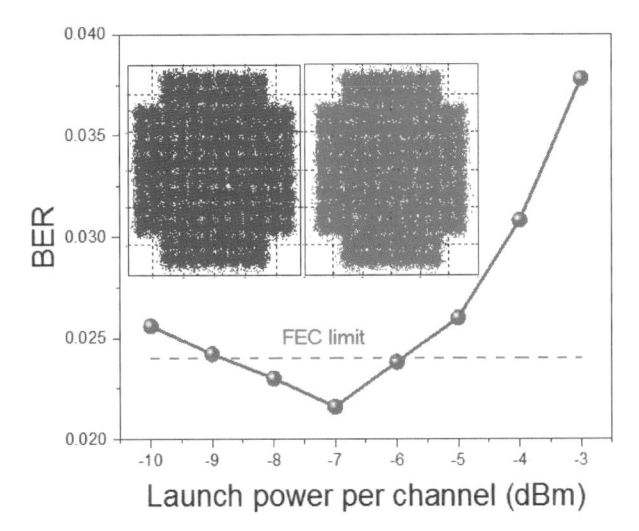

Fig. 5.18. Measured BER versus launch power.

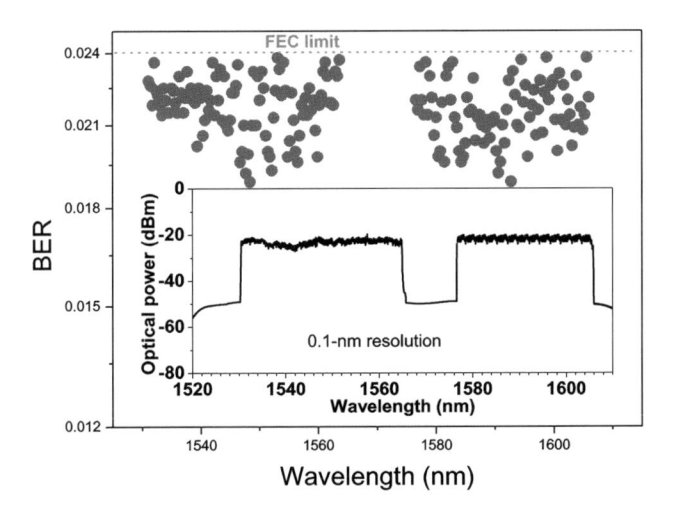

Fig. 5.19. Measured BERs of all 182 channels.

versus different channel spacing of C-band channel, 58 in WDM OBTB configuration at BER of 2.4×10^{-2} with results shown in Fig. 5.19. Significant OSNR penalty appears when channel spacing is less than 41 GHz, which is mainly caused by the side lobes of adjacent channels. The OSNR penalty is 1 dB with 42-GHz channel spacing.

5.4.2. Scheme summary

Here, we demonstrated 182 WDM channels with 40-GHz PDM-128QAM-OFDM signals at 42-GHz channel spacing achieving spectral efficiency of 10.77 bit/s/Hz over 2 × 50-km SMF. By using optimized pre-equalization and DD-LMS post-equalizer, we achieved "the 20% SD-FEC-defined" error-free transmission for all 182 channels, each loaded with 400 Gb/s data rate.

5.5. Summary

In this chapter, we introduce the hot techniques to overcome high-frequency power attenuation and reduce the PAPR for achieving 400+ Gb/s or 1+ Tb/s bit rate per wavelength transmission in the CO-OFDM system. In Section 5.2, DFT-spread and pre-equalization techniques are introduced and then we experimentally compared the performance of one/multiple bands DFT-spread 16QAM-OFDM, conventional 16QAM-OFDM and pre-equalized 16QAM-OFDM on resisting high frequency power attenuation and nonlinear effects in fiber link in Section 5.3. Finally, we have successfully transmitted 182 WDM channels with 40-GHz PDM-128QAM-OFDM signals at 42-GHz channel spacing achieving spectral efficiency of 10.77 bit/s/Hz over 2 × 50-km SMF in Section 5.4.

References

[1] B. Zhu, S. Chandrasekhar, X. Liu and D. W. Peckham, Transmission performance of a 485-Gb/s CO-OFDM superchannel with PDM-16QAM sub-carriers over ULAF and SSMF-based links, *IEEE Photon. Technol. Lett.* **23** (2011) 1400–1402.

[2] X. Liu, S. Chandrasekhar, B. Zhu, P. Winzer, A. Gnauck and D. Peckham, 448-Gb/s reduced-guard-interval CO-OFDM transmission over 2000 km of ultra-large-area fiber and five 80-GHz-Grid ROADMs, *J. Lightw. Technol.* **29** (2010) 483–490.

[3] S. L. Jansen, I. Morita, T. C. W. Schenk and H. Tanaka, 121.9-Gb/s PDM-OFDM transmission with 2 b/s/Hz spectral efficiency over 1000 km of SSMF, *J. Lightwave Technol.* **27** (2009) 177–188.

[4] M. -F. Huang, S. Zhang, E. Mateo, D. Qian, F. Yaman, T. Inoue, Y. Inada and T. Wang, EDFA-only WDM 4200-km transmission of OFDM-16QAM and 32QAM, *IEEE Photon. Technol. Lett.* **24** (2012) 1466–1468.

[5] Y. Fang, L. Liu, C. Wong, S. Zhang, T. Wang, G. Liu and X. Xu, Silicon IQ modulator based 480 km 80 × 453.2 Gb/s PDM-eOFDM transmission on 50 GHz grid with SSMF and EDFA-only link, in *OFC* 2015, paper M3G.5.

[6] S. Zhang, Y. Zhang, M.-F. Huang, F. Yaman, E. Mateo, D. Qian, L. Xu, Y. Shao and I. Djordjevic, Transoceanic transmission of 40 × 117.6 Gb/s PDM-OFDM-16QAM over hybrid large-core/ultralow-loss fiber, *J. Lightwave Technol.* **31** (2013) 498–505.

[7] F. Li, X. Li and J. Yu, Performance Comparison of DFT-spread and pre-equalization for 8×244.2-Gb/s PDM-16QAM-OFDM, *J. Lightwave Technol.* **33** (2015) 227–233.

[8] D. Qian, M. Huang, E. Ip, Y. Huang, Y. Shao, J. Hu and T. Wang, 101.7 Tb/s (370 × 294 Gb/s) PDM-128QAM-OFDM transmission over 3 × 55 km SSMF using pilot-based phase noise mitigation, in *OFC* 2011, paper PDPB5.

[9] T. Omiya, K. Toyoda, M. Yoshida and M. Nakazawa, 400 Gbit/s frequency-division-multiplexed and polarization-multiplexed 256 QAM-OFDM transmission over 400 km with a spectral efficiency of 14 bit/s/Hz, in *OFC* 2012, paper OM2A.7.

[10] D. Qian, E. Ip, M.-F. Huang, M.-J. Li and T. Wang, 698.5-Gb/s PDM-2048QAM transmission over 3 km multicore fiber, in *Proc. 39th ECOC* 2013, paper Th.1.C.5.

[11] C. Xie, B. Zhu and E. Burrows, Transmission performance of 256-Gb/s PDM-16QAM with different amplification schemes and channel spacings, *J. Lightwave Technol.* **32** (2014) 2324–2331.

[12] J. Zhang and H. C. Chien, A novel adaptive digital pre-equalization scheme for bandwidth limited optical coherent System with DAC for signal generation, in *OFC* 2014, paper W3K.4.

[13] Z. Dong, H.C. Chien, Z. Jia and X. Li, Joint digital preequalization for spectrally efficient super Nyquist-WDM signal, *J. Lightwave Technol.* **31** (2013) 3237–3242.

[14] F. Li, Z. Cao, J. Zhang, X. Li and J. Yu, Transmission of 8 × 520 Gb/s Signal based on single band/λ PDM-16QAM-OFDM on a 75-GHz grid, in *OFC 2016*, paper Tu3A.3.

[15] D. Chang, F. Yu, Z. Xiao, N. Stojanovic, F. N. Hauske, Y. Cai, C. Xie, L. Li, X. Xu and Q. Xiong, LDPC convolutional codes using layered decoding algorithm for high speed coherent optical transmission, in *OFC* 2012, paper OW1H.4.

[16] W.-R. Peng, T. Tsuritani and I. Morita, Simple carrier recovery approach for RF-pilot-assisted PDM-CO-OFDM systems, *J. Lightwave Technol.* **31** (2013) 2555–2564.

[17] F. Li, X. Li, J. Yu and L. Chen, Optimization of training sequence for DFT-spread DMT signal in optical access network with direct detection utilizing DML, *Opt. Express* **22** (2014) 22962–22967.

[18] S. J. Savory, Digital filters for coherent optical receivers, *Opt. Express* **16** (2008) 804–817.

[19] S. L. Jansen, I. Morita, T. C. W. Schenk, N. Takeda and H. Tanaka, Coherent optical 25.8-Gb/s OFDM transmission over 4160-km SSMF, *J. Lightwave Technol.* **26** (2008) 6–15.

[20] W. Shieh, H. Bao and Y. Tang, Coherent optical OFDM: Theory and design, *Opt. Express* **16** (2008) 841–859.
[21] R. Kudo, T. Kobayashi, K. Ishihara, Y. Takatori, A. Sano and Y. Miyamoto, Coherent optical single carrier transmission using overlap frequency domain equalization for long-haul optical systems, *J. Lightwave Technol.* **27** (2009) 3721–3728.

Chapter 6

Digital Signal Processing for Dual/Quad Subcarrier OFDM Coherent Detection

6.1. Introduction

Optical orthogonal frequency division multiplexing (OFDM) has attracted lots of attention due to its high spectral efficiency (SE) and robustness to transmission impairments enabled by digital signal processing (DSP) [1–22]. In the traditional coherent OFDM transmission system, the frequency offset estimation (FOE), channel estimation, equalization, and phase recovery are implemented with training sequence (TS) and pilot tones [2–4]. As the TS and pilot tones are critical in the frequency-domain equalization scheme, the number of subcarriers in the OFDM modulation/demodulation with inverse fast Fourier transform/fast Fourier transform (IFFT/FFT) is usually larger than 64 in order to reduce the overhead including pilot tones and TSs and acquire more accurate channel estimation. Unfortunately, an OFDM signal with a large IFFT/FFT size has high peak-to-average power ratio (PAPR) values [8]. We have discussed the perniciousness of high PAPR in OFDM in the previous chapter, so the PAPR of the OFDM signal must be reduced in the CO-OFDM system. In order to mitigate the impairments induced by high PAPR of the OFDM signals, many techniques have been proposed to reduce the PAPR of the OFDM signal [6, 9, 10] such as companding algorithm in the previous chapter. While among these schemes, the PAPR of the OFDM signal is still very high. A more straightforward approach is to reduce the number of subcarriers in the OFDM modulation with IFFT. The PAPR of the OFDM signal can be reduced quickly with the reduction of the number of subcarriers, while this will cause a dramatic increase in overhead and the channel estimation based on TS in frequency domain cannot effectively work. To overcome these problems, one blind equalization in the time domain to recover the received signal in the few subcarrier CO-OFDM system is introduced in [20–22].

A 2-subcarrier dual-polarization QPSK-OFDM signal transmission system with blind equalization in time domain is introduced in Section 6.2 of this chapter [20]. Two subcarrier Quadrature Phase Shift Keying OFDM (QPSK-OFDM) signal demonstrates as a 9-ary quadrature amplitude modulation (9-QAM) signal in the time domain [15] and therefore can be blindly equalized with cascaded multi-modulus algorithm (CMMA) equalization method in the time domain [11]. With the blind equalization in the time domain, the FOC, channel estimation and phase recovery can be implemented without TS and pilot tones. The overhead that existed in the traditional optical OFDM transmission system can be completely removed in the two subcarrier optical OFDM transmission system with blind equalization which leads to higher spectral efficiency. Experimental results in this section show that the bit error ratio (BER) of 48-Gb/s dual-polarization 2-subcarrier OFDM signal is less than the pre-forward-error-correction (pre-FEC) threshold of 3.8×10^{-3} after 5600-km single-mode fiber-28 (SMF-28) transmission, while the 32.1-Gbit/s dual-polarization traditional 256-subcarrier OFDM signal which is created for comparison can be only transmitted 3500 km under the pre-FEC threshold with frequency domain equalization based on TS and pilot tones. The nonlinear effect resistance and transmission distance of 2-subcarrier OFDM with blind equalization can be enhanced compared with traditional OFDM transmission system based on frequency equalization with TS.

Quad-Carrier QPSK-OFDM signal transmission and reception is introduced with blind equalization in the time domain in Section 6.3 [21–23]. Compared to two-subcarrier OFDM scheme [8], 4-subcarrier OFDM scheme is much more flexible in power allocation and pre-equalization as bandwidth of each subcarrier is smaller. The main difference between our scheme with 4-subcarrier all optical OFDM is the generation and detection. For 4-subcarrier all optical OFDM signal generation, we need to generate four frequency-locked subcarriers. The channel spacing between four subcarriers should be equal to the baud rate of each subchannel in order to make the 4-subcarrier orthogonal. So, this generation is usually complicated compared to our scheme. At receiver, after optical to electrical (O/E) conversion, a digital filter is used to separate the 4-subcarrier, and then DSP is applied for each subcarrier [12]. 4-subcarrier OFDM signal shows as a 25-QAM signal in the time domain, and it can be blindly equalized with cascaded multi-modulus algorithm (CMMA) equalization in the time domain [11, 13–15]. With the blind equalization, channel estimation and equalization, FOE, and phase recovery can be implemented without TS and pilot tones. The overhead existing in the traditional optical OFDM transmission system can be completely eliminated in the 4-subcarrier optical OFDM transmission system with blind equalization. In this section, transmission and reception of 48 Gbit/s dual-polarization quad-carrier quadrature-phase-shift-keying OFDM (QPSK-OFDM) signal are demonstrated. In the offline DSP, the FOE should be done with 25-QAM signal before four subcarriers are separated with FFT. Compared to the traditional OFDM signal with 256 subcarriers, the PAPR of quad-carrier QPSK-OFDM signal with blind equalization is decreased dramatically from 14.4 to 6.4 dB at the probability of 1×10^{-4}. There is no penalty after 80-km single-mode fiber-28 (SMF-28) transmission.

In Section 6.4, we introduced how to transmit and detect dual-subcarrier 16QAM-OFDM signal with blind equalization in the time domain [23]. In Sections 6.2 and 6.3, the transmission and reception of dual-carrier QPSK-OFDM signal and quad-carrier QPSK-OFDM signal are introduced, respectively. Compared to the traditional scheme based on frequency-domain equalization, the SE is improved adopting CMMA blind equalization, but the SE is still limited as the modulation format is QPSK. To further increase the SE of the system, we investigate the transmission of dual-subcarrier 16QAM-OFDM signal in this section. Dual-subcarrier 16QAM-OFDM signal demonstrates as a 49QAM signal in the time domain, and therefore it can also be blindly equalized with the CMMA equalization method in the time domain. The transmitter and receiver of our scheme are simpler compared with dual-subcarrier all optical 16QAM OFDM. In this section, 16Gbaud PDM dual-subcarrier 16QAM-OFDM signal transmission and reception are introduced. The experimental results show that there is no power penalty observed between optical back to back (OBTB) and after 80-km SMF-28. We also measured the optical signal-to-noise ratio (OSNR) penalty versus the bandwidth of the channel and the experimental results show that a 0.6 dB OSNR penalty is observed when the bandwidth of channel is set at 25 GHz.

6.2. Transmission and Reception of Dual-Polarization 2-Subcarrier Coherent QPSK-OFDM Signals

6.2.1. System principle

The IFFT size during the OFDM modulation is N and the time length of one OFDM symbol is T. After IFFT, the OFDM signal can be expressed as

$$s(t) = \sum_{k=0}^{N-1} c_k \exp(j2\pi f_k t) \quad (1 \leq t \leq T) \tag{6.1}$$

where k is the index of subcarriers, and f_k is the frequency of the kth subcarrier and can be expressed as

$$f_k = k\Delta f = \frac{k}{T}. \tag{6.2}$$

In the 2-subcarrier scheme, the N and the time length of OFDM symbol are both only 2 when only two-subcarriers are used in the OFDM modulation and demodulation, and the expression can be simplified as

$$\begin{aligned} s(t) &= \frac{1}{\sqrt{2}} \sum_{k=0}^{1} c_k \exp(j2\pi f_k t) \\ &= \frac{1}{\sqrt{2}} \left(c_0 + c_1 \exp\left(j2\pi \frac{t}{2}\right) \right) \quad (1 \leq t \leq 2) \end{aligned} \tag{6.3}$$

where c_0 and c_1 represent the data modulated onto two subcarriers, respectively. After IFFT, the OFDM symbols are generated and one OFDM symbol includes two samples. Two samples are denoted by time slot 1 and time slot 2 in the following part and can be expressed as

$$\begin{aligned} s(0) &= \frac{1}{\sqrt{2}}(c_0 - c_1) \\ s(1) &= \frac{1}{\sqrt{2}}(c_0 + c_1) \end{aligned} \tag{6.4}$$

Table I in Fig. 6.1 shows the data before and after 2-point IFFT. Data on 2-subcarrier in frequency domain represent data before IFFT, while two time slots in one OFDM symbol indicate data after IFFT. The data on 2-subcarrier are obtained under the QPSK mapping rules, and two-time slots in one OFDM symbol can be calculated via Eq. (6.4). The constellations of data on two subcarriers demonstrate as 4-QAM signal with blue in Fig. 6.1,

Table I

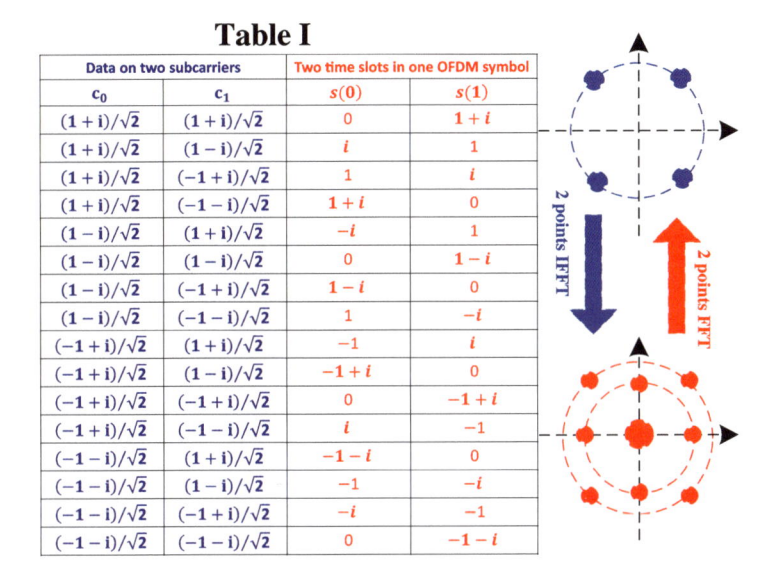

Data on two subcarriers		Two time slots in one OFDM symbol	
c_0	c_1	$s(0)$	$s(1)$
$(1+i)/\sqrt{2}$	$(1+i)/\sqrt{2}$	0	$1+i$
$(1+i)/\sqrt{2}$	$(1-i)/\sqrt{2}$	i	1
$(1+i)/\sqrt{2}$	$(-1+i)/\sqrt{2}$	1	i
$(1+i)/\sqrt{2}$	$(-1-i)/\sqrt{2}$	$1+i$	0
$(1-i)/\sqrt{2}$	$(1+i)/\sqrt{2}$	$-i$	1
$(1-i)/\sqrt{2}$	$(1-i)/\sqrt{2}$	0	$1-i$
$(1-i)/\sqrt{2}$	$(-1+i)/\sqrt{2}$	$1-i$	0
$(1-i)/\sqrt{2}$	$(-1-i)/\sqrt{2}$	1	$-i$
$(-1+i)/\sqrt{2}$	$(1+i)/\sqrt{2}$	-1	i
$(-1+i)/\sqrt{2}$	$(1-i)/\sqrt{2}$	$-1+i$	0
$(-1+i)/\sqrt{2}$	$(-1+i)/\sqrt{2}$	0	$-1+i$
$(-1+i)/\sqrt{2}$	$(-1-i)/\sqrt{2}$	i	-1
$(-1-i)/\sqrt{2}$	$(1+i)/\sqrt{2}$	$-1-i$	0
$(-1-i)/\sqrt{2}$	$(1-i)/\sqrt{2}$	-1	$-i$
$(-1-i)/\sqrt{2}$	$(-1+i)/\sqrt{2}$	$-i$	-1
$(-1-i)/\sqrt{2}$	$(-1-i)/\sqrt{2}$	0	$-1-i$

Fig. 6.1. Data and constellations before and after 2-point FFT for the QPSK signal.

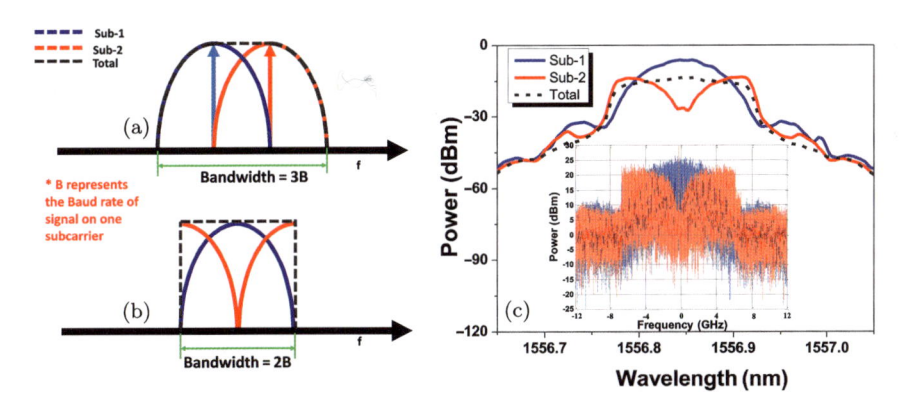

Fig. 6.2. Principle of 2-subcarrier OFDM implemented (a) with all optical approach and (b) with 2-point IFFT/FFT. (c) Optical and electrical spectra of 2-subcarrier OFDM signal implemented with 2-point IFFT/FFT.

while after IFFT, the constellations of two time slots of the OFDM symbols display as 9-QAM signal with red in Fig. 6.1.

2-subcarrier OFDM signal generated by 2-point IFFT in electrical domain in this letter is not the same as all optical OFDM with only 2-subcarrier implemented in [8, 16, 17]. Figure 6.2 shows these differences from the view of the spectra. Assume B represents the baud rate of signal on only one subcarrier. Figure 6.2(a) shows the 2-subcarrier OFDM scheme [8, 16, 17] implemented with all optical approach and Fig. 6.2(b) shows the 2-subcarrier OFDM scheme implemented in electrical domain with 2-point IFFT. There are some differences between these two 2-subcarrier OFDM schemes. The first difference is that the arrangement of two-subcarriers is not the same, the second difference is that the spectral shape of the signal after the combination of two-subcarriers is different, and the most important difference is that the bandwidth of optical 2-subcarrier OFDM is 3B, while the bandwidth of 2-subcarrier OFDM signal generated in electrical domain is only 2B. The other main difference between our scheme and optical OFDM with 2-subcarrier is the generation and detection. For 2-carrier optical OFDM signal generation, we need to generate two frequency-locked subcarriers. The channel spacing between the two subcarriers should be equal to the baud rate of signal on each subcarrier in order to make the 2-subcarrier orthogonal. The generation is usually complicated compared to our scheme. At the receiver, after optical-to-electrical (O/E) conversion, a digital filter is used to separate the two subcarriers, and then DSP is done for each subcarrier [12]. According to the analysis, the two 2-subcarrier OFDM schemes are different and in this letter, we propose a new scheme where the implementation of electrical-to-optical OFDM can be realized without any overhead for channel estimation and equalization. The optical and electrical spectra of the 2-subcarrier OFDM signal in the experiment with 2-point IFFT/FFT for generation and demodulation are shown in Fig. 6.2(c). The blue and red solid lines represent the components of subcarriers 1 and 2, respectively. The black dashed line is the optical spectrum of 2-subcarrier OFDM signal after the combination of subcarriers 1 and 2 with 0.01-nm resolution. The electrical spectra of 2-subcarrier in the OFDM signal are also shown in the inset of Fig. 6.2(c). The arrangement of 2-subcarrier both in optical domain and electrical domain agrees well with the principle.

6.2.2. Experimental setup and results

Figure 6.3 shows the experimental setup of the coherent optical OFDM (CO-OFDM) transmission system [20]. In the transmitter, an external cavity laser (ECL) is modulated by an in-phase/quadrature (I/Q) modulator driven by an electrical baseband OFDM signal at 12-GSa/s sample rate. Here, two types of the OFDM signal transmission are employed in the

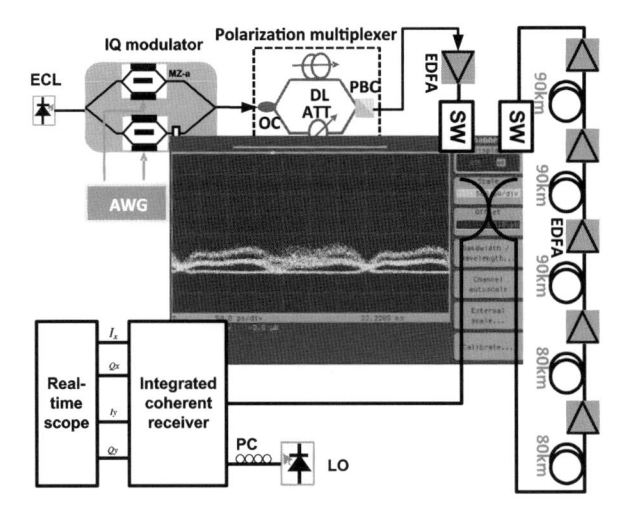

Fig. 6.3. Experimental setup of QPSK CO-OFDM transmission system.

system. One is the traditional OFDM signal containing 256-subcarrier with frequency equalization via TS and 32-sample cyclic prefix (CP) is added to the 256 samples, while the other has only two subcarriers without additional CP and TS and the signal is equalized with CMMA blind equalization method. For optical OFDM modulation, two parallel Mach–Zehnder modulators (MZMs) in the I/Q modulator are both biased at the null point and the phase difference between the upper and lower branches of the I/Q modulator is controlled at $\pi/2$. The polarization multiplexing is realized by polarization multiplexer, comprising a polarization-maintaining optical coupler (OC) to halve the signal into two branches, an optical delay line (DL) to remove the correlation between X-polarization and Y-polarization by providing a time delay, an optical attenuator to balance the power of two branches and a polarization beam combiner (PBC) to recombine the signal. The generated signal is boosted via an erbium-doped fiber amplifier (EDFA) before getting launched into optical recirculating loop. The optical recirculating loop consisted of two spans of 80 km and three spans of 90 km SMF-28 and five EDFAs with 5 dB noise figure. The output signal is then injected into the integrated coherent receiver to implement O/E detection.

In the 256-subcarrier scheme, 200 subcarriers are employed to convey data, the first subcarrier is set to zero for DC-bias and the remaining 55 null subcarriers at the edge are reserved for oversampling and 8 pilot tones

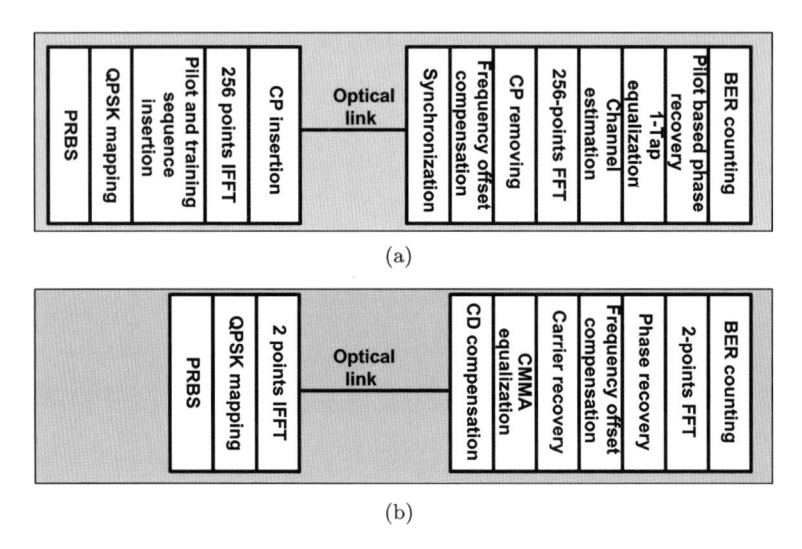

Fig. 6.4. Offline DSP: (a) traditional OFDM signal and (b) 2-subcarrier OFDM signal.

are reserved for phase recovery. The offline DSP of 256-subcarrier scheme is shown in Fig. 6.4(a). In the transmitter, the pseudo-random binary sequence (PRBS) is firstly mapped to QPSK, and then TS and pilot tones are added for frequency domain equalization. A 256-point IFFT is applied to convert the signal into time domain, and finally CP is added to the 256 samples. After optical link, in the receiver of the traditional OFDM signal with 256-subcarrier, time synchronization is realized by the conjugate symmetric TS in time domain placed in the front of the frame at the transmitter, and FOC is implemented with the aid of the TS. After CP removal, FFT is applied to transform the OFDM into frequency domain and channel estimation is implemented by a pair of TS orthogonal in time domain at the transmitter in two polarizations. After channel estimation, demultiplexing can be realized in order to minimize crosstalk between two branches. One-tap zero-forcing equalization is used to equalize the signal, and the phase noise cancellation in two branches is implemented with the pilot tones inserted in each OFDM symbol. At last, the BER is obtained via bit error counting. The total bit rate of 256-subcarrier OFDM signal is 32.1 Gbit/s after removing overhead which include CP, TS, pilot tone and virtual subcarriers. The offline processing for the 2-subcarrier scheme is shown in Fig. 6.4(b). In the transmitter, after QPSK mapping, the data on 2-subcarrier are converted to time domain via 2-point IFFT. The time-domain signal demonstrates as a 9-QAM signal with three level according to the previous analysis in this

Digital Signal Processing for Dual/Quad Subcarrier OFDM Coherent Detection   141

chapter and the optical eye diagram of 9-QAM signal is inserted as inset in Fig. 6.3. After optical link, the 9-QAM signal can be equalized with CMMA method without additional overhead compared to traditional OFDM signal with frequency-domain equalization. In the receiver, the oversampling ratio is 2 and a $T/2$-spaced time-domain finite-impulse-response (FIR) filter is firstly used for chromatic dispersion (CD) compensation, where the filter coefficients are calculated from the known fiber CD transfer function using the frequency-domain truncation method. Secondly, the CMMA is used to retrieve the modulus of the PDM-9QAM signal and realize polarization demultiplexing. The subsequent step is to realize the FOC and phase recovery [11]. After these procedures, 9-QAM signal in time domain is converted into QPSK signal in frequency domain with 2-point FFT and then the BER can also be obtained with the BER counting. As blind equalization is applied for 2-subcarrier OFDM signal, there is no overhead and the capacity is 48 Gbit/s.

Figure 6.5 shows the measured back-to-back (BTB) BER of two types of OFDM signal. Compared to OFDM signal with 2-subcarrier, there is about 0.3-dB optical signal-to-noise ratio (OSNR) penalty at the BER of 3.8×10^{-3} for 256-subcarrier OFDM signal. This penalty is mainly induced by the high PAPR of 256-subcarrier OFDM signal, and it is obvious that

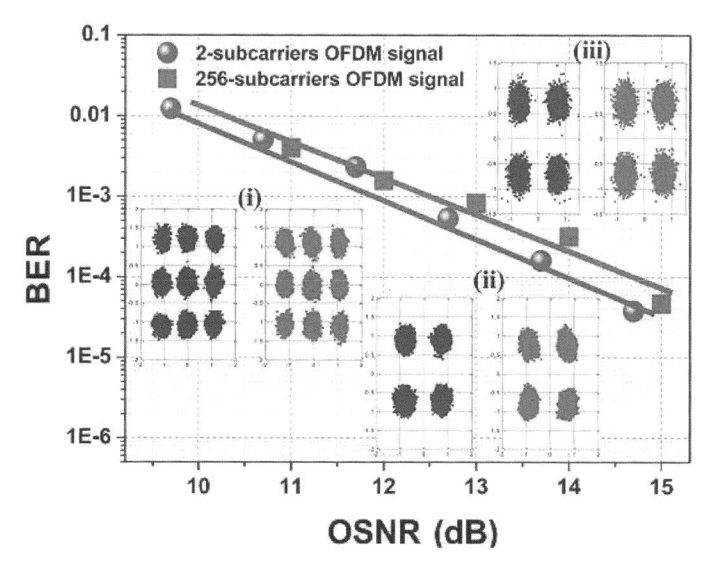

Fig. 6.5. Measured BTB BER of two types of OFDM signal versus OSNR and constellations.

the PAPR of the OFDM signal with 256-subcarriers is much higher than the three-level OFDM signal with 2-subcarrier. The PAPR of 2-subcarrier QPSK-OFDM signal is only a constant and it is only 1 dB, while the probability of PAPR of the 256-subcarrier OFDM signal exceeding 10 dB is 1×10^{-4}, which means the nonlinear effect will be significantly reduced if 2-subcarrier QPSK-OFDM is transmitted in the system. Nonlinear distortion is serious for signal with high PAPR during digital-to-analog convertor (DAC) with arbitrary waveform generator (AWG) and analog-to-digital convertor (ADC) in real-time scope. As the resolutions of DAC in the AWG and ADC in real-time scope are 10 bits and 8 bits, respectively, the nonlinear distortion is not very serious when the number of subcarriers is increased to 256. The constellations of dual polarization 2-subcarrier OFDM signal after phase recovery with OSNR @17dB are shown in inset (i) of Fig. 6.5, and the signal is converged into 9QAM in time domain. The 9QAM can be resorted into QPSK after FFT and the constellations of dual polarization after this procedure are shown in inset (ii) of Fig. 6.5, while for the 256-subcarrier OFDM signal, the constellations of dual-polarization signal with OSNR @17dB after pilot-based phase recovery are shown in Fig. 6.5(iii). Compared to the constellations of the OFDM signal with 256-subcarrrier at the same OSNR, these OFDM signals with 2-subcarrier are more concentrated due to the low PAPR.

Figure 6.6 shows the measured BER versus transmission fiber length for two types of OFDM signal. A 48 Gbit/s dual-polarization 2-subcarrier OFDM signal can be transmitted over 5600-km SMF-28, while the

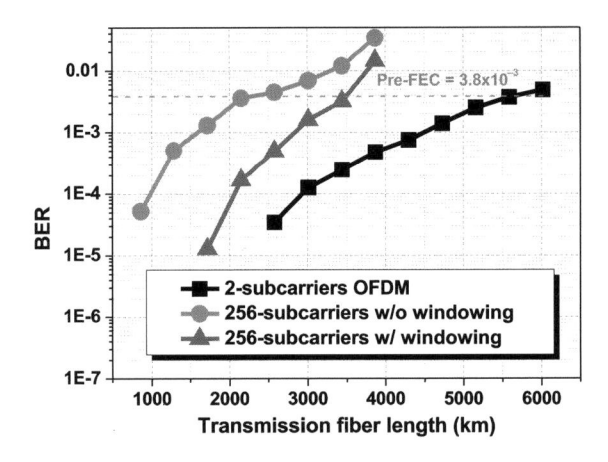

Fig. 6.6. Measured BER versus transmission fiber length for two types of OFDM signal.

32.1 Gbit/s dual-polarization traditional 256-subcarrier OFDM signal can be only transmitted 2300 km under the pre-FEC threshold of 3.8×10^{-3}. In the traditional OFDM transmission system, window shaping can be used to eliminate the impact of sampling error and residual frequency offset [18], and the transmission distance can be extended to 3500 km. But it is still much shorter than that of OFDM with 2-subcarrier. The nonlinear distortion in the fiber transmission for 2-subcarrier OFDM signal is relatively less than 256-subcarrier traditional OFDM signal as the PAPR is very low. Thus, the nonlinear effect resistance and transmission distance of 2-subcarrier OFDM with blind equalization can be enhanced compared with traditional OFDM transmission system based on frequency equalization with TS.

Figure 6.7(a) shows the measured BER versus input power of 2-subcarrier OFDM signal with 3440-km fiber transmission. The optimal input power is 1 dBm and the constellations of dual polarizations signal after phase recovery and after FFT are both inserted in Fig. 6.7(a) as insets (i) and (ii), respectively. The measured BER versus input power of 256-subcarrier traditional OFDM signal with 3010-km fiber transmission is shown in Fig. 6.7(b). The optimal input power is 0 dBm and the constellations of dual polarization after phase recovery are inserted as inset in Fig. 6.7(b). The 2-subcarrier OFDM signal with blind equalization has better nonlinear performance due to its lower PAPR and higher SE as no additional overhead is needed for the equalization.

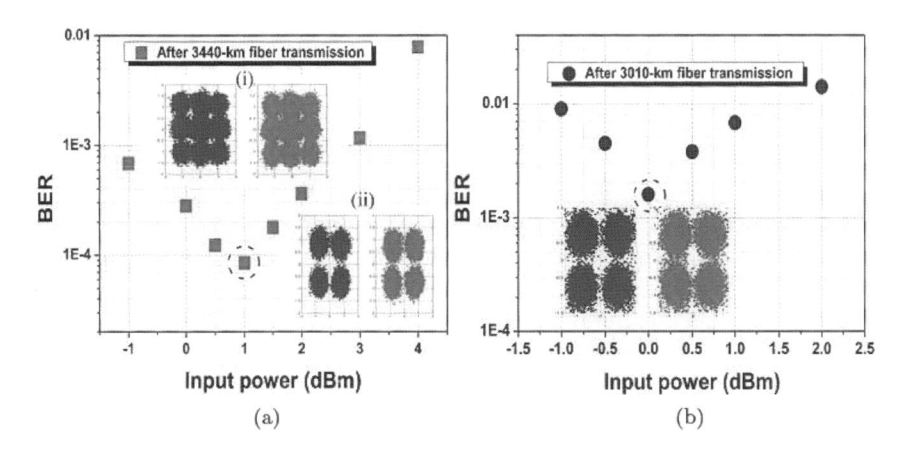

Fig. 6.7. Measured BER of OFDM signal versus input power and constellations: (a) 2-subcarriers and (b) 256-subcarriers.

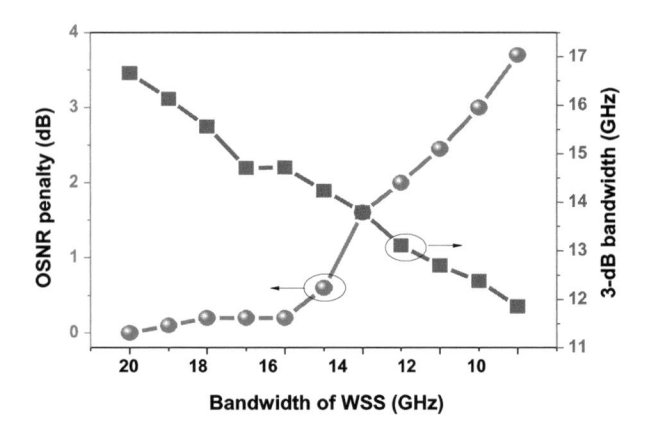

Fig. 6.8. Measured BTB OSNR penalty and 3-dB bandwidth versus bandwidth of WSS.

In the BTB case, a wavelength selective switch (WSS) is used as an optical tunable filter to determine the minimum bandwidth for the 48 Gbit/s 2-subcarrier OFDM signal transmission. Figure 6.8 shows the measured BTB OSNR penalty and 3-dB bandwidth versus bandwidth of WSS. The BTB OSNR penalty can be neglected when the bandwidth of WSS is set to 15 GHz and the 3-dB bandwidth is 14.3 GHz.

Summary: In this section, we experimentally demonstrate a 48-Gb/s 2-subcarrier dual-polarization QPSK-OFDM signal transmission system. The signal is blindly equalized with cascaded multi-modulus algorithm (CMMA) equalization method in the time domain [11]. Compared with the traditional 256-subcarrier QPSK-OFDM signal based on frequency domain equalization, the advantages of 2-subcarrier QPSK-OFDM signal adopting CMMA blind equalization include higher spectral efficiency, lower PAPR, and longer transmission distance.

6.3. Transmission and Reception of Quad-Carrier QPSK-OFDM Signal

6.3.1. *System principle*

In 4-subcarrier OFDM signal the OFDM signal can be expressed as

$$s(t) = \frac{1}{2}\left(c_0 + c_1 \exp\left(j2\pi\frac{t}{4}\right) + c_2 \exp\left(j2\pi\frac{t}{2}\right)\right.$$
$$\left. + c_3 \exp\left(j2\pi\frac{3t}{4}\right)\right) \quad (0 \leq t \leq 3) \tag{6.5}$$

where c_0, c_1, c_2 and c_3 represent the QPSK symbols modulated onto 4-subcarrier, respectively. After IFFT, the OFDM symbols are generated and one OFDM symbol includes four samples in time domain. Four samples can be expressed as

$$s(0) = \frac{1}{2}(c_0 + c_1 + c_2 + c_3), \quad s(1) = \frac{1}{2}(c_0 + jc_1 - c_2 - jc_3)$$
$$s(2) = \frac{1}{2}(c_0 - c_1 + c_2 - c_3), \quad s(3) = \frac{1}{2}(c_0 - jc_1 - c_2 + jc_3) \tag{6.6}$$

where $s(0)$, $s(1)$, $s(2)$ and $s(3)$ denote the symbols after IFFT. $s(0)$ in Eq. (4.6) is obtained by the addition of four QPSK symbols, and four 2-level $(-\sqrt{2}/2, \sqrt{2}/2)$ real/imaginary components are combined to get one 5-level $(-\sqrt{2}, -\sqrt{2}/2, 0, \sqrt{2}/2, \sqrt{2})$ real/imaginary component. The real and imaginary components of the other three symbols $s(1)$, $s(2)$ and $s(3)$ are also 5-level according to the Eq. (4.6). After 4-point IFFT, QPSK symbol on 4-subcarriers in the frequency domain becomes 25-QAM signal in the time domain.

In this section, 4-subcarrier OFDM signal is generated by 4-point IFFT and the spectral distribution of the subcarriers in quad-carrier QPSK-OFDM signal with 12Gbaud rate is shown in Fig. 6.9(a). Assume B represents the baud rate of signal on only one subcarrier, and the total bandwidth of quad-carrier QPSK-OFDM signal generated in electrical domain is only 4B. The constellations of QPSK and 25-QAM and conversion conditions are shown in Fig. 6.9(b). In the dual-polarization quad-carrier QPSK-OFDM signal transmission system, DSP algorithms are required to realize demultiplexing, FOE, channel estimation and phase recovery. In the traditional optical OFDM system, channel estimation and equalization are implemented in the frequency domain with known TSs and pilot tones (i.e., time-interleaved TSs for demultiplexing, FOE, and channel estimation, and the pilot tones for phase recovery). If the frequency-domain equalization-based DSP algorithms are applied in the dual-polarization quad-carrier QPSK-OFDM signal transmission system, the SE will decrease dramatically as the overhead occupies a large portion of the total data. In order to avoid such overhead, we propose to utilize time-domain blind equalization to recover dual-polarization quad-carrier QPSK-OFDM signal as a 25-QAM signal. In the blind equalization, the CMMA algorithm is used to implement the polarization demultiplexing and channel estimation. As for the FOE, the fourth power method is applied to estimate the frequency offset between the signal and the LO. The fourth power method can be performed on either 25-QAM signal before 4-point FFT

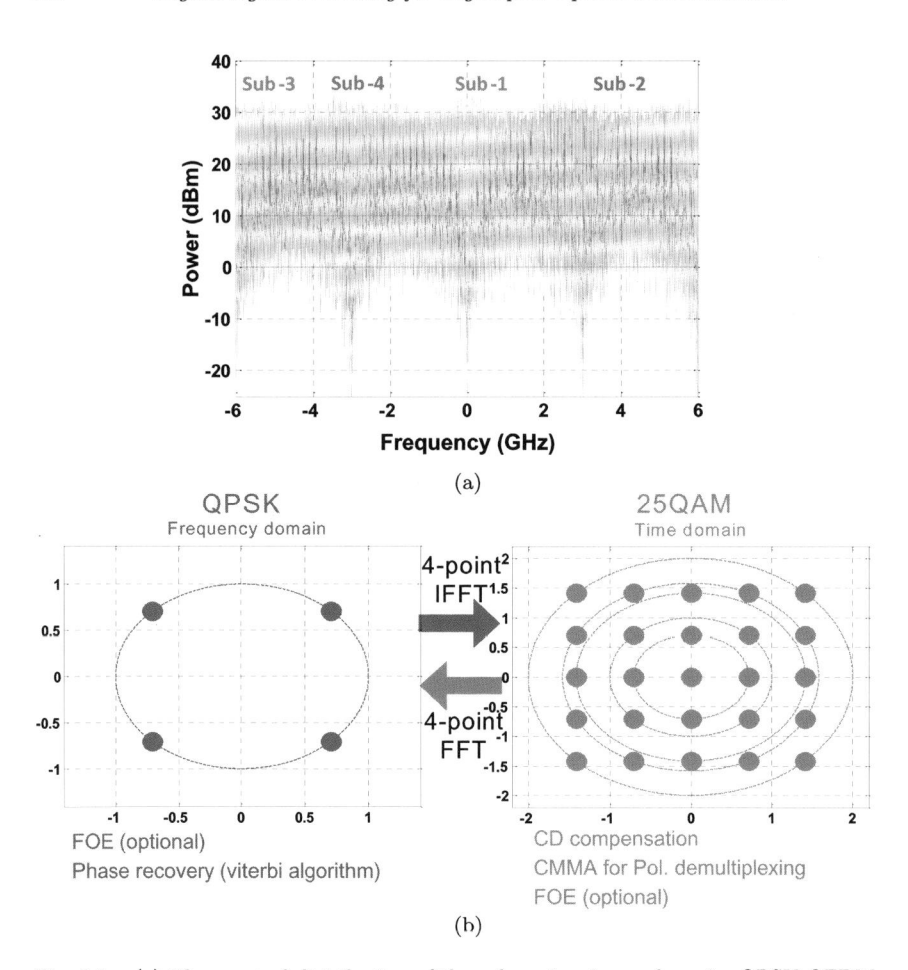

Fig. 6.9. (a) The spectral distribution of the subcarriers in quad-carrier QPSK-OFDM signal and (b) signal conversion between QPSK and 25-QAM.

or QPSK signal after 4-point IFFT, which we will discuss later in this chapter. For the phase recovery, Viterbi algorithm is utilized to cancel the phase noise of QPSK signal after 4-point FFT. For the CMMA algorithm, we only select the inner three rings/radii for the error signal calculation to increase equalizer robustness [13, 14], which is the same as CMMA algorithm for 9-QAM signal [11, 15]. We also analyze the PAPR of quad-carrier QPSK-OFDM signal and traditional QPSK-OFDM signal. The PAPR performance is evaluated by the complementary cumulative distribution function (CCDF). CCDF presents the probability distribution in

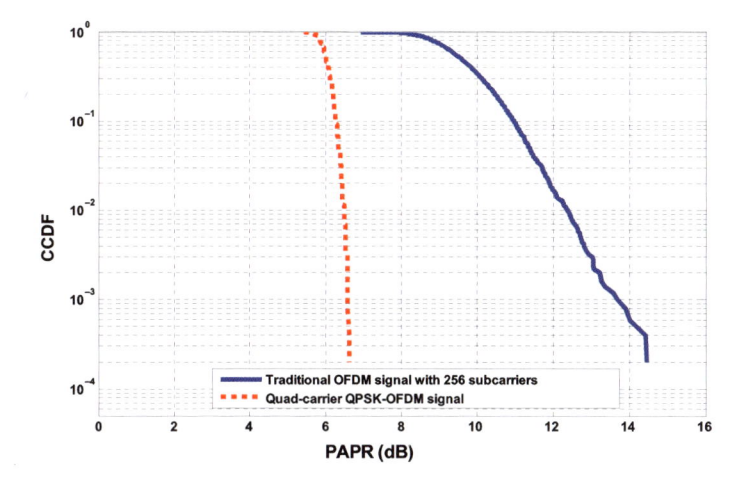

Fig. 6.10. The CCDFs of the traditional QPSK-OFDM signal with 256 subcarriers and quad-carrier QPSK-OFDM signal.

which the PAPR of current OFDM symbol is higher than certain threshold. CCDF curves of PAPR for traditional QPSK-OFDM signal with 256 subcarriers and quad-carrier QPSK-OFDM signal are shown in Fig. 6.10. The PAPR of quad-carrier QPSK-OFDM signal outperforms the traditional OFDM and there is an 8-dB PAPR improvement at the probability of 1×10^{-4}.

6.3.2. Experimental setup and results

Figure 6.11 shows the experimental setup of quad-carrier QPSK-OFDM signal transmission system. The optical transmitter is quite similar to that in Section 4.2.2. In this section, the quad-carrier QPSK-OFDM signal is without additional CP and TS and the signal is equalized with CMMA blind equalization method. The generated signal is boosted via an EDFA before getting launched into 80 km SMF-28. The output signal is then injected into the integrated coherent receiver to implement optical to electrical detection. After integrated coherent receiver, the signal is captured by the real-time oscilloscope with 50GSa/s sample rate.

The optical eye diagram of quad-carrier QPSK-OFDM signal is inserted as inset (a) in Fig. 6.11. The optical spectra before and after 80-km SMF-28 transmission with 0.1-nm resolution are shown in Fig. 6.11(b) and

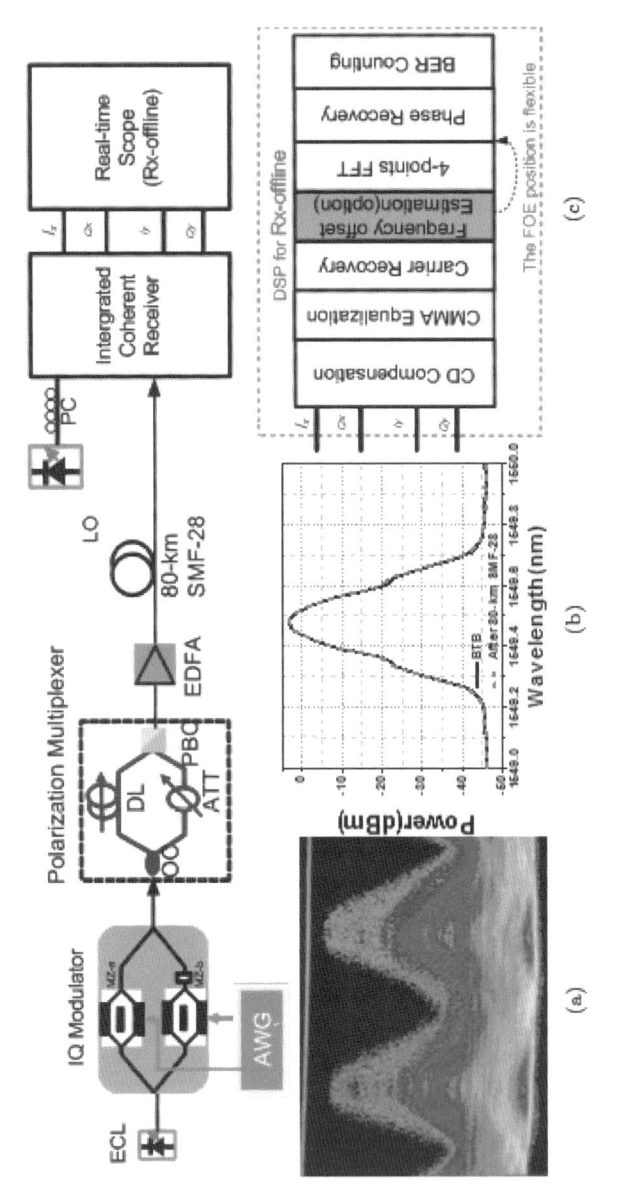

Fig. 6.11. Experimental setup.

there is no optical signal-to-noise ratio (OSNR) degradation observed after 80-km SMF-28 transmission. The DSP for receiver offline processing of the Quad-Carrier QPSK-OFDM signal is shown in Fig. 6.11(c). At the receiver, the 25-QAM quad-carrier QPSK-OFDM signal can be equalized with the CMMA method without additional overhead compared to traditional OFDM signal with frequency domain equalization. After optical link, four signal components are first captured by the real-time oscilloscope with 50 GSa/s sample rate. Secondly, a $T/2$-spaced time-domain finite-impulse-response (FIR) filter is firstly used for chromatic dispersion (CD) compensation, where the filter coefficients are calculated from the known fiber CD transfer function using the frequency-domain truncation method. Thirdly, the CMMA is used to retrieve the modulus of the polarization-division-multiplexed 25-QAM signal and realize polarization demultiplexing. The subsequent step is to realize the FOE, and here we have to claim the position of FOE is flexible and it can also be done after 4-point FFT. After these procedures, 4-point FFT is applied to convert the 25-QAM signal in time domain into QPSK signal in frequency domain and then the bit error ratio (BER) can also be obtained with the BER counting after QPSK signal phase recovery. As blind equalization is applied for quad-carrier QPSK-OFDM signal, there is no overhead and the capacity is 48 Gbit/s. In this experiment, the BER is counted over 10×10^6 bits (10 data sets, and each set contains 10^6 bits). The optical spectra of different subcarriers are shown in Fig. 6.12 with 0.01 nm resolution, and it can be seen that the distribution

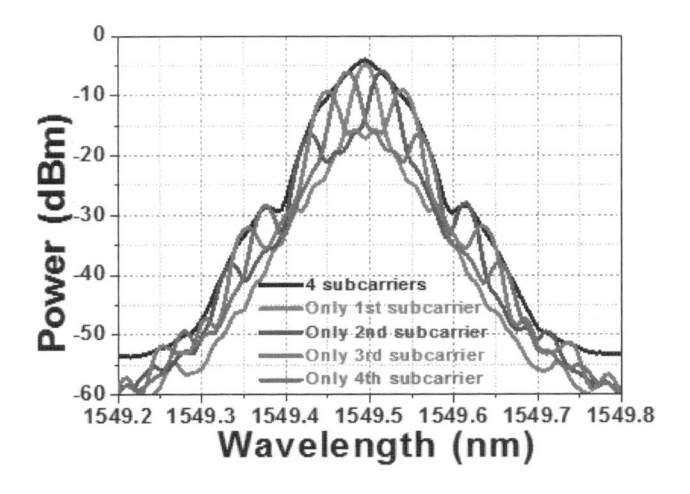

Fig. 6.12. The optical spectra of different subcarriers.

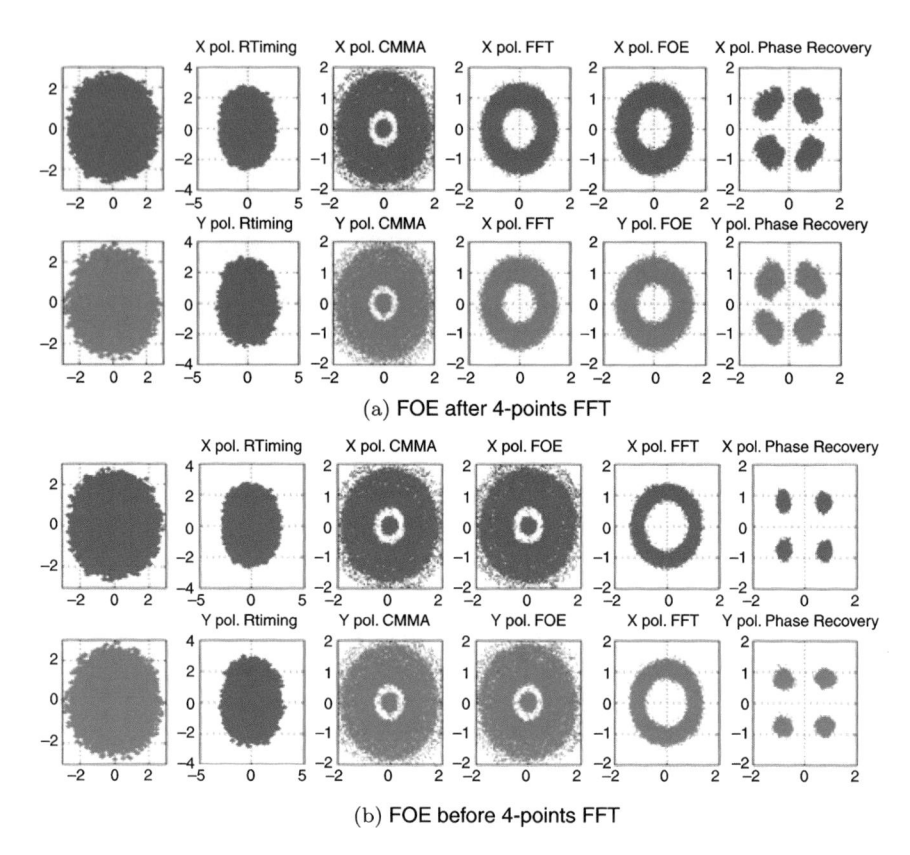

Fig. 6.13. Constellations in different stages of DSP: (a) FOE after 4-point FFT and (b) FOE before 4-point FFT.

of subcarriers in optical domain is the same as that in the electrical domain demonstrated in Fig. 6.9(a).

Figure 6.13 shows the constellations of quad-carrier QPSK-OFDM signal with OSNR of 20 dB in different stages of the offline DSP. In Fig. 6.13(a), the FOE is done after 4-point FFT in the receiver offline DSP, while in Fig. 6.13(b), FOE is completed before 4-point FFT. Compared to the constellations after phase recovery in Fig. 6.13(a), those in Fig. 6.13(b) are converged much better, which means the FOE should be done with 25-QAM signal before 4-point FFT. As FFT causes the spread of noise induced by frequency offset [6], it should be better to finish the FOE before the FFT in the time domain.

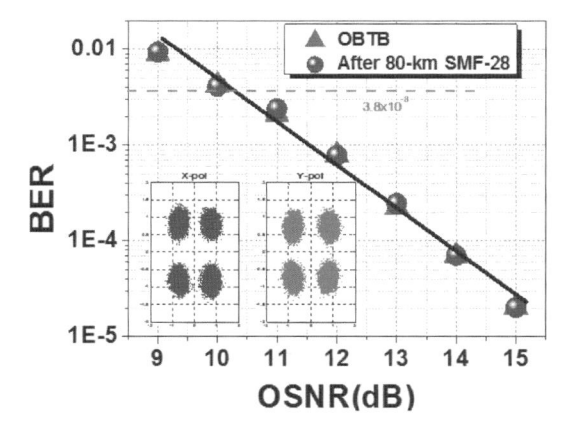

Fig. 6.14. Measured BER of quad-carrier QPSK-OFDM signal versus OSNR.

Figure 6.14 shows the measured BER of quad-carrier QPSK-OFDM signal versus OSNR. There is nearly no OSNR penalty observed after 80-km SMF-28 transmission. The BER for the 48-Gbit/s dual-polarization quad-carrier QPSK-OFDM signal is less than the pre-forward-error-correction (7% overhead) threshold of 3.8×10^{-3} when the OSNR is higher than 10 dB after 80-km SMF-28 transmission. The constellations of dual-polarization quad-carrier QPSK-OFDM signal after phase recovery with OSNR of 17 dB after 80-km SMF-28 transmission are shown in the inset of Fig. 6.14.

In the back-to-back (BTB) case, we adjust the receiver bandwidth by changing the bandwidth of the real-time oscilloscope to determine the minimum bandwidth for the 48-Gbit/s quad-carrier QPSK-OFDM signal transmission. Figure 6.15 shows measured BER versus receiver bandwidth. Compared to the situation when the receiver bandwidth is larger than 6 GHz, there is less than 0.3 dB OSNR penalty when the bandwidth of the receiver is set to 6 GHz according to the relationship between OSNR and BER in Fig. 6.14. The electrical spectra of the obtained signal with different receiver bandwidths are inserted as insets (i)–(v) in Fig. 6.15.

The receiver bandwidth of real-time oscilloscope can only be changed by integral interval, and it has been found that the signal cannot be recovered with 5-GHz receiver bandwidth in the experiment. The OSNR penalty and BER versus receiver bandwidth when the bandwidth is set between 5 and 6 GHz are measured and shown in Fig. 6.16. The fractional change of

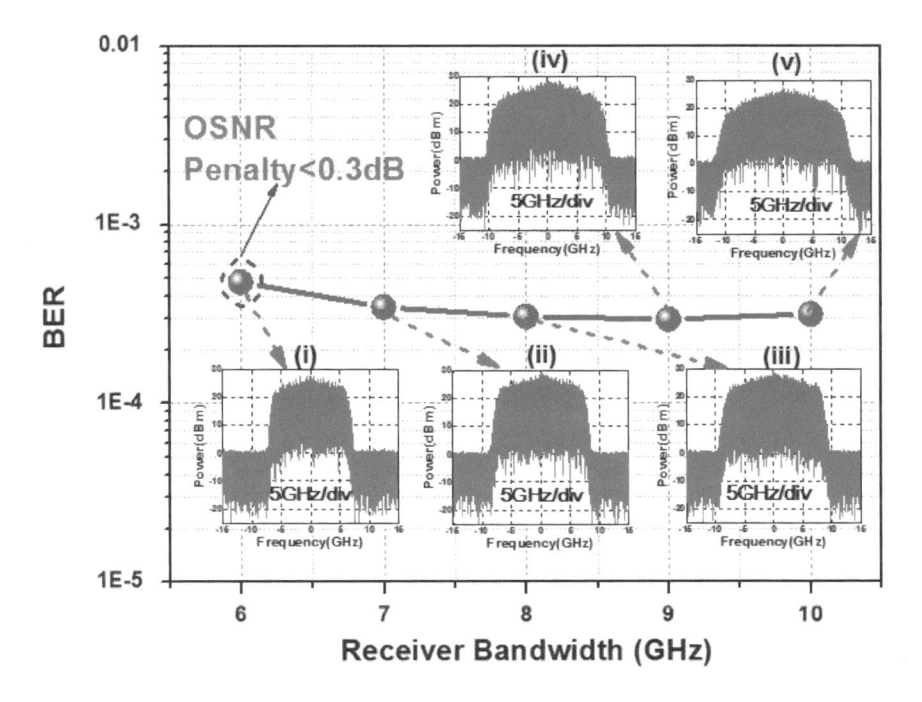

Fig. 6.15. Measured BER versus receiver bandwidth.

the receiver bandwidth is virtually realized via offline low pass filter (LPF), and during this virtual realization, the frequency offset of two lasers in the experiment should be close to 0. The received sample for the test is the data obtained with 10 GHz bandwidth receiver. The signal cannot be recovered if the bandwidth of receiver is smaller than 5.2 GHz because in this case, some useful spectral components are filtered out due to the inadequate bandwidth. There is 3-dB OSNR penalty when the bandwidth of the LPF is only 5.2 GHz and the BER is 1.02×10^{-2}.

Summary: In this section, transmission and reception of 48 Gbit/s dual-polarization quad-carrier QPSK-OFDM signal is successfully demonstrated in CO-OFDM system. The signal is blindly equalized with CMMA equalization method in the time domain. The phase recovery can be implemented with simple Viterbi algorithm and the FOE should be done before 4 sub-carriers are separated with FFT. Using these techniques, we successfully generate and transmit 48-Gbit/s quad-carrier QPSK-OFDM signal over 80-km SMF-28 without penalty.

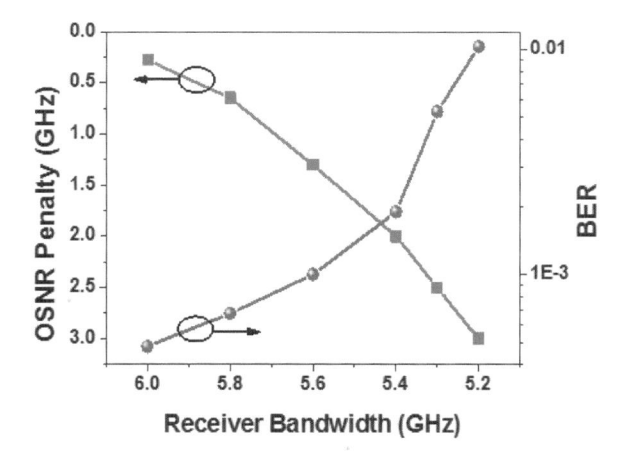

Fig. 6.16. Measured OSNR penalty and BER when the receiver bandwidth is set between 5 and 6 GHz.

6.4. Transmission and Reception of PDM Dual-Subcarrier Coherent 16QAM-OFDM Signals

6.4.1. *System principle*

In Sections 6.2 and 6.3, transmission and reception of dual-polarization dual-carrier and quad-carrier QPSK-OFDM signals are successfully demonstrated in the CO-OFDM system, respectively. Adopting the CMMA blind equalization method in time domain can avoid extra overhead such as the insertion of pilots and training sequence which can improve system spectral efficiency. In this section, we investigate the transmission of higher order modulation format signal in few subcarrier schemes. To determine blind equalization scheme of different modulation formats, we must understand the time-domain characteristic of signal. Equations (6.4) and (6.6) give the sample values of dual-subcarrier QPSK-OFDM and quad-subcarrier QPSK-OFDM, respectively. According to the transformational rule from frequency-domain signal to time-domain sample in the two equations, output time-domain sample signal can be determined when the modulation format of frequency-domain signal added on subcarriers is fixed.

Table 6.1 shows the input data and output data after 2-point IFFT and 4-point IFFT. The QPSK in the frequency domain is transformed to 9-QAM and 25-QAM in time domain after 2-point and 4-point IFFT, respectively. 21QAM and 88QAM can be obtained with 2-point and 4-point IFFT of rectangle 8QAM. 16QAM input data is changed to 49QAM and

Table 6.1. The output data of dual-subcarrier and quad-subcarrier OFDM.

Input data	Dual-subcarrier OFDM	Quad-subcarrier OFDM
QPSK	9QAM	25QAM
8QAM	21QAM	88QAM
16QAM	49QAM	169QAM
32QAM	109QAM	401QAM
64QAM	225QAM	841QAM

169QAM in the dual-subcarrier and quad-subcarrier scheme. The formats of output data of dual-subcarrier and quad-subcarrier OFDM with 32QAM in the time domain are 109QAM and 401QAM, respectively. 225QAM and 841QAM can be obtained with 2-point and 4-point IFFT of 64QAM. In this section, we only discuss the transmission and reception of dual-subcarrier 16QAM OFDM, the dual-subcarrier 16QAM OFDM can be equalized as 49QAM signal in time domain with the CMMA equalization method.

6.4.2. *Experimental setup and results*

Figure 6.17 shows the experimental setup of the dual-subcarrier coherent optical 16QAM OFDM (CO-16QAM-OFDM) transmission system. The transmitter is similar to that described in Section 4.2.2. The dual-carrier

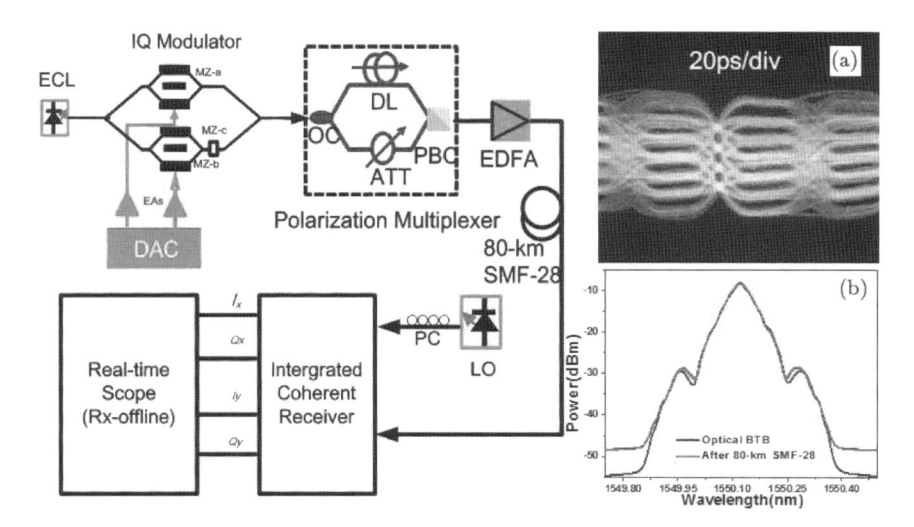

Fig. 6.17. Experimental setup of QPSK CO-OFDM transmission system: (a) electrical eye diagram and (b) optical spectra.

16QAM OFDM signal is generated by commercial DAC. The generated dual-carrier 16QAM OFDM signal is boosted via an EDFA before getting launched into 80-km SMF-28. The 80-km SMF-28 has 18-dB average loss and 17-ps/km/nm CD at 1550 nm without optical dispersion compensation. The output signal is then injected into an integrated coherent receiver to implement optical-to-electrical detection. The output signal is then injected into the integrated coherent receiver to implement optical-to-electrical detection. After integrated coherent receiver, the signal is captured by the real-time oscilloscope with 80 GSa/s sample rate. The electrical eye diagram of the in-phase component of 16Gbaud Dual-subcarrier 16QAM-OFDM signal is inserted as inset (a) in Fig. 6.17. The level of this signal is 7. The optical spectra before and after 80-km SMF-28 transmission with 0.1-nm resolution are shown in Fig. 6.17(b) and there is no optical signal-to-noise ratio (OSNR) degradation observed after 80-km SMF-28 transmission.

The DSP for receiver (Rx)-offline processing of the dual-subcarrier 16QAM-OFDM signal is shown in Fig. 6.18. At the receiver, the dual-subcarrier 16QAM OFDM can be equalized with CMMA method like 49QAM without any additional overhead. After the integrated receiver, four signal components are first captured by the real-time scope with 80 GSa/s sample rate. Secondly, a $T/2$-spaced time-domain FIR filter is

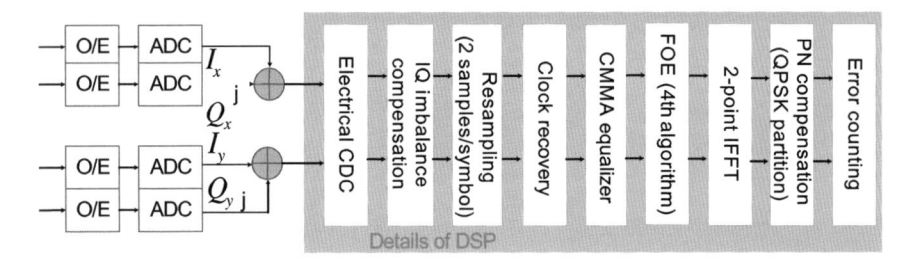

Fig. 6.18. The DSP for receiver offline processing of the dual-subcarrier 16QAM-OFDM signal.

firstly used for chromatic dispersion compensation (CDC), where the filter coefficients are calculated from the known fiber CD transfer function using the frequency-domain truncation method. Thirdly, the 10-radius CMMA is used to retrieve the modulus of the PDM dual-subcarrier 16QAM OFDM like a 49-QAM signal and realize polarization demultiplexing. The subsequent step is to realize the FOE, with a fourth algorithm. After these procedures, 2-point FFT is applied to convert the time domain 49-QAM signal into two 16QAM signal in frequency domain. QPSK partition algorithm is used to realize the CPE [11, 19] and then the bit error ratio (BER) can also be obtained with the BER counting. As blind equalization is applied for dual-subcarrier 16QAM-OFDM signal, there is no overhead and the capacity is 128 Gbit/s. In this experiment, the BER is counted over 20×10^6 bits (20 data sets, and each set contains 10^6 bits).

In the single polarization dual-subcarrier 16QAM-OFDM transmission and reception, we set the optical source in the transmitter and receiver to be the same ECL and the linewidth of this ECL is 400 Hz, which means that there is no frequency offset and negligible phase noise during the reception of the single polarization dual-subcarrier 16QAM-OFDM signal. We can see the 49QAM signal after retiming and CMMA, which is shown in Fig. 6.19. The constellations rotate due to the sight phase noise.

Figure 6.20 shows the constellations in different stages of DSP for PDM dual-subcarrier 16QAM-OFDM with OSNR of 28 dB in different stages of the offline DSP, which is described in detail in Fig. 6.18. FOE should be implemented before 2-point FFT according to the previous section as FFT may lead to the spread of noise induced by frequency offset [6]. The object of phase noise estimation is 16QAM signal because the phase noise is after FFT. QPSK partition algorithm is used to remove the phase noise in dual-subcarrier 16QAM signal.

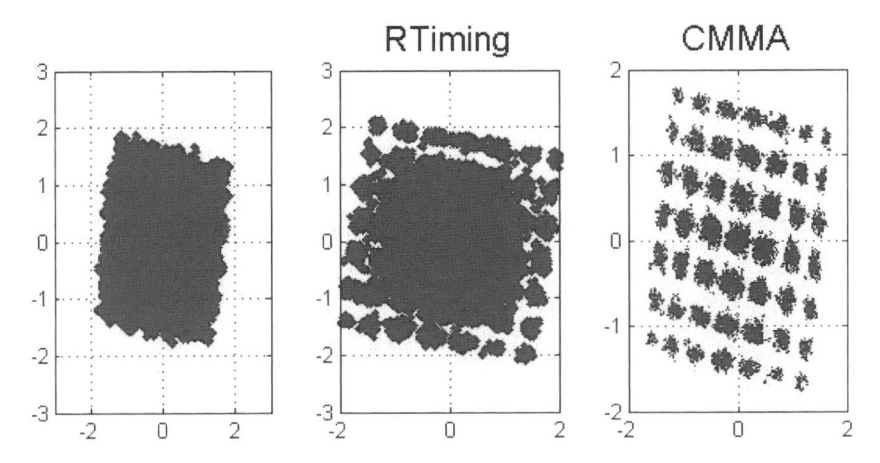

Fig. 6.19. Constellations in different stages of DSP for single polarization dual-subcarrier 16QAM-OFDM.

Figure 6.21 shows the measured BER of 16 Gbaud dual-subcarrier 16QAM-OFDM signal versus OSNR. There is nearly no OSNR penalty observed after 80-km SMF-28 transmission. The BER for the 128-Gbit/s PDM dual-subcarrier 16QAM-OFDM signal is less than the pre-forward-error-correction (7% overhead) threshold of 3.8×10^{-3} when the OSNR is higher than 18 dB after 80-km SMF-28 transmission. The constellations of 16 Gbaud PDM dual-subcarrier 16QAM-OFDM signal after phase recovery with OSNR of 19 dB after 80-km SMF-28 transmission are shown in the inset of Fig. 6.21.

In the OBTB case, we insert a wavelength selective switch (WSS) to adjust the bandwidth of channel to find the OSNR penalty for the 128-Gbit/s dual-subcarrier 16QAM-OFDM signal transmission with different optical channel bandwidth. Figure 6.22 shows measured BTB OSNR requirement and OSNR penalty versus bandwidth of WSS. OSNR penalty increases rapidly when the optical channel bandwidth decreases. We think the dual-subcarrier 16QAM-OFDM cannot overcome the high frequency attenuation due to the sufficient bandwidth as it demonstrates a 49QAM in the time domain, such a high-order QAM is vulnerable to high-frequency attenuation. A 0.6-dB OSNR penalty is observed when the bandwidth of the channel is set at 25 GHz.

Summary: In this section, 16-Gbaud PDM dual-subcarrier 16QAM-OFDM signal transmission and reception in the CO-OFDM system are successfully demonstrated with blind equalization like a 49QAM signal. Therefore,

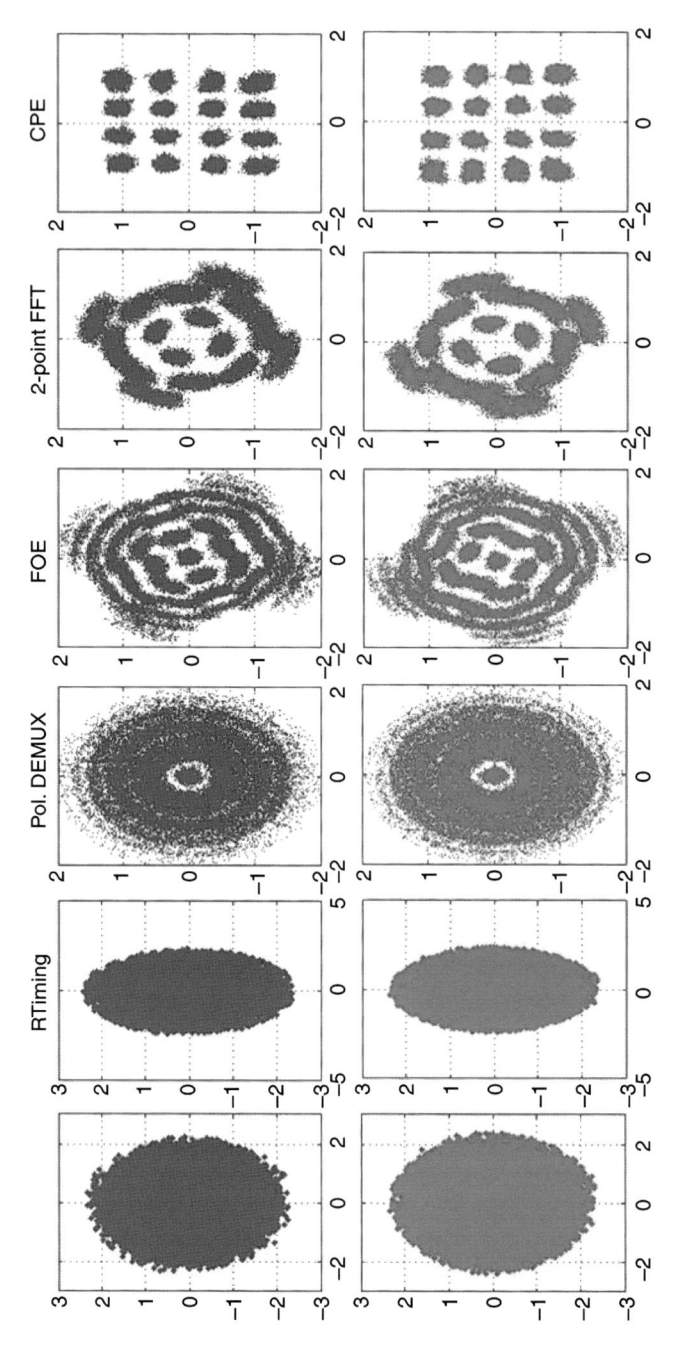

Fig. 6.20. Constellations in different stages of DSP for PDM dual-subcarrier 16QAM-OFDM.

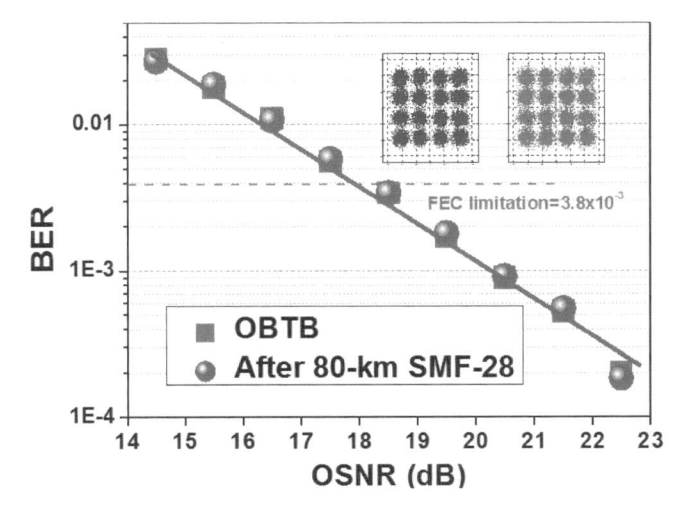

Fig. 6.21. Measured BER versus transmission fiber length for two types of OFDM signal.

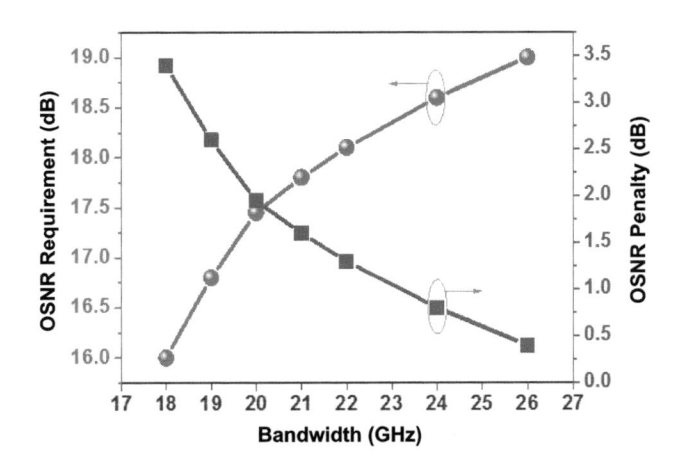

Fig. 6.22. Measured BTB OSNR requirement and OSNR penalty versus bandwidth of WSS.

signal can be recovered adopting CMMA blind equalization in time domain. A 128-Gbit/s PDM dual-subcarrier 16QAM-OFDM signal is successfully transmitted over 80-km SMF-28 without penalty and 0.6 dB optical signal-to-noise ratio (OSNR) penalty is observed when the bandwidth of the channel is set at 25 GHz.

6.5. Summary

In this chapter, we discuss blind equalization in the time domain to recover the received signal in the few subcarriers CO-OFDM system. In Section 6.2, a 2-subcarrier dual-polarization QPSK-OFDM signal transmission system is successfully demonstrated with blind equalization in time domain. Quad-carrier QPSK-OFDM signal transmission and reception is successfully demonstrated with blind equalization in the time domain in Section 6.3. Finally, the transmission and reception of dual-subcarrier 16QAM-OFDM signal with blind equalization in the time domain are realized in Section 6.4.

References

[1] J. Armstrong, OFDM for optical communications, *J. Lightwave Technol.* **27**(3) (2009) 189–204.
[2] S. L. Jansen, I. Morita, T. C. W. Schenck, N. Takeda and H. Tanaka, Coherent optical 25.8-Gb/s OFDM transmission over 4160-km SSMF, *J. Lightwave Technol.* **26**(1) (2008) 6–15.
[3] Z. Cao, J. Yu, W. Wang, L. Chen and Z. Dong, Direct-detection optical OFDM transmission system without frequency guard band, *IEEE Photon. Technol. Lett.* **22**(11) (2010) 736–738.
[4] W.-R. Peng, T. Tsuritani and I. Morita, Simple carrier recovery approach for RF-pilot-assisted PDM-CO-OFDM systems, *J. Lightwave Technol.* **31**(15) (2013) 2555–2564.
[5] A. J. Lowery, Improving sensitivity and spectra efficiency in direct-detection optical OFDM systems, in *Proc. OFC'08* (2008), paper OMM4.
[6] L. Tao, J. Yu, Y. Fang, J. Zhang, Y. Shao and N. Chi, Analysis of noise spread in optical DFT-S OFDM systems, *J. Lightwave Technol.* **30**(20) (2012) 3219–3225.
[7] W.-R. Peng, H. Takahashi, I. Morita and T. Tsuritani, Per-symbol-based digital back-propagation approach for PDM-CO-OFDM transmission systems, *Opt. Express* **21**(2) (2013) 1547–1554.
[8] T. Kobayash, A. Sano, E. Yamada, E. Yoshida and Y. Miyamoto, Over 100 Gb/s electro-optically multiplexed OFDM for high capacity optical transport network, *J. Lightwave Technol.* **27**(16) (2009) 3714–3720.
[9] H. Wang, Y. Li, X. G. Yi, D. M. Kong, J. Wu and J. T. Lin, APSK modulated CO-OFDM system with increased tolerance toward fiber nonlinearities, *IEEE Photon. Technol. Lett.* **24** (2012) 1085.
[10] C. Li, Q. Yang, T. Jiang, Z. He, M. Luo, C. Li, X. Xiao, D. Xue and X. Yi, Investigation of coherent optical multi-band DFT-S OFDM in long haul transmission, *IEEE Photon. Technol. Lett.* **24** (2012) 1704.
[11] J. Zhang, J. Yu, N. Chi, Z. Dong, J. Yu, X. Li, L. Tao and Y. Shao, Multi-modulus blind equalizations for coherent quadrature duobinary

spectrum shaped PM-QPSK digital signal processing, *J. Lightwave Technol.* **31** (2013) 1073.
[12] J. Yu, Z. Dong and N. Chi, 1.96 Tb/s (21 × 100 Gb/s) OFDM optical signal generation and transmission over 3200-km fiber, *IEEE Photon. Technol. Lett.* **23**(15) (2011) 1061–1063.
[13] X. Zhou, J. Yu, M.-F. Huang, Y. Shao, T. Wang, L. Nelson, P. Magill, M. Birk, P. I. Borel, D. W. Peckham, R. Lingle and B. Zhu, 64-Tb/s, 8 b/s/Hz, PDM-36QAM transmission over 320 km using both pre- and post-transmission digital signal processing, *J. Lightwave Technol.* **29**(4) (2011) 571–577.
[14] X. Zhou, L. E. Nelson, P. Magill, R. Isaac, B. Zhu, D.W. Peckham, P. I. Borel and K. Carlson, PDM-Nyquist-32QAM for 450-Gb/s per-channel WDM transmission on the 50 GHz ITU-T grid, *J. Lightwave Technol.* **30**(4) (2012) 553–559.
[15] J. Zhang, B. Huang and X. Li, Improved quadrature duobinary system performance using multi-modulus equalization, *IEEE Photon. Technol. Lett.* **25**(16) (2013) 1630–1633.
[16] A. Sano, E. Yamada, H. Masuda, E. Yamazaki, T. Kobayashi, E. Yoshida, Y. Miyamoto, R. Kudo, K. Ishihara and Y. Takatori, No-guard-interval coherent optical OFDM for 100-Gb/s long-haul WDM transmission, *J. Lightwave Technol.* **27** (2009) 3705.
[17] Y. Huang, D. Qian, R. E. Saperstein, P. N. Ji, N. Cvijetic, L. Xu and T. Wang, Dual-polarization 2×2 IFFT/FFT optical signal processing for 100-Gb/s QPSK-PDM all-optical OFDM, in *Optical Fiber Communication Conference and National Fiber Optic Engineers Conference*, OSA Technical Digest (CD) (Optical Society of America, 2009), paper OTuM4.
[18] L. Tao, J. Yu, J. Zhang, Y. Shao and N. Chi, Reduction of intercarrier interference based on window shaping in OFDM RoF systems, *IEEE Photon. Technol. Lett.* **25** (2013) 851.
[19] I. Fatadin, D. Ives and S. J. Savory, Laser linewidth tolerance for 16-QAM coherent optical systems using QPSK partitioning, *IEEE Photon. Technol. Lett.* **22**(9) (2010) 631–633.
[20] F. Li, J. Zhang, J. Yu and X. Li, Blind equalization for dual-polarization two-subcarrier coherent QPSK-OFDM signals, *Opt. lett.* **39**(2) (2014) 201–204.
[21] F. Li, J. Zhang, J. Xiao and X. Li, Transmission and reception of quad-carrier QPSK-OFDM signal with blind equalization, in *Optical Fiber Communication Conf.*, (2014) Th2A. 64.
[22] F. Li, J. Zhang, Z. Cao, J. Yu, X. Li, L. Chen, Y. Xia and Y. Chen, Transmission and reception of quad-carrier QPSK-OFDM signal with blind equalization and overhead-free operation, *Opt. Express* **21** (2013) 30999–31005.
[23] F. Li, J. Zhang, J. Yu and X. Li. Transmission and reception of PDM dual-subcarrier coherent 16QAM-OFDM signals, *Optical Fiber Technol.* **26** (2015) 201–205.

Chapter 7

DSP for MIMO OFDM Signal

7.1. Introduction

Recently, the demand for ultra-high data rate optical transmission has been continuously growing. Social multimedia, mobile front-haul services and other bandwidth-intensive services are constantly posing challenges. Facing these problems, direct-detection (DD) optical transmission is considered as a more attractive and feasible solution in terms of system construction cost, computation complexity and low power consumption [1–4]. Currently, a 100-Gb/s/λ system is strongly desired for short-reach applications with a DD-type optical link, which can use single sideband (SSB) or vestigial sideband (VSB) to combat the power-fading impairment in typical double-sideband (DSB) systems. A 100-G/λ DD transmission has been demonstrated by discrete multi-tone (DMT) SSB modulation with an 80-km standard single-mode fiber (SSMF) using an I/Q modulator [4]. To further improve the bandwidth efficiency, 200-G/λ twin-SSB transmission over a 160-km SSMF with discrete Fourier transform spread orthogonal frequency division multiplexing (DFT-S OFDM) using joint image Cancellation and nonlinearity mitigation has been presented [5]. However, it uses the I/Q modulator, which contains two Mach–Zehnder modulators (MZMs). References [6, 8] show the feasibility using only one dual-driver MZM (DD-MZM) to achieve 100-G/λ and 200-G/λ twin-SSB transmission with bit-loading DMT modulation. The improvement in papers [6–8] is based on more advanced equipment and a frequency-domain multiple-input multiple-output (MIMO) array algorithm. However, the improvement from using the frequency-domain MIMO-array algorithm is only 20% [8].

In this chapter, we discuss a time-domain MIMO-Volterra algorithm to overcome both the interference and nonlinearity penalty of a twin-SSB signal in a DD-MZM modulator, and some experimental results will be presented to show this scheme by applying twin SSB can double the spectral efficiency, and the MIMO-Volterra algorithm can improve the data rate by more than 45%.

7.2. Principles

7.2.1. Generation of the twin-SSB signal

Figure 7.1(f) shows the principles of an SSB signal generation process based on a DD-MZM. The output of the DD-MZM can be simplified as $I+jQ$ [7]. We set the electrical signal I as a real signal x and the signal Q as its Hilbert pair $\hat{x}$. The output of $x + j\hat{x}$ is the analytic signal of x and is a right-band SSB signal. Then, the optical domain expression is

$$E_{\text{out}} = E_{\text{in}}(x + j\hat{x}) \tag{7.1}$$

From (7.1), we observe that it is an optical right-band SSB signal. Thus, the left-band SSB signal is expressed as:

$$E_{\text{out}} = E_{\text{in}}(y - j\hat{y}) \tag{7.2}$$

The twin-SSB signal is [6]

$$\begin{aligned} E_{\text{out}} &= E_{\text{in}}(x_r + j\hat{x}_r) + E_{\text{in}}(x_l - j\hat{x}_l) \\ &= E_{\text{in}}[(x_r + x_l) + j(\hat{x}_r - \hat{x}_l)] \end{aligned} \tag{7.3}$$

Two independent DFT-S OFDM signals x_r and x_l are combined to drive the upper arm of the DD-MZM, whereas their Hilbert pairs are subtracted to drive the lower arm of the DD-MZM. The bias of the two parallel phase modulators (PMs) in the DD-MZM is driven with a bias difference of $V_\pi/2$.

However, the simplified output of the DD-MZM ignores the nonlinearity and interference. The simplified process is expressed as

$$\begin{aligned} E_{\text{out}} &= \frac{\sqrt{2}}{2} E_{\text{in}} \{ e^{j[\frac{\pi}{V_\pi}I(t) - \frac{\pi}{2}]} + e^{j[\frac{\pi}{V_\pi}Q(t)]} \} \\ &= \frac{\sqrt{2}}{2} E_{\text{in}} \{ -je^{j[\frac{\pi}{V_\pi}I(t)]} + e^{j[\frac{\pi}{V_\pi}Q(t)]} \} \end{aligned} \tag{7.4}$$

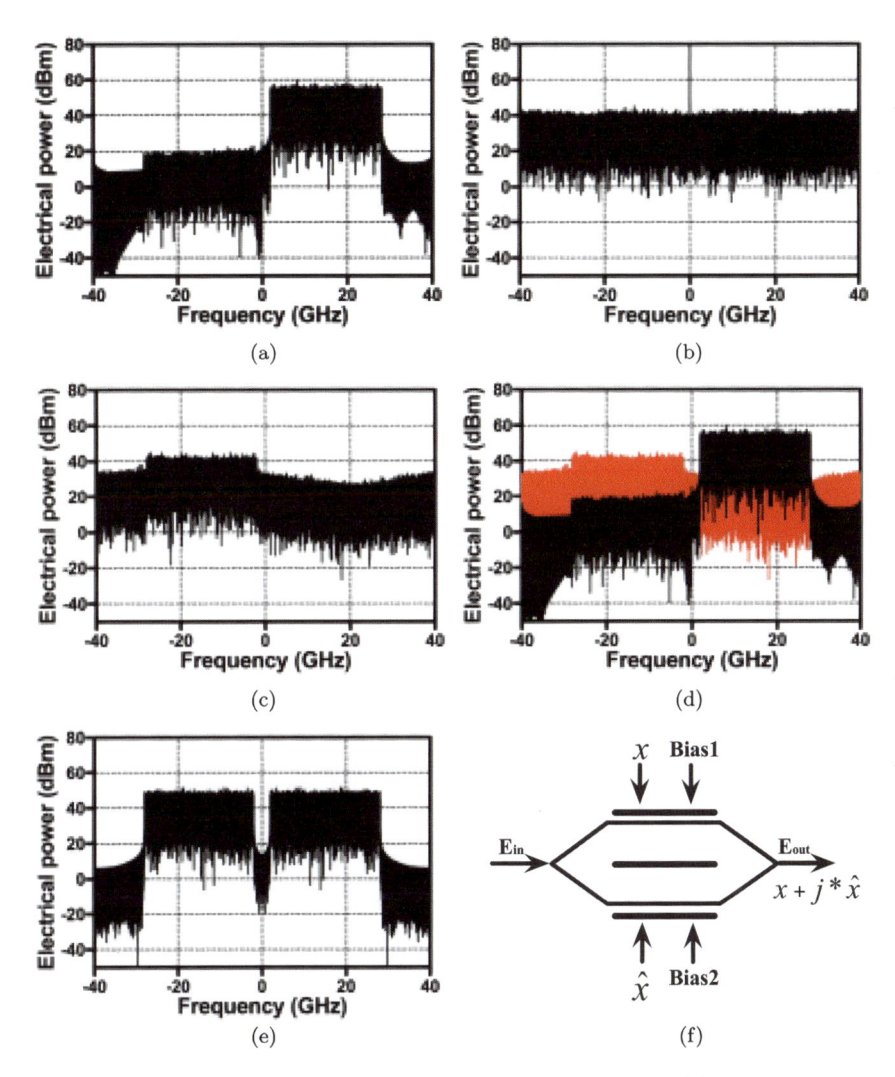

Fig. 7.1. Simulated electrical spectrum of different order terms: (a) first-order term; (b) second-order term; (c) third-order term; (d) both first- (black) and third- (red) order terms; (e) twin-SSB; and (f) principles of an optical SSB signal generation process based on a DD-MZM.

$$\approx \frac{\sqrt{2}}{2} E_{in} \left\{ -j \left[1 + j\frac{\pi}{V_\pi} I(t) \right] + \left[1 + j\frac{\pi}{V_\pi} Q(t) \right] \right\}$$

$$= \frac{\sqrt{2}}{2} E_{in} \left\{ \frac{\pi}{V_\pi} [I(t) + j \times Q(t)] + 1 - j \right\} \quad (7.5)$$

Equation (7.4) is the output of the DD-MZM in the optical domain. The simplified process from (7.4) to (7.5) uses the Taylor expansion of e^x. Reference [7] only uses the first-order term to approximate the linear conversion. Considering the higher order terms, the output is expressed as

$$E_{\text{out}} \approx \frac{\sqrt{2}}{2} E_{\text{in}} \left\{ \begin{array}{l} -j\left[1+j\frac{\pi}{V_\pi}I(t)\right] + \left[1+j\frac{\pi}{V_\pi}Q(t)\right] \\[6pt] -j\dfrac{[1+j\frac{\pi}{V_\pi}I(t)]^2}{2!} + \dfrac{[1+j\frac{\pi}{V_\pi}Q(t)]^2}{2!} \\[6pt] -j\dfrac{[1+j\frac{\pi}{V_\pi}I(t)]^3}{3!} + \dfrac{[1+j\frac{\pi}{V_\pi}Q(t)]^3}{3!} + \cdots \end{array} \right\} \quad (7.6)$$

$$= \frac{\sqrt{2}}{2} E_{\text{in}} \left\{ \begin{array}{l} \dfrac{\pi}{V_\pi}[I(t)+jQ(t)] \\[6pt] +\dfrac{\pi^2}{2V_\pi^2}[jI(t)^2 - Q(t)^2] \\[6pt] +\dfrac{\pi^3}{6V_\pi^3}[I(t)^3 - jQ(t)^3] \\[6pt] +1-j \end{array} \right\} \quad (7.7)$$

If we set the electrical signal I as a real signal x and signal Q as its Hilbert pair $\hat{x}$, the first-order term in (7.7) is a right-band SSB signal, which is expressed as

$$\begin{aligned} I(t) &= x(t) \\ Q(t) &= x(t)h(t) \end{aligned} \quad (7.8)$$

$$H(jw) = \begin{cases} -j, & w > 0 \\ j, & w < 0 \end{cases} \quad (7.9)$$

$$f\{I(t)+jQ(t)\} = X(jw) + j \cdot H(jw)X(jw)$$
$$= \begin{cases} 2X(jw), & w > 0 \\ 0, & w > 0 \end{cases} \quad (7.10)$$

$$f\{jI(t)^2 - Q(t)^2\} = j \cdot X(jw)X(jw) - [H(jw)X(jw)][H(jw)X(jw)]$$

$$= \begin{cases} (j-1)\sum_{n=-\infty}^{\infty} X(jn)X[j(w-n)] \\ \quad +2\sum_{n=-\infty}^{w} X(jn)X[j(w-n)], & w > 0 \\ (j-1)\sum_{n=-\infty}^{\infty} X(jn)X[j(w-n)], & w = 0 \\ (j-1)\sum_{n=-\infty}^{\infty} X(jn)X[j(w-n)] \\ \quad +2\sum_{n=w}^{0} X(jn)X[j(w-n)], & w < 0 \end{cases} \quad (7.11)$$

where H is the frequency response of the Hilbert transform and f is the Fourier transform. Equation (7.11) shows the frequency response of the first order, which is the reason why we can achieve the SSB signal. The expansion of the second order is shown in (7.11), where there is the large direct-current (DC) component and nonlinearity noise in the frequency domain. To obtain a better picture of the noise, the simulated electrical spectra of different order terms are shown in Fig. 7.1. The signal is a 26-Gbaud electrical signal. The first-order term is the right-band SSB signal in Fig. 7.1(a), whereas the second-order term is the direct-current (DC) component and background noise in the output (7.7), which is shown in Fig. 7.1(b). The third-order term is the left-band interference in Fig. 7.1(c). In Fig. 7.1(d), we combine the first- and third-order terms, which are shown in black and red, respectively. Compared with the theoretical twin-SSB signal in Fig. 7.1(e), there is large interference from the right-band SSB signal in the left band. If we continue considering the higher order terms of the Taylor expansion, we find that the orders of 3, 7, 11,... are the interference in the left band, the orders of 5, 9, 13,... are the nonlinearity in the right band, and all even-order terms are the background nonlinearity noise. Without considering the nonlinearity from the receiver such as a photodetector (PD) or an electronic amplifier (EA), the nonlinearity of the fifth-order term has less effect on the performance than the interference of the third-order term. Thus, in [6], the data rate only achieved a 10% increase using twin SSB with nonlinearity equalization (145 Gb/s with twin SSB; 133 Gb/s with conventional SSB in the back-to-back (BTB) case). In [8], the increase is only 20% with the frequency-domain MIMO method (210 Gb/s without MIMO and 260 Gb/s with MIMO). Similarly, the left-band SSB also introduces the interference in the right band and nonlinearity in the left band.

7.3. MIMO-Volterra Equalization Algorithm

The time-domain joint image cancellation (IC) and nonlinearity equalization (NE) were proposed in [5]. However, the MIMO effect is only considered in the linear domain. Considering the interference and nonlinearity in this twin-SSB system, we first separate them. The interference between two band signals can be canceled using the MIMO algorithm [9], which regards these two bands as multiple inputs. For the nonlinearity effect, the Volterra-series-based equalizer appears to be a better choice. Volterra series can be used to simultaneously estimate the response of a nonlinear system and capture the memory effect of devices or fibers [10]. Both linear and nonlinear parts are considered as the interference from the other band signal. The construction of our proposed time-domain MIMO-Volterra equalizer is shown in Fig. 7.2(a).

The Volterra series expansion includes a linear term and nonlinear terms. Considering the tradeoff between computation complexity and equalization performance, we only regard the interference as the linear term, and only the second-order term is used in the calculation. Therefore, the output of the equalizer is expressed as

$$\begin{aligned}
y_l(n) = &\sum_{i=0}^{N-1} h_{ll}(n)x_l(n-i) + \sum_{i=0}^{N-1} h_{lr}(n)x_r(n-i) \\
&+ \sum_{k=0}^{L-1}\sum_{i=k}^{L-1} w_{ll}(n)x_l(n-k)x_l(n-i) \\
&+ \sum_{k=0}^{L-1}\sum_{i=k}^{L-1} w_{lr}(n)x_r(n-k)x_r(n-i) \\
y_r(n) = &\sum_{i=0}^{N-1} h_{rr}(n)x_r(n-i) + \sum_{i=0}^{N-1} h_{rl}(n)x_l(n-i) \\
&+ \sum_{k=0}^{L-1}\sum_{i=k}^{L-1} w_{rr}(n)x_r(n-k)x_r(n-i) \\
&+ \sum_{k=0}^{L-1}\sum_{i=k}^{L-1} w_{rl}(n)x_r(n-k)x_r(n-i)
\end{aligned} \quad (7.12)$$

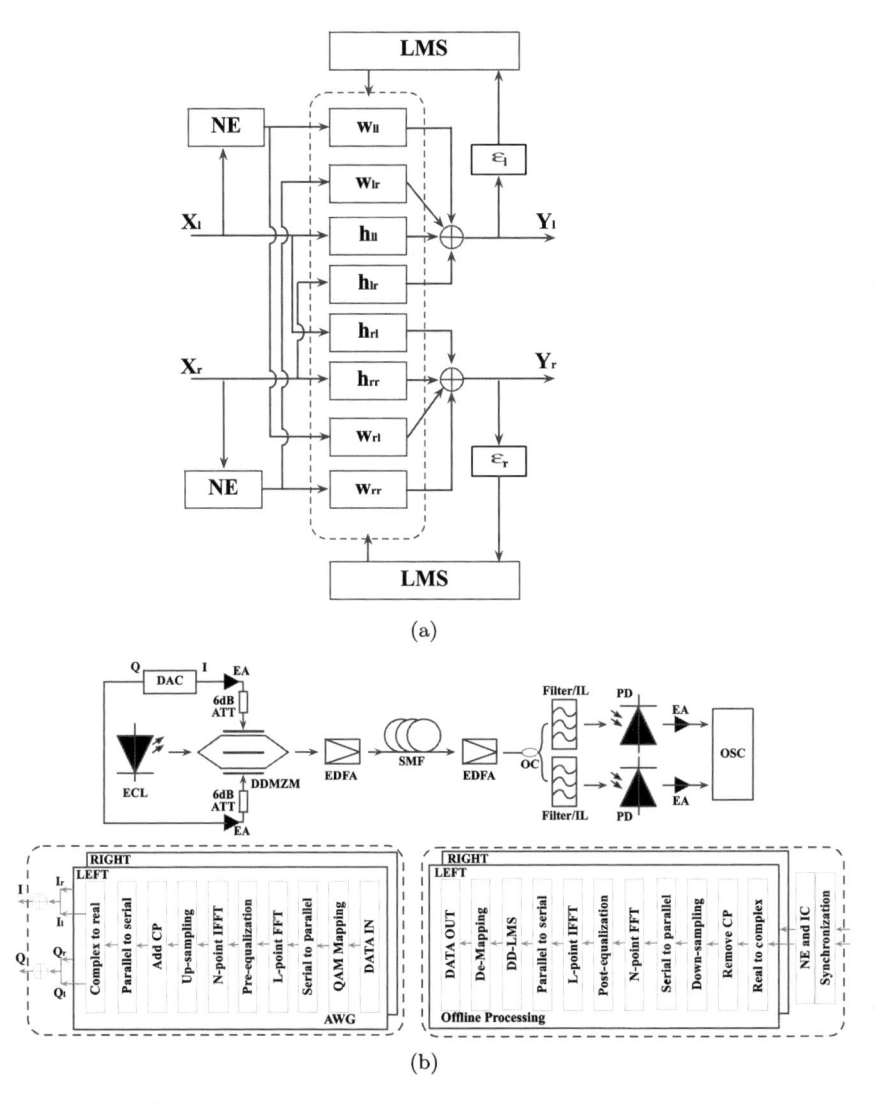

Fig. 7.2. (a) Structure of the MIMO-Volterra equalizer and (b) experimental setup of the twin-SSB system. ECL: external cavity laser, EA: electronic amplifier, ATT: attenuator, DAC: digital-to-analog converter, DD-MZM: dual-driver Mach–Zehnder modulator, EDFA: erbium doped fiber amplifiers, SMF: single mode fiber, OC: optical coupler, IL: interleave, PD: photodetectors, OSC: oscilloscope.

Here, N and L are the tap numbers of the linear and nonlinear equalizers, respectively. The four terms are the linear term, linear interference term, nonlinearity term and nonlinearity interference term. Training sequences are used to update the weight coefficients according to the Least Mean Square (LMS) error function. As a quasi-static transmission system, the updated weight coefficients can be used for a long time after finishing training.

7.4. Experimental Setup and Results

Figure 7.2(b) shows the experimental setup of 208-Gb/s/λ DFT-S OFDM transmission over a 40-km SSMF using twin-SSB modulation. We generated the drive signals using an 80-GSa/s digital-to-analog converter (DAC) with a 20-GHz bandwidth and an offline Matlab® program. Before driving the upper and lower arms of the DD-MZM, the signals were amplified by electrical amplifiers (EA, 32-GHz bandwidth and 20-dB gain), and 6-dB electrical attenuators were used to fit the linear region of the modulator. A continual wave (CW) light at 1549.76 nm was fed into a DD-MZM with a 25-GHz optical bandwidth and a 1.8 V driving voltage. Before and after the 40-km SSMF fiber transmission, two erbium-doped fiber amplifiers (EDFA) were used to boost the optical signal. An optical coupler (OC) and an interleave (IL) were used to separate the left-band and right-band optical signals before they were detected by two 50-GHz photodetectors (PDs). Finally, the signals were sampled by a digital real-time oscilloscope with an 80-GSa/s sampling rate and a 36-GHz electrical bandwidth.

At the transmitter, the data were first mapped into complex symbols of 16-QAM or 32-QAM. Then, a 2048-point FFT was used to generate the DFT-S signal, and IFFT was used to generate the OFDM signal with 2048 subcarriers. A cyclic prefix (CP) was added to alleviate the intersymbol interference (ISI) incurred by the CD. After the parallel-to-serial (P/S) conversion, we used subcarrier modulation to generate the real-value DFT-S OFDM. In this experiment, the bandwidth of the OFDM signal was 24–30 GHz, and offline digital signal processing (DSP) was applied to demodulate the signal sampled by the OSC.

During the offline process, two data streams of receivers were first processed by the MIMO-Volterra equalization algorithm after synchronization. Then, the data required OFDM demodulation before DFT-S demodulation.

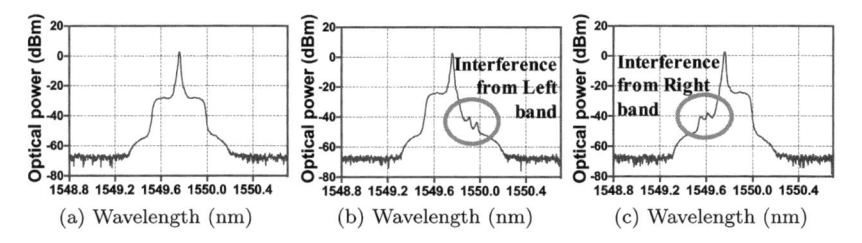

Fig. 7.3. Optical spectra of (a) twin SSB, (b) only left-band SSB, and (c) only right-band SSB.

The bit error ratio (BER) performance of the final data was measured after the demapping process.

Figure 7.3 illustrates the optical spectra of the twin SSB, left-band SSB and right-band SSB. The twin SSB was generated by two independent SSBs using equation (7.3), and (b) and (c) show the conventional SSB using (7.1) and (7.2). There is obvious interference from the left- and right-band signals. If we do not handle this part, it will destroy the performance of the twin-SSB system.

Figure 7.4 shows the BER performance versus data rate for the left- and right-band SSB signal in the BTB case. According to the constellations in Fig. 7.4, we find that the MIMO interference cancellation method has a much larger effect on the BER performance than the nonlinearity equalization, as we analyzed in Section 7.2. There is a small difference in performance between the left- and right-band SSBs when they are processed without MIMO IC because the IL does not correctly match the wavelength. However, if the MIMO-Volterra algorithm is used, the BER performances of both band SSB signals are almost identical.

Figure 7.5(a) shows the BER performance between the twin SSB with different methods and the conventional SSB DFT-S OFDM signal at different bit rates. In this figure, the conventional SSB performance decreases more sharply than that of the twin SSB because when the data rate is notably high, the conventional SSB generated by (1) requires more bandwidth (e.g., a 45-GHz bandwidth, 16-QAM) or higher order modulation (e.g., a 36-GHz bandwidth, 32-QAM), both of which reach the limit of this system. However, for the twin SSB, even at 240 Gbit/s, the bandwidth is only 30 GHz, and the modulation format is only 16 QAM. Using the MIMO-Volterra algorithm, we achieve 240 Gb/s/λ of twin-SSB DFT-S OFDM in the BTB case, whereas the twin SSB without the MIMO algorithm is only 165 Gb/s/λ, the twin SSB with the Joint method [5] is only

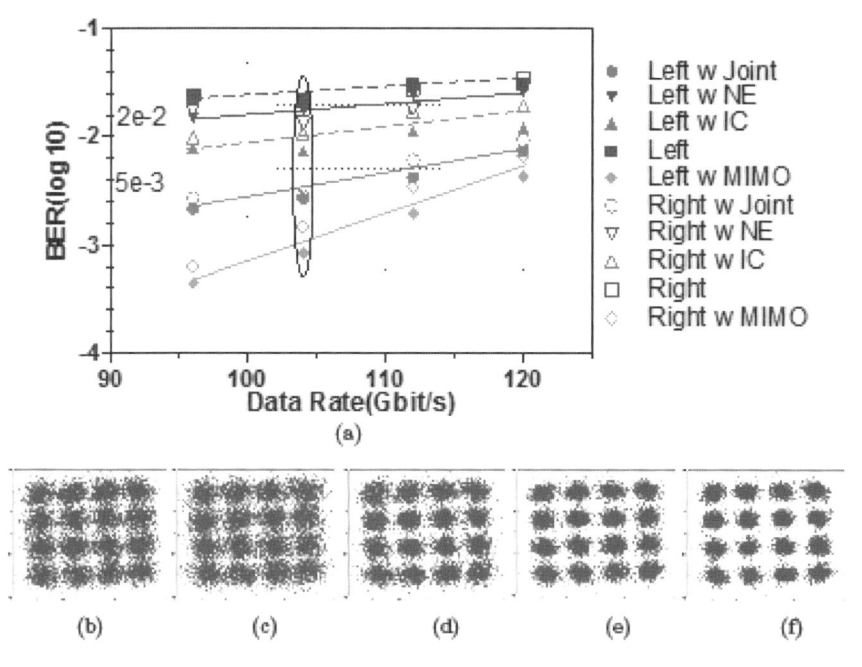

Fig. 7.4. BER versus data rate of the (a) left-band SSB and right-band SSB; constellations of the (b) left band without IC and NE, (c) left band with only NE, (d) left band with only IC, (e) left band with joint IC and NE, and (f) left band with MIMO-Volterra.

224 Gb/s/λ, and the conventional SSB can only achieve 164 Gb/s/λ with a BER less than 5×10^{-3}. Thus, the net data rate is 224 Gb/s/λ using the MIMO algorithm with a BER less than 5×10^{-3}, which can be converted to error-free transmission using only 7% overhead single-BCH FEC coding and decoding [11]. The resulting data rate is improved by more than 45%, whereas the improvement is only more than 20% in [8] (gross data rate: 210 Gb/s without the MIMO method and 260 Gb/s with the MIMO method in the BTB case with a BER of 4.5×10^{-3}).

Thus, we have experimentally demonstrated 208-Gb/s/λ DFT-S OFDM transmission over a 40-km SSMF without CD compensation at the FEC threshold of 2×10^{-2} as shown in Fig. 7.5(b). The BERs after 40 km transmission for the left- and right-band SSB using the MIMO method are 1.138×10^{-2} and 1.229×10^{-2} respectively. Hence, the net data rate after the 40-km transmission is 173 Gb/s/λ assuming the use of 20% overhead FEC. The performance of the left- and right-band SSB signal remains identical after the fiber transmission.

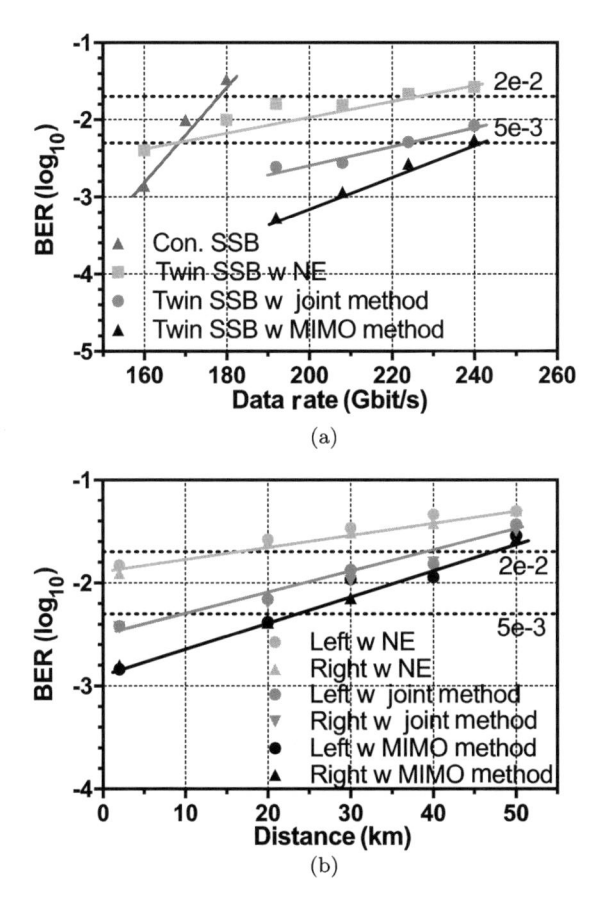

Fig. 7.5. (a) Capacity comparison of the twin-SSB signal with only NE, with the joint method and with the MIMO-Volterra method and the conventional SSB signal and (b) BER versus distance of SSMF transmission for a twin-SSB signal.

7.5. Conclusions

With the proposed time-domain MIMO-Volterra equalization algorithm, 208-Gb/s/λ DFT-S OFDM transmission over a 40-km SSMF without CD compensation is experimentally demonstrated. Compared with other methods, we achieve 240 Gb/s/λ of twin-SSB DFT-S OFDM in the BTB case, whereas the twin SSB without the MIMO algorithm can only reach 165-Gb/s/λ, the twin SSB with the joint method can only reach 224 Gb/s/λ, and the conventional SSB can only achieve 164 Gb/s/λ with a BER less than 5×10^{-3}. The results show the feasibility that the twin SSB generated by a commercial DD-MZM can double the spectral efficiency, and the MIMO-Volterra algorithm can improve the data rate by more than 45%.

References

[1] J. C. Rasmussen, T. Takahara, T. Tanaka, Y. Kai, M. Nishihara, T. Drenski, L. Li, W. Yan and Z. Tao, 2014 *European Conf. Optical Communication (ECOC)*, (IEEE, 2014).

[2] W. Yan, L. Li, B. Liu, H. Chen, Z. Tao, T. Tanaka, T. Takahara, J. Rasmussen and D. Tomislav, *Optical Fiber Communication Conf.*, (Optical Society of America, 2014).

[3] S. Randel, D. Pilori, S. Chandrasekhar, G. Raybon and P. Winzer, *2015 European Conf. Optical Communication (ECOC)*, (IEEE, 2015).

[4] Y. Wang, J. Yu and N. Chi, Demonstration of 4 × 128-Gb/s DFT-S OFDM signal transmission over 320-km SMF with IM/DD, *IEEE Photon. J.* **8**(2) (2016) 1–9.

[5] Y. Wang, J. Yu, H. C. Chien, X. Li and N. Chi, *Proc. VDE: 42nd European Conf. Optical Communication (ECOC)* (2016).

[6] L. Zhang, T. Zuo, Q. Zhang, E. Zhou, G. N. Liu and X. Xu, *2015 European Conf. Optical Communication (ECOC)*, (IEEE, 2015).

[7] L. Zhang, E. Zhou, Q. Zhang, G. N. Liu and T. Zuo, *Optical Fiber Communication Conf.*, (Optical Society of America, 2015).

[8] L. Zhang, T. Zuo, Q. Zhang, J. Zhou, E. Zhou and G. N. Liu, Single lane 150-Gb/s, 100-Gb/s and 70-Gb/s 4-PAM transmission over 100-m, 300-m and 500-m MMF using 25-G class 850 nm VCSEL, *Proc. VDE: 42nd European Conf. Optical Communication (ECOC)*, (2016), pp. 1–3.

[9] P. R. King and S. Stavrou, Low elevation wideband land mobile satellite MIMO channel characteristics, *IEEE Trans. Wireless Commun.* **6**(7) (2007) 2712–2720.

[10] Y. Wang, L. Tao, X. Huang, J. Shi and N. Chi, 8-Gb/s RGBY LED-based WDM VLC system employing high-order CAP modulation and hybrid post equalizer, *IEEE Photon. J.* **7**(3) (2015) 1–7.

[11] M. Li, Z. Xiao, F. Yu, N. Stojanovic, I. B. Djaordjevic, X. Shi and L. Li, *Optical Fiber Communication Conf.*, (Optical Society of America, 2015), Th3E. 2.

Chapter 8

DSP Implementation in OFDM Signal Systems

8.1. Introduction

Recently, the emergence of bandwidth-hungry services, such as High-definition television (HDTV), video call and cloud computing, has driven the speed of optical communication systems to higher and higher bit rate. It requires ultra-short reach (<10 km) optical fiber systems in data centers and interconnect applications and metropolitan (Metro) networks to operate at a bit rate of 100 Gb/s. Optical orthogonal frequency division multiplexing (OFDM) with high spectrum efficiency (SE) is a competitive candidate to realize high-speed data transmission [1–10]. In the optical OFDM schemes with cyclic prefix (CP), the inter-symbol interference induced by chromatic dispersion (CD) and polarization-mode dispersion (PMD) can be compensated utilizing simple frequency domain channel estimation and equalization [1–5, 11–17]. High bit rate optical OFDM transmission systems are demonstrated with direct detection [1–10] or coherent detection [11–18]; direct detection OFDM scheme can be applied to realize high capacity ultra-short reach optical fiber systems and Metro networks can be implemented with coherent detection OFDM. Most of these works regarding these two aspects are realized with offline digital signal processing (DSP) approaches. Therefore, computational complexity and computing resource allocation still need to be discussed and effectively verified.

Intensity modulation and direction detection (IM/DD) is considered as a promising candidate with low cost especially when the directly modulated laser (DML) is used to realize electrical-to-optical conversion (E/O) [8, 9, 19–21]. Advanced modulation formats including half-cycle

16-ary quadrature-amplitude modulation (16QAM), [19] discrete multi-tone modulation (DMT) [8, 9, 20, 21] and carrierless amplitude phase modulation (CAP) [22] are applied in recently reported 100 Gb/s signal generation and transmission in short reach optical fiber systems. DMT is a multicarrier scheme which derives from OFDM with effective data only on positive frequency bins. Negative frequency bins are filled to stratify Hermitian conjugate symmetry with positive frequency bins to generate real-value signal. We classify DMT to be a special form of OFDM as it inherits all advantages of the OFDM signal, such as transparency to modulation formats and robustness to chromatic dispersion and polarization mode dispersion. Among these three types of technology, DMT is the most practical solution, as it is easy to upgrade to high throughout with higher level modulation formats with the same equalization method. Unfortunately, all these 100 Gb/s transmission in short reach optical fiber systems are realized offline [19–22]. A 50-Gb/s real-time direct detection optical DMT transmission system with 64QAM modulation format was the highest capacity prior to this work [5]. Here, transmission and reception of real-time direct detection optical DMT (DDO-DMT) system with 16QAM operating at a new record rate of 100 Gb/s are successfully demonstrated. A 100-Gb/s 16QAM-DMT signal is carried on dual optical subcarriers generated by two DMLs, which means the capacity for each optical carrier is 50 Gb/s. After transmission of 20-km LEAF, no received optical power penalty is observed after vestigial sideband modulation (VSB) enabled by the narrow bandwidth optical filter is adopted to resist the CD-induced frequency fading and the output bit error rate (BER) is lower than 2×10^{-2}, the soft-decision pre-FEC limit. Rather than inserting several TSs, an intra-symbol frequency-domain averaging (ISFA) filter [23] is utilized to reduce ASE noise during the channel estimation with only one TS.

In Metro networks, transmission distance of fiber is usually more than 100 km, it is difficult to realize such a long distance transmission for 100 Gb/s OFDM with direct detection. In the optical OFDM system with coherent detection, CD-induced frequency fading effect does not exist anymore. Moreover, SE and receiver sensitivity can be both improved compared to direct detection [15–18]. Coherent optical OFDM (CO-OFDM) with offline processing has been successfully demonstrated beyond 100 Gb/s, [13, 14] while real-time implementation of CO-OFDM reception beyond 100 Gb/s is reported [17] with multi-band transmission. The generation and separation of multi-carrier for multi-band OFDM signal in the transceiver are complicated and the guard band will be required in the real application.

Moreover, only QPSK format is used in the reception of real-time OFDM signal beyond 100 Gb/s [17] and the (SE) is low. In this chapter, we study the key DSP techniques and mainly discuss the simplification of time synchronization and frequency synchronization in the DSP of real-time receiver. The single-band CO-OFDM signal is successfully received and recovered after 200-km standard single mode fiber (SSMF) transmission under BER of 3.8×10^{-3}. To the best of our knowledge, it is the first time to transmit 100-Gb/s OFDM signal on only one optical carrier. The bit rate per one optical carrier is the highest so far in the demonstration of real-time optical OFDM systems.

8.2. Details of DSP in Receiver

Figure 8.1 gives out the details of DSP in direct detection receiver. Frequency offset does not exist in the direct detection system and phase noise in ultra-short reach with direct detection can be ignored, so the DSP in direct detection receiver is very simple. It can be divided into six steps: (1) signal resampling, (2) timing synchronization, (3) N-point FFT, (4) channel estimation and equalization, (5) symbol hard decision and (6) error counting. Details of DSP in polarization–division–multiplexing (PDM)–16QAM–OFDM with coherent receiver are shown in Fig. 8.2. The DSP can be divided into nine stages: (1) signal resampling, (2) timing synchronization, (3) frequency synchronization, (4) N-point FFT, (5) polarization demultiplexing and channel estimation utilizing Jones matrix, (6) phase noise estimation with the aid of pilot tones, (7) L-point IFFT, (8) symbol hard decision, and (9) error counting. The DSP in coherent detection receiver is much more complicated than that in direct detection receiver. Algorithms for time synchronization and frequency offset estimation in coherent receiver are resource-intensive, they should be

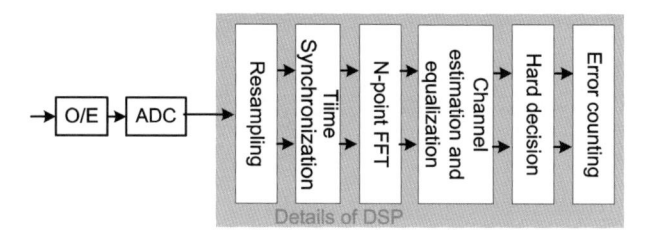

Fig. 8.1. Details of DSP in direct detection receiver.

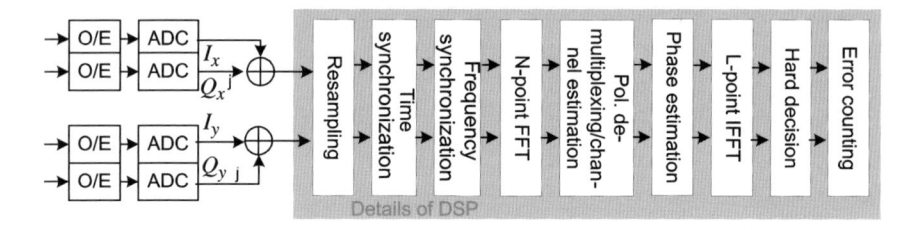

178   *Digital Signal Processing for High-Speed Optical Communication*

Fig. 8.2.   Details of DSP in coherent detection receiver.

simplified during real-time implementation. Some stages of signal processing are discussed in detail in the following sections.

8.2.1. *Signal resampling*

In the dual optical carrier 100-Gb/s 16QAM-DMT system with direct detection, the sampling rates in digital-to-analog converter (DAC) in the transmitter and analog-to-digital converter (ADC) in the receiver are 33.544 and 41.93 GSa/s, respectively. While in the single optical carrier 100-Gb/s PDM-16QAM-OFDM system with coherent detection, the sampling rates in DAC in the transmitter and ADC in the receiver are 62.895 and 41.93 GSa/s, respectively. In this stage, the signal is resampled to one sample/symbol with Lagrange interpolation in field-programmable gate arrays (FPGAs).

8.2.2. *Timing synchronization*

Timing synchronization in both the systems is realized with the aid of training sequence (TS). In the dual optical carrier 100-Gb/s 16QAM-DMT system with direct detection, TS including random specific length "0" and "1" sequences in the time domain is applied for low computational complexity synchronization, in the receiver, synchronization is realized by comparing the received symbols after decision with symbols in TS. In the single optical carrier 100-Gb/s PDM-16QAM-OFDM system with coherent detection, the processing of synchronization should be more complicated due to the frequency offset.

In the TS for synchronization, only real-value data ("1" or "−1") is filled on the even subcarriers in frequency domain, a conjugate symmetry time domain sequence is generated after IFFT in the transmitter. We can use this feature to achieve the timing synchronization with autocorrelation. As the autocorrelation is resource-consuming due to too many multiplication operations, we have to simplify the autocorrelation. During the autocorrelation, we are only concerned with the sign bit of every sample to avoid multiplication operations. The start point of the signal can be obtained as follows:

$$M(p) = \left| \sum_{k=1}^{N/2-1} r(p + N/2 + N_{\text{CP}} + 1 + k) \right.$$

$$\left. \times r(p + N/2 + N_{\text{CP}} + 1 - k) \right| \quad (8.1)$$

where $M(p)$ is the timing metric, N is the number of the sub-carriers, N_{CP} is the number of the CP and $|\cdot|$ represents the modulo operation. As we are only concerned with the sign bit of the received symbols, the timing metric can be expressed as

$$M(p) = \left| \sum_{k=1}^{N/2-1} \text{sgn}(r(p + N/2 + N_{\text{CP}} + 1 + k)) \right.$$

$$\left. \times \text{sgn}(r(p + N/2 + N_{\text{CP}} + 1 - k)) \right| \quad (8.2)$$

where $\text{sgn}(\cdot)$ represents the sign function and the output of this function can be $1+j, -1+j, 1-j$, or $-1-j$. The results after the multiplication of two sign functions should be $2, -2, 2j$ or $-2j$. We have to point out that $2, -2, 2j$ and $-2j$ can be obtained when the phase offset between the symbol and the symbol after $N/2$ point is $3\pi/2, \pi/2, 0$ and π, respectively. These multiplications in the FPGA can be realized with lookup table the given in Table 8.1 and the outputs are simplified to $1, -1, j$ and j in the table. The timing metric is obtained by the sum of two sign function multiplication. When the probe point p is just the exact start point, the result of two sign function multiplication of two samples with $N/2$ interval after coherent detection should be the same even though there is a frequency offset. The real and the imaginary parts after sum of two sign function multiplication are assumed to be $A(p)$ and $B(p)$, and Eq. (8.2) can be simplified as

$$M(p) = |A(p) + iB(p)|$$
$$= \sqrt{(A(p))^2 + (B(p))^2} \quad (8.3)$$

Table 8.1. Lookup table for timing metric acquisition.

Sign of symbol	Sign of symbol after $N/2$ interval	Result of multiplication
$1+j$	$1+j$	j
$1+j$	$1-j$	1
$1+j$	$-1+j$	-1
$1+j$	$-1-j$	$-j$
$1-j$	$1+j$	1
$1-j$	$1-j$	$-j$
$1-j$	$-1+j$	j
$1-j$	$-1-j$	-1
$-1+j$	$1+j$	-1
$-1+j$	$1-j$	j
$-1+j$	$-1+j$	$-j$
$-1+j$	$-1-j$	1
$-1-j$	$1+j$	$-j$
$-1-j$	$1-j$	-1
$-1-j$	$-1+j$	1
$-1-j$	$-1-j$	j

The calculation of timing metrics is simplified to obtain the module of the autocorrelation of sign bit. In the FPGA implantation, it is still very difficult as squaring operation should be realized with multiplication. As mentioned above, either $A(p)$ or $B(p)$ will be very large only when the probe point is the exact start point, otherwise $A(p)$ and $B(p)$ will be both very small. According to this feature, we further simplify the calculation of timing metric

$$M(p) = |A(p)| + |B(p)| \qquad (8.4)$$

in which the timing metric can be obtained by just the addition operation. When the probe point p is not the exact start point, $\sqrt{(A(p))^2 + (B(p))^2}$ and $|A(p)| + |B(p)|$ are both very small, where $\sqrt{(A(p))^2 + (B(p))^2}$ can be approximated as $|A(p)| + |B(p)|$, that is $\sqrt{(A(p))^2 + (B(p))^2} \approx |A(p)| + |B(p)|$ can be obtained according to Taylor formula when the probe point p is the exact start point.

8.2.3. Frequency synchronization

As mentioned above, no frequency offset exists in the system with direct detection. Frequency offset between signal laser and local laser must be estimated and compensated before further processing in the single optical carrier 100-Gb/s PDM-16QAM-OFDM system with coherent detection. Frequency offset is also estimated in the processing of first type of TS. Frequency offset leads to the cyclic shift of subcarriers in the frequency domain. According to this theory, the frequency offset can be easily found with the cross-correlation of transmitted and received sequences in the frequency domain. The frequency offset metrics can be calculated as

$$F(k) = \left| \sum_{k=1}^{M} \mathrm{sgn}(R(k)) \, \mathrm{sgn}(T(k)) \right| \qquad (8.5)$$

where M is the number of data subcarriers, $T(k)$ is the training symbols carried on 1 or -1 in the frequency domain for frequency offset estimation, and $R(k)$ is the received symbols with frequency offset. As frequency offset and I/Q imbalance exist, the $R(k)$ will not be the only real value. In order to simplify the calculation of the frequency offset metrics, we only extract the sign of real part for cross-correlation, the frequency offset metrics can be simplified as

$$F(k) = \sum_{k=1}^{M} \mathrm{sgn}(\mathrm{real}(R(k))) \, \mathrm{sgn}(T(k)) \qquad (8.6)$$

In Eq. (8.6), the multiplications in the FPGA can also be realized with XNOR gate to save resources.

8.2.4. Channel estimation

Channel estimation is a critical procedure in optical OFDM transmission systems. Physical impairments of the fiber transmission link such as CD and PMD can be obtained with channel estimation. Hence, the subsequent channel equalization can be performed to restore the signal quality. In this paper, zero forcing (ZF) algorithm is used for channel estimation due to its lower computational complexity. In the dual optical carrier 100-Gb/s 16QAM-DMT system with direct detection, channel estimation is implemented with the aid of another type of TS. The modulation format for this TS is QPSK and the DSP for TS in the transmitter is exactly the same for the OFDM signal symbols. The channel response obtained by TS is used to equalize OFDM signal symbols in the frame. The ISFA algorithm is applied in this paper to improve the accuracy of channel estimation.

In the single optical carrier 100-Gb/s PDM-16QAM-OFDM system with coherent detection, polarization demultiplexing should also be realized in the channel estimation. In order to realize polarization, we should first get the Jones matrix. A pair of time-interleaved TSs is transmitted in the two independent tributaries [13, 14] to estimate this Jones matrix H at the receiver. ZF algorithm is used to obtain the Jones matrix H. The ISFA algorithm is applied in this paper to suppress the channel noise during Jones matrix H estimation.

8.2.5. Phase estimation

In the optical OFDM transmission system with direct detection, phase noise can be ignored when the fiber transmission distance is very short. While in the coherent optical OFDM system, phase noise will destroy the orthogonality among data subcarriers and lead to inter-carrier interference (ICI). Phase noise has to be removed in the DSP of coherent detection receiver. In the single optical carrier 100-Gb/s PDM-16QAM-OFDM system with coherent detection, pilot tones are inserted in every OFDM signal symbol to obtain and compensate phase noise.

8.3. Experimental Setup and Results

Figure 8.3 shows the experimental setup of real-time transmitter and receiver in the DDO-16QAM-DMT transmission system. At the transmitter, there are two DFB-based DML at 1537.02 nm and 1537.9 nm with ~25-MHz linewidth used as dual optical carrier. The continuous-wavelength (CW) lightwaves from DMLs are modulated by electrical baseband DMT signal from two sets of DAC. The samples generated offline with Matlab are fed to a 6-bit resolution DAC with 3-dB bandwidth of larger than 13 GHz running with 33.544 GSa/s sample rate. The DAC in the transmitter are shown in Fig. 8.3(a). Two low pass filters (LPFs) are cascaded after DACs to remove the residual sideband DMT and then two linear amplifiers are applied to boost the DMT signal to 2.4 V. Here, the FFT size for DMT generation is 256, in which 200 subcarriers are employed with 100 conveying data in the positive frequency bins. 16-QAM is taken on all the 100 information-bearing subcarriers. An 8-sample cyclic prefix is added to the 256 samples, giving 264 samples per OOFDM symbol. The bit rate on each optical carrier in the system is 50.82 Gb/s ($33.544 \times 100/264 \times 4$ Gb/s ≈ 50.82 Gb/s) and the total bit rate of the dual carrier is 101.64 Gb/s, the bandwidth of each DMT signal is 1.56 GHz

Fig. 8.3. Experimental setup of real-time DDO-16QAM-DMT transmission system: (a) transmitter, the optical spectrum (0.02-nm resolution) of combined optical signal, (b) before and (c) after 20-km LEAF transmission, (d) the optical spectrum of selected optical carrier at 1537.9 nm, (e) the ADCs and FPGAs in the receiver, (f) electrical spectrum of received 16QAM-DMT signal and (g) constellations of received 16QAM-DMT signal.

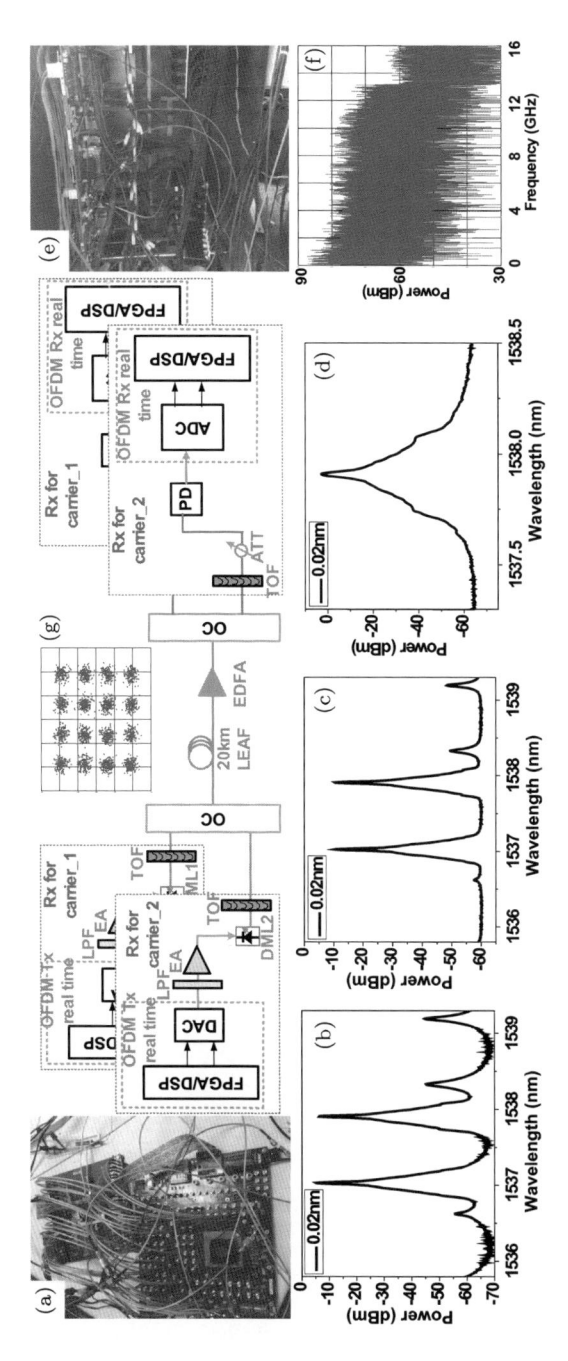

($100/256 \times 33.544\,\text{GHz} \approx 13.1\,\text{GHz}$). For optical DMT modulation, two commercial DMLs with 10-GHz bandwidth are biased at 52 mA to produce 6.3-dBm average output power separately. Two modulated optical carriers are combined by an optical coupler (OC). The total power of combined signal is 9.2 dBm. The optical spectrum (0.02-nm resolution) of combined optical signal before and after 20-km LEAF transmission is inserted as insets (b) and (c) in Fig. 8.3, respectively.

At the receiver, the received optical OFDM signal is coupled to an erbium-doped fiber amplifier (EDFA) to compensate the loss of the LEAF and then an OC is used to divide the optical signal into two branches. In each branch, a tunable optical filter (TOF) with 0.9 nm bandwidth is used to select the optical carrier and suppress the ASE noise and the optical spectrum of selected optical carrier-2 at 1537.9 nm is shown in Fig. 8.3(d). An optical attenuator is applied in the system to adjust the optical power into photodiode (PD). An optical receiver with 3 dB bandwidth of 13 GHz converts the received optical signal to the electrical domain. The converted electrical DMT signal is fed into a 6-bit resolutions ADC that operates at 41.93 GS/s with 3-dB bandwidth of 16 GHz. Finally, the digital samples from two ADCs are fed into other two FPGAs, which perform the real-time DSP on the received symbols and calculate the BER. The ADCs and FPGAs in the receiver are shown in Fig. 8.3(e) in the receiver. Detailed descriptions of the DSP within the real-time OOFDM transceivers are discussed in Section 8.2. In order to verify the feasibility of this system, the electrical spectrum and constellations are tested and shown in insets (f) and (g) of Fig. 8.3, respectively.

The BER in the optical back-to-back (OBTB) case and after 20-km LEAF transmission for optical carrier-2 at 1537.9 nm is shown in Fig. 8.4(a). Rather than inserting several TSs, ISFA technique with 11 taps is utilized to suppress channel noise during the channel estimation with only one TS. It can be seen that after 20-km LEAF transmission, the BER performance degrades significantly. The main reason for the performance degradation is the fading effect caused by CD. In order to remove this CD-induced fading effect, a narrow bandwidth filter with 10 GHz minimum pass band is applied to convert the double sideband (DSB) modulation, which is vulnerable to the CD, to the VSB modulation, which is robust to the CD to some extent. BER versus received optical power for two carriers after 20-LEAF with VSB-DMT modulation by the narrow filter is shown in Fig. 8.4(b). The BERs of 16-QAM-DMT signal on two carriers are both lower than 2×10^{-2}. Compared to the BTB case in Fig. 8.4(a), the penalty is eliminated with the

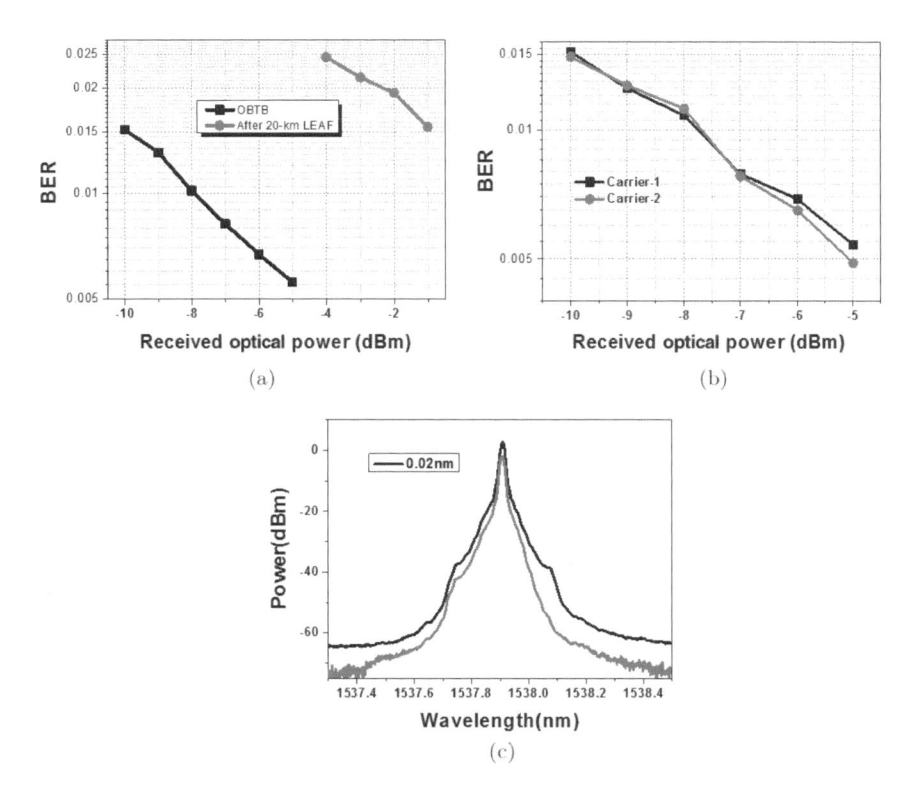

Fig. 8.4. (a) BER versus received optical power for carrier-2, (b) BER versus received optical power for two carriers after 20-LEAF with VSB-DMT, and (c) optical spectrum of 16QAM-DMT signal on carrier-2 before and after narrow band filter.

dispersion-immune VSB-DMT signal. The optical spectrum of carrier-2 at 1537.9 nm before and after narrow band optical filter is shown in Fig. 8.4(c), and the electrical spectra of received DMT signal with and without narrow band optical filter are shown in Figs. 8.5(a) and 8.5(b), respectively. It can be seen that the fading in the high frequencies has been significantly improved.

8.4. Single Optical Carrier 100-Gb/s PDM-16QAM-OFDM Transmission and Real-Time Reception with Direct Detection

Figure 8.6 shows the experimental setup for dual polarization (DP) 16QAM-CO-OFDM transmission with real-time reception. At the transmitter,

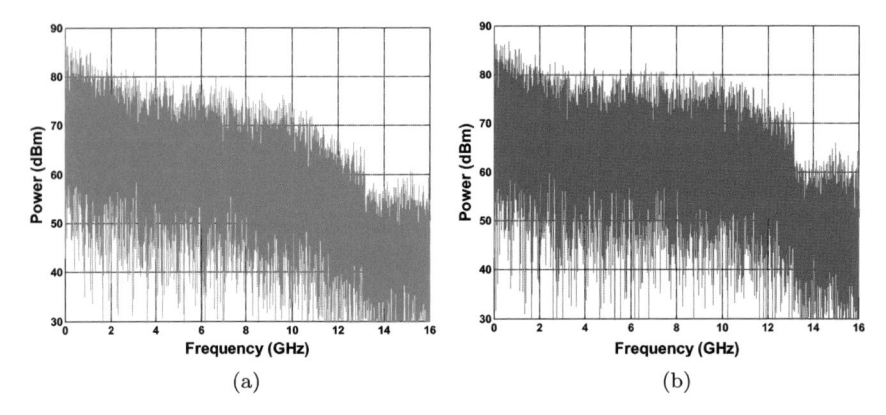

Fig. 8.5. Optical spectrum of 16QAM-DMT signal with (a) DSB and (b) VSB modulation.

there is an external cavity laser (ECL) at 1548.53 nm with less than 100-kHz linewidth and maximum output power of 14.5 dBm. The continuous-wavelength (CW) lightwave from ECL is modulated by in-phase/quadrature (I/Q) modulator driven by an electrical baseband OFDM signal. The OFDM signal is generated by a DAC shown in Fig. 8.6(a) and the sampling rate is set to 62.985 Gsa/s and then boosted by two linear electrical amplifiers (EAs). In the OFDM modulation, the inverse fast Fourier transform (IFFT) size is 1024. Among the 1024 subcarriers, 256 subcarriers are allocated for data transmission with 16QAM, eight subcarriers are filled with pilots for phase estimation, the first subcarrier is set to zero for DC-bias and the remaining 759 null subcarriers at the edge are reserved for oversampling. An addition 256-point DFT is added to realize DFT-spread and this can help the reallocation of SNR among all data and reduce the PAPR of the OFDM signal [18, 24]. Each 16-QAM data symbol is spread to all data that carried 256-subcarriers after DFT-spread, so the signal becomes robust to high-frequency power attenuation and narrowband interface. After IFFT, 25-sample CP is added in the OFDM symbol. Two types of TSs are added in the front of the data stream which includes 27 OFDM symbols. The first type TS with subcarriers is used for time and frequency synchronization, while the other one is a pair of time-interleaved TSs used for demultiplexing and channel estimation. For optical OFDM modulation, the two parallel Mach–Zehnder modulators (MZMs) in I/Q modulator are both biased at the null point and the phase difference between the upper and lower branches of I/Q

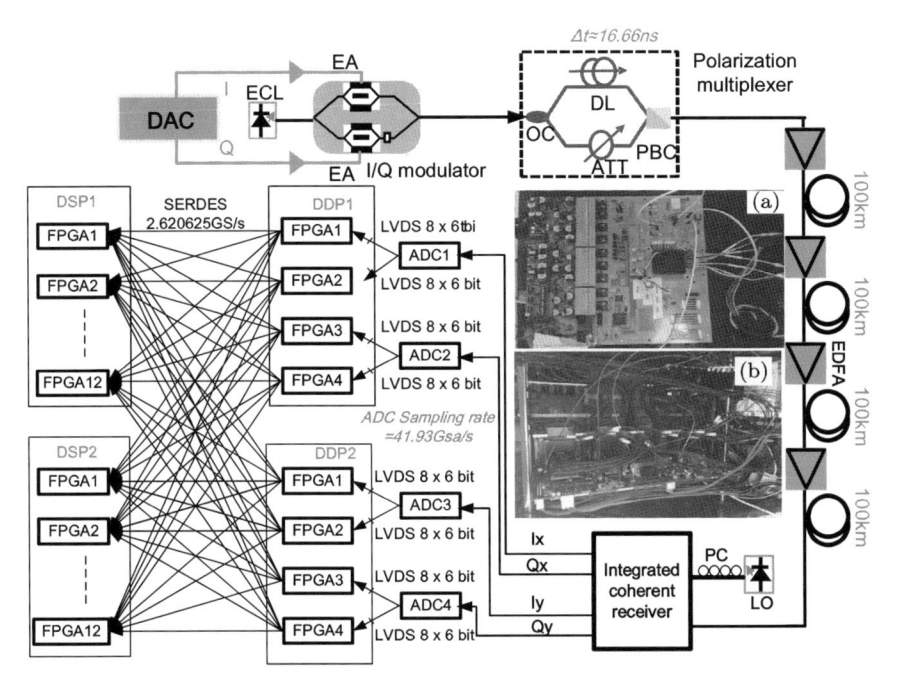

Fig. 8.6. Experimental setup: (a) DAC and (b) Integrated coherent receiver, ADC and FPGAs.

modulator is controlled at $\pi/2$. The polarization multiplexing is realized by polarization multiplexer, comprising a polarization-maintaining optical coupler (OC) to halve the signal into two branches, an optical delay line (DL) to remove correlation between X-polarization and Y-polarization by delaying exactly. one-symbol $(1/62.895\text{G} \times (1024 + 25))\text{s} \approx 16.66\,\text{ns}$), an optical attenuator to balance the power of two branches and a polarization beam combiner (PBC) to recombine the signal. After Y-polarization is delayed by exactly one OFDM symbol compared to X-polarization after polarization multiplexer, a pair of time-interleaved TSs for demultiplexing can be constructed.

The generated signal is launched into four spans of 100-km SSMF, which has 21-dB loss and chromatic dispersion of 17-ps/km/nm CD at 1550 nm. EDFAs with 5-dB noise figure are used to compensate for the fiber loss. The optical spectrum of optical OFDM signal in the OBTB case and after 200-km SSMF transmission are shown in Fig. 8.7(a) with 0.02-nm resolution. The output signal is then injected into the integrated coherent

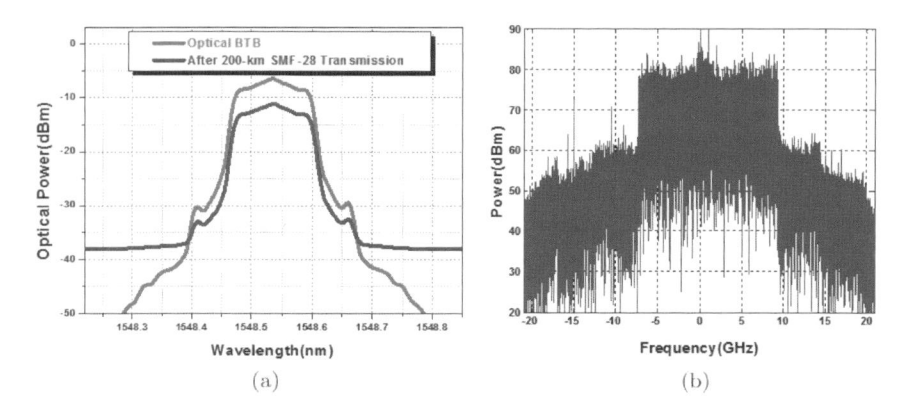

Fig. 8.7. (a) Optical spectrum (0.02 nm) of the OFDM signal and (b) electrical spectrum of received OFDM signal.

receiver to implement optical to electrical detection. Then, the detected RF signals are sampled by four high-speed ADCs at 41.93 GSa/s. The integrated receiver, DSP and FPGAs modules are integrated and inserted in Fig. 8.2 as inset (b). The resolutions of DAC and ADC in this paper are 8 bits and 6 bits, respectively. The bandwidth of the 16QAM-OFDM signal is 16.28 GHz $((256 + 8 + 1)/1024 \times 62.895 \text{ GHz} \approx 16.28 \text{ GHz})$ and the spectrum of the OFDM signal after optical to electrical detection is inserted in Fig. 8.7(b). The effective data rate after removing all overheads of the OFDM signals is 107.05 Gbit/s $(256/(1024 + 24) \times 4 \times 2 \times 27/31 \times 62.895 \text{ Gbit/s} \approx 107.05 \text{ Gbit/s})$. The four ADCs produce four outputs Ix, Qx, Iy, and Qy separately. The four ADC outputs are then transmitted to two data preprocessing (DPP) boards: Ix and Qx are transmitted into DPP1, while Iy and Qy are transmitted into DPP2. Each DPP board has four Altera EP4S100G FPGAs. In the two DPP boards, the ADC outputs will be calibrated to eliminate the random delays caused by ADC sampling and then reorganized with FPGAs before DSP processing. The preprocessing of data in DPP boards is essential for DSP processing later. There are two DSP boards following DPP boards, and DSP boards and DPP boards are connected by a high-speed backboard. The FPGAs in DSP boards are much more powerful than FPGAs in DPP boards as high computational complexity algorithms need to be implemented in the DSP boards. Each DSP board has 12 Xilinx 6VSX475 FPGAs, and the total 24 Xilinx 6VSX475 FPGAs can fulfill the huge amount of calculations of the algorithms. Both the Altera EP4S100G FPGAs and the Xilinx 6VSX475

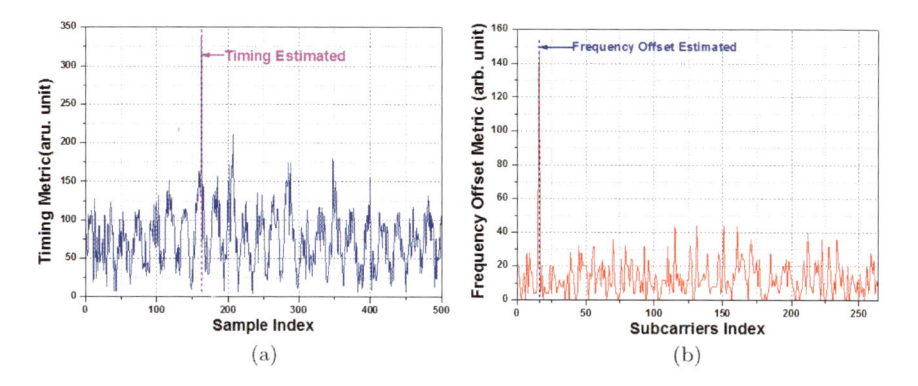

Fig. 8.8. (a) Timing metric and (b) frequency offset metric.

FPGAs have high-speed SERDES interfaces, and the signals transmitted from DPP FPGAs to DSP FPGAs pass through the SERDES interfaces. The transmission speed can be up to 6.5 Gb/s and used at 2.620625 Gb/s in this project. Each FPGA in the DSP boards contains eight parallel calculation channels, and the calculation clock frequency inside the FPGAs can be up to 296 MHz. The recovered data after the DSP processing and the original transmitted data will be compared by the FPGAs of the DSP boards and the error can be counted. The DSP processing and all processing results can be controlled and monitored by PC through JTAG interface and ChipScope Pro debugging modules.

The timing metric of real-time received data according to Eq. (8.4) is shown in Fig. 8.8(a), the high peak in the timing metric is the exactly estimated start point of the frame of the OFDM signal. Figure 8.8(b) demonstrates the estimated frequency offset metric with the first TS according to Eq. (8.5). The estimated Jones matrix with ZF algorithm is displayed in Fig. 8.8 as blue line. As only one pair of TSs is used to implement the channel estimation, the channel response estimated by TS is in dramatic fluctuation mode, which will not appear in the real channel response as it should be smooth and have less fluctuation. As mentioned above, the ISFA algorithm is applied in the FPGAs in order to make the estimated channel response much more smooth and the size of subcarriers used for ISFA is set to 13. The ISFA algorithm-based estimation of the elements in the Jones matrix becomes smooth, which is shown in Fig. 8.9 as red line.

Figure 8.10 shows the BER performance versus OSNR of COOFDM signal at back-to-back transmission. Single-polarization (SP) and DP CO-OFDM are both measured. Every BER point in this figure is obtained

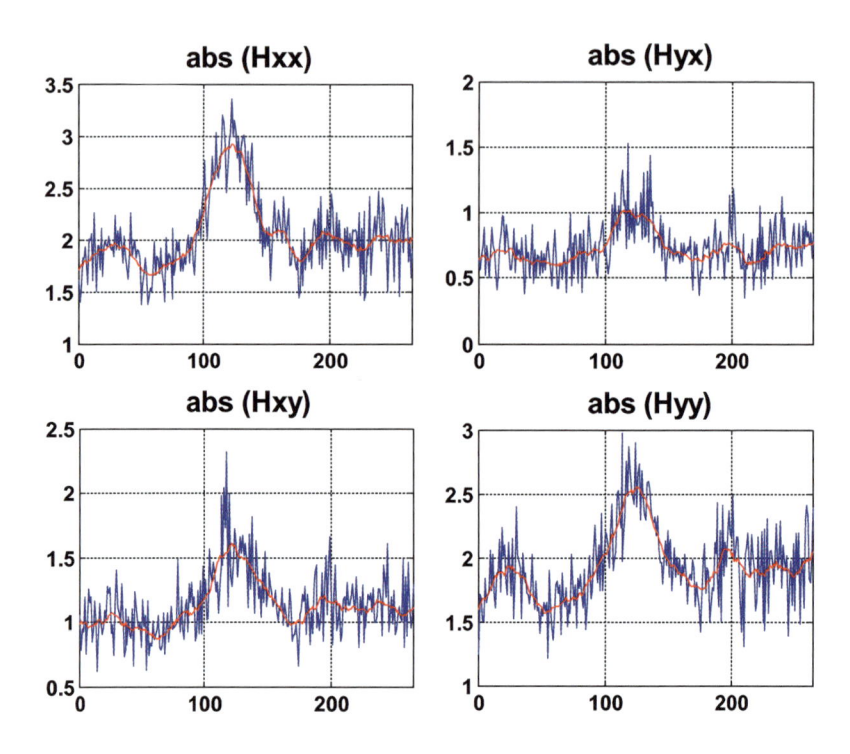

Fig. 8.9. Jones matrix.

by averaging over 270 OFDM symbols. A BER of 3.8×10^{-3} which is the FEC threshold with 7% overhead can be obtained at an OSNR of 14.3 dB and 17.2 dB for SP-16QAM and DP-16QAM OFDM signals, respectively. The inset shows the constellation diagrams for the DP 16QAM-CO-OFDM signal with an OSNR of 21 dB. After removing the FEC overhead, the effective data rate becomes 100.12 Gbit/s. Figure 8.11 shows the measured BER versus input power of DP 16QAM-CO-OFDM signal with 100-km SSMF transmission. The optimal input power is 2.5 dBm, we can see that the 6-dBm launch power demonstrates the optimal BER performance. When the launch power is larger than 2.5 dBm, the BER performance is degraded due to the nonlinear effect in the fiber. The constellations of dual polarizations after 100-km SSMF transmission with 2.5 dB input optical power are both inserted in Fig. 8.11.

Figure 8.12 shows the BER performance versus transmission fiber distance for the DP 16QAM-CO-OFDM signal. As the sampling rate of DAC

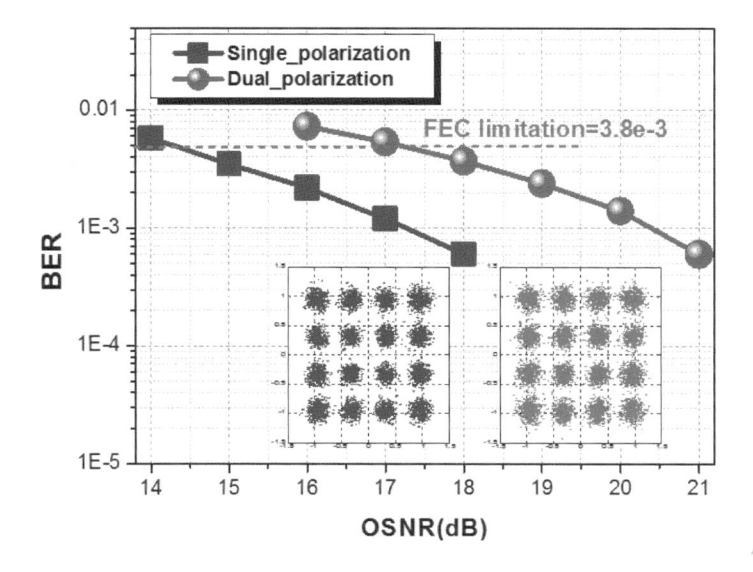

Fig. 8.10. Measured BER versus OSNR.

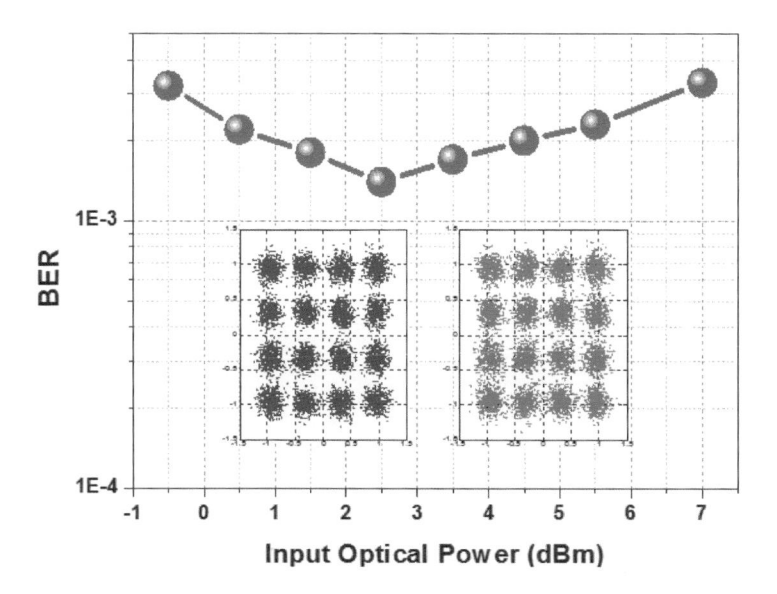

Fig. 8.11. Measured BER versus input optical power.

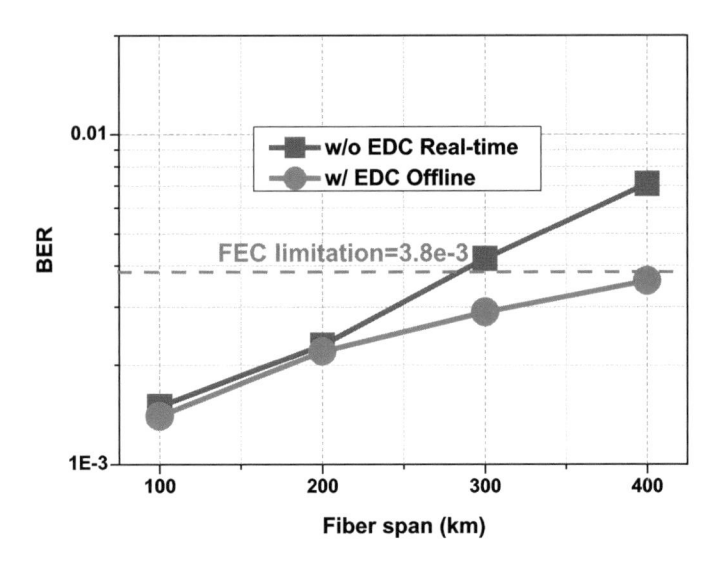

Fig. 8.12. Measured BER versus fiber distance.

is very high and the length of CP is relatively short, the inter-symbol interference (ISI) induced by the CD after long fiber span transmission leads to the BER performance degradation. In addition, the ISI causes inaccurate timing synchronization. In this paper, windowing technique is applied to alleviate the impact of inaccurate timing synchronization. As the electrical dispersion compensation (EDC) is really resource-intensive, we abandoned this in the real-time implementation with FPGAs. The offline test results are added with EDC for comparison to investigate the impact of CD in fiber transmission. The measured BER versus fiber span shown in Fig. 8.12 indicates that the 100 Gbit/s DP 16QAM-CO-OFDM signal with real-time and offline implementation under the BER of 3.8×10^{-3} can be transmitted over 200 km and 400 km, respectively. We can see that there is almost no penalty between real-time implementation without EDC and offline implementation with EDC in Fig. 8.12 when the fiber span is less than 200 km and the penalty between implementations with and without EDC becomes obvious with more than 200-km SSMF transmission. This means the CD-induced ISI after more than 200-km SSMF transmission becomes irresistible with 25-sample CP and windowing technique and should be overcome by EDC.

8.5. Conclusions

Here, we demonstrate a fully real-time dual optical carrier DDO-16QAM-OFDM system running at 101.64 Gb/s. After 20-km LEAF, the BER is lower than 2.7×10^{-2}. This was implemented and achieved not only by using the high-rate DACs and ADCs with FPGAs but also by using the dispersion-immune VSB-DMT signal. The results prove DMT with direct detection is a competitive candidate for the high throughput data centers and inter-connect applications. Also, we have experimentally demonstrated the reception of 100-Gbit/s real-time DP 16QAM-CO-OFDM signal carried on only one optical carrier for the first time. The computational complexity of the acquisition of timing metric and frequency offset metric is significantly reduced. After 200-km SSMF without EDC, the BER of DP 16QAM-CO-OFDM signal is lower than 3.8×10^{-3}. The results show that commercial Metro networks with high capacity can be implemented with CO-OFDM.

References

[1] W.-R. Peng, I. Morita, H. Takahashi and T. Tsuritani, Transmission of high-speed (>100 Gb/s) direct-detection optical OFDM superchannel, *J. Lightwave Technol.* **30**(12) (2012) 2025–2034.

[2] Z. Cao, J. Yu, W. Wang, L. Chen and Z. Dong, Direct-detection optical OFDM transmission system without frequency guard band, *IEEE Photon. Technol. Lett.* **22**(11) (2010) 736–738.

[3] Z. Li, X. Xiao, T. Gui, Q. Yang, R. Hu, Z. He, M. Luo, C. Li, X. Zhang, D. Xue, S. You and S. Yu, 432-Gb/s direct-detection optical OFDM superchannel transmission over 3040-km SSMF, *IEEE Photon. Technol. Lett.* **25**(15) (2013) 1524–1526.

[4] F. Li, J. Yu, Y. Fang, Z. Dong, X. Li and L. Chen, Demonstration of DFT-spread 256QAM-OFDM signal transmission with cost-effective directly modulated laser, *Opt. Express* **22**(7) (2014) 8742–8748.

[5] X. Q. Jin, R. P. Giddings, E. Hugues-Salas and J. M. Tang, Real time demonstration of 128-QAM-encoded optical OFDM transmission with a 5.25 bit/s/Hz spectral efficiency in simple IMDD systems utilizing directly modulated DFB lasers, *Opt. Express* **17**(22) (2009) 20484–20493.

[6] M. Chen, J. He, J. Tang, X. Wu and L. Chen, Experimental demonstration of real-time adaptively modulated DDO-OFDM systems with a high spectral efficiency up to 5.76 bit/s/Hz transmission over SMF links, *Opt. Express* **22**(15) (2014) 17691–17699.

[7] M. Chen, J. He, J. Tang, X. Wu and L. Chen, A real-time 10.4-Gb/s single-band optical 256/64/16QAM receiver for OFDM-based PON systems, *IEEE Photon. Technol. Lett.* **26**(20) (2014) 2012–2015.

[8] F. Li, X. Xiao, X. Li and Z. Dong, Real-time demonstration of DMT based DDO-OFDM transmission and reception at 50 Gb/s, in *Proc. ECOC* (2013), paper P.6.13.

[9] X. Xiao, F. Li, J. Yu, Y. Xia and Y. Chen, Real-time demonstration of 100 Gbps class dual-carrier DDO-16QAM-DMT transmission with directly modulated laser, in *Proc. OFC* (2014), paper M2E.6.

[10] S. L. Jansen, I. Morita, T. C. W. Schenk, N. Takeda and H. Tanaka, Coherent optical 25.8-Gb/s OFDM transmission over 4160-km SSMF, *J. Lightwave Technol.* **26** (2008) 6–15.

[11] S. L. Jansen, I. Morita, N. Takeda and H. Tanaka, 20-Gb/s OFDM transmission over 4,160-km SSMF enabled by RF-pilot tone phase noise compensation, in *Proc. OFC/NFOEC 2007*, Anaheim, CA (March 25–29, 2007), paper PDP15.

[12] S. L. Jansen, I. Morita, T. C. W. Schenk and H. Tanaka, 121.9-Gb/s PDM-OFDM transmission with 2-b/s/Hz spectral efficiency over 1000 km of SSMF, *J. Lightwave Technol.* **27**(2) (2009) 177–188.

[13] Q. Yang, Y. Tang, Y. Ma and W. Shieh, Experimental demonstration and numerical simulation of 107-Gb/s high spectral efficiency coherent optical OFDM, *J. Lightwave Technol.* **27**(2) (2009) 168–176.

[14] Q. Yang, N. Kaneda, X. Liu, S. Chandrasekhar, W. Shieh and Y. K. Chen, real-time coherent optical OFDM receiver at 2.5-GS/s for receiving a 54-Gb/s multi-band signal, in *Proc. OFC/NFOEC*, San Diego, CA (March 2009), paper PDPC5.

[15] N. Kaneda, Q. Yang, X. Liu, S. Chandrasekhar, W. Shieh and Y. Chen, Real-time 2.5 GS/s coherent optical receiver for 53.3-Gb/s sub-banded OFDM, *J. Lightwave Technol.* **28**(4) (2010) 494–501.

[16] S. Chen, Q. Yang, Y. Ma and W. Shieh, Multi-gigabit real-time coherent optical OFDM receiver, in *Proc. OFC/NFOEC*, San Diego, CA (March 2009), paper OTuO4.

[17] X. Xiao, F. Li, J. Yu, X. Li, Y. Xia and Y. Chen, 100-Gb/s Single-band Real-time coherent optical DP-16QAM-OFDM transmission and reception, in *Proc. OFC* (2014), paper Th5C.6.

[18] C. Li, Q. Yang, T. Jiang, Z. He, M. Luo, C. Li, X. Xiao, D. Xue and X. Yi, Investigation of coherent optical multiband DFT-S OFDM in long haul transmission, *IEEE Photon. Technol. Lett.* **24**(19) (2012) 1704–1707.

[19] A. S. Karar and J. C. Cartledge, Generation and detection of a 56 Gb/s signal using a DML and half-cycle 16-QAM Nyquist-SCM, *IEEE Photon. Technol. Lett.* **25**(8) (2013) 757–760.

[20] T. Tanaka and M. Nishihara, Experimental investigation of 100-Gbps transmission over 80-km single mode fiber using discrete multi-tone modulation, in *SPIE 8646N*, (2013).

[21] W. Yan, T. Tanaka, B. Liu, M. Nishihara, L. Li, T. Takahara, Z. Tao, J. C. Rasmussen and T. Drenski, 100 Gb/s optical IM-DD transmission with 10G-class devices enabled by 65 GSamples/s CMOS DAC core, in *Proc. OFC* 2013, paper OM3H.1.

[22] M. I. Olmedo, T. Zuo, J. B. Jensen, Q. Zhong, X. Xu and I. T. Monroy, Towards 400GBASE 4-lane solution using direct detection of multiCAP signal in 14 GHz bandwidth per lane, in *Proc. OFC* 2013, paper PDP5C.10.

[23] J. Zhao and H. Shams, Fast dispersion estimation in coherent optical 16QAM fast OFDM systems, *Opt. Express* **21**(2) (2013) 2500–2505.

[24] Y. Tang, W. Shieh and B. S. Krongold, DFT-spread OFDM for fiber nonlinearity mitigation, *IEEE Photon. Technol. Lett.* **22**(16) (2010) 1250–1252.

Chapter 9

Advanced DSP for Free-Space Optical Communication

9.1. Introduction

Free-space optical (FSO) communication is a novel communication technology that uses visible or invisible light beam emitting from a laser or LED for data transmission and reception. As this technology is cost-effective, license-free, electromagnetic interference-free, and can provide high available bandwidth and data security, it has garnered increasing attention and can be expected to serve as complimentary solution for future access network [1].

Figure 9.1 gives the schematic diagram of point-to-point FSO communication system, which consists of transmitter, free-space channel and receiver. At the transmitter side, the original binary streams are firstly sent to coding and modulation models and then modulated on light waves. At the receiver side, the captured optical signals are firstly converted to electrical signals and sent for demodulation, post-equalization and decoding models. The modulated optical signals are delivered to the receiver side via free-space channel. Additionally, considering the divergence of light beams, beam formation and focus are needed in this system.

However, there are two main problems in this intensity modulation with direct detection (IM/DD)-based FSO system. One problem is the relatively low bandwidth of the whole system, which is mainly limited by the optical and electrical devices. The other one is the strong nonlinearity induced by the nonlinear characteristics of optical device and square-law detection. In this chapter, these two problems are discussed and several possible solutions are proposed, including pre-equalization,

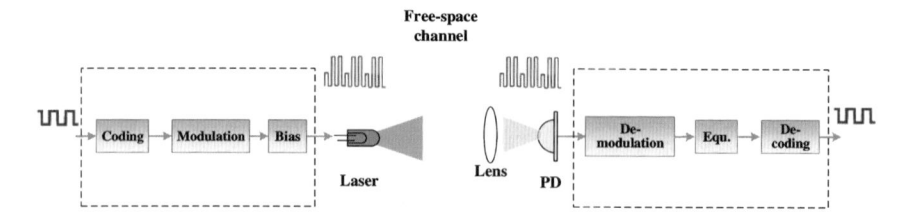

Fig. 9.1. Schematic diagram of FSO communication.

spectral-efficient modulation with Nyquist and super-Nyquist shaping, signal-to-signal beating noise elimination algorithm and hybrid time and frequency equalization algorithm.

9.2. Pre-equalization

As mentioned above, the end-to-end system bandwidth is mainly decided by the implemented independent optic-electronic devices, such as power amplifier, directly modulated laser (DML), light-emitting diode (LED) and photodiode. The relatively low 3-dB bandwidth of these devices, especially the LED and DML, will seriously limit the whole system available bandwidth, which will introduce inter-symbol interference and signal-to-noise ratio (SNR) reduction. Figure 9.2 gives the channel frequency response of white LED-based free-space optical communication system. From the measured frequency response, we can find that the 3-dB bandwidth of this system is limited to several MHz and the 20-dB bandwidth is still limited to 20 MHz, which are too low to support Gb/s or beyond Gb/s high-speed data communication. From the fitting frequency response, we can find the channel gains are in exponential reduction at high-frequency part.

Therefore, the pre-equalization technique is a promising solution for bandwidth-limitation impairment compensation. It can be realized by hardware circuit or software algorithm. In this section, we will focus on the software algorithm. Basically speaking, the realization of pre-equalization can be divided into two steps: attain the channel information and then employ equalization at the transmitter side by the attained channel knowledge. The principle of the pre-equalization technology is depicted in Fig. 9.3. Figure 9.3(a) gives the flowchart of the channel matrix estimation. First, the original signals $S(w)$ without pre-equalization are sent and passed through the free-space channel $H(w)$.

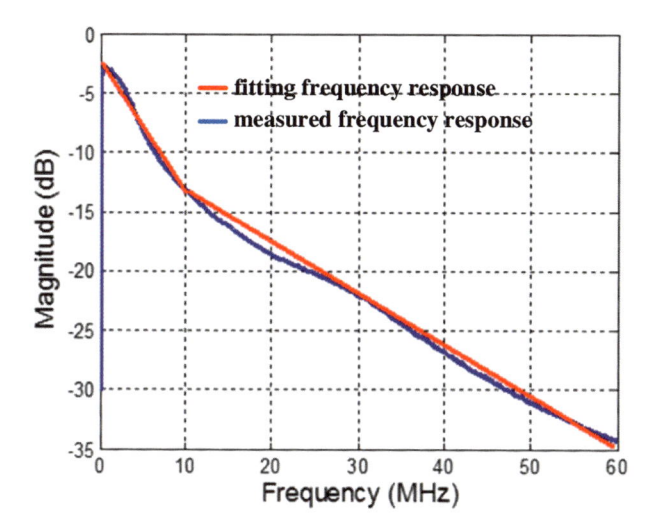

Fig. 9.2. Frequency response of free-space optical communication system.

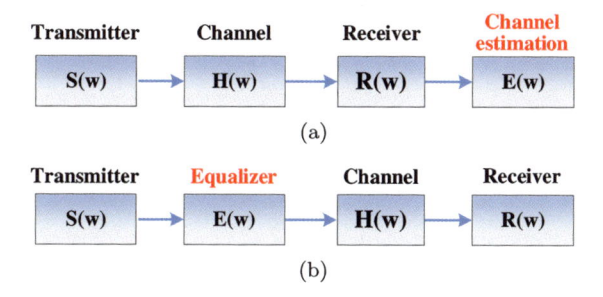

Fig. 9.3. Principle of the proposed pre-equalization technique: (a) channel matrix estimation and (b) pre-equalization at the transmitter side.

Assuming the channel is additive white Gaussian noise (AWGN) type, the received signals $R(w)$ at the receiver side can be expressed as

$$R(w) = S(w) \times H(w) + N(w) \qquad (9.1)$$

These signals are represented in the frequency domain, and $N(w)$ is the AWGN. Neglecting the noise, the equalized item can be written as

$$E(w) = S(w)/R(w) \qquad (9.2)$$

Once we obtain the channel information and equalized item, the pre-equalization can be accomplished at the transmitter side. The detailed steps are given in Fig. 9.3(b), and the equalized signals $S'(w)$ can be expressed as

$$S'(w) = S(w) \times E(w). \tag{9.3}$$

In these ways, the pre-equalization algorithm can be realized. In this section, the pre-equalization is implemented in frequency domain, and it can also be realized by FIR filter in time domain. In different modulation formats and various application scenarios, additional interpolation may be needed.

Let us take a widely used modulation format orthogonal frequency division multiplexing (OFDM) for example [2]. Figures 9.4(a)–(d) show the all the steps in detail. In this example, the FFT length of OFDM is 64. First, original signals without pre-equalization should be sent to attain the channel matrix. For accuracy estimation, multiple symbols are suggested to be transmitted at the same time and averaging performed at the receiver end. Moreover, smooth frequency may also be needed to eliminate the fluctuation. The detailed channel information estimation is depicted in

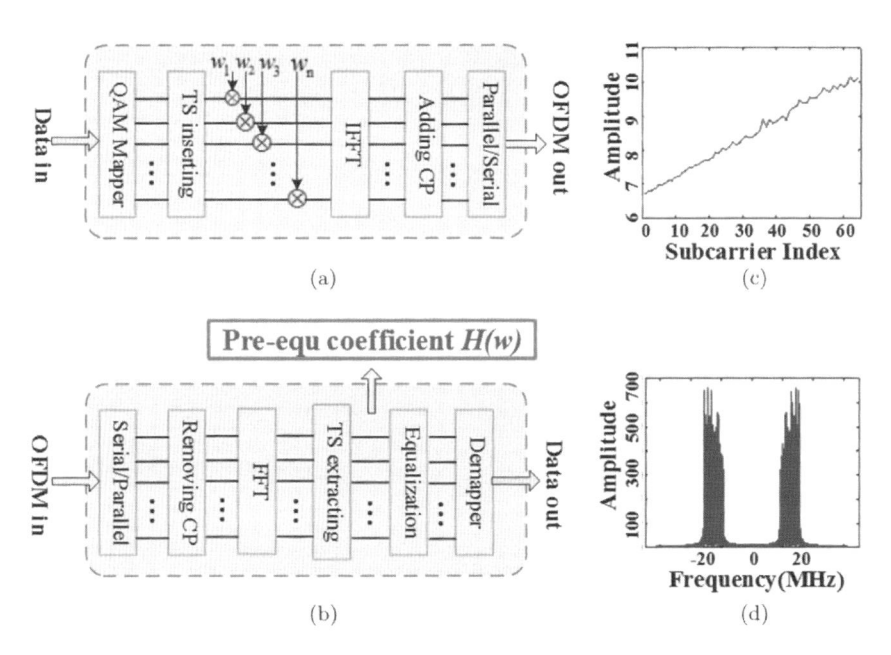

Fig. 9.4. Pre-equalization for the OFDM modulation format: (a) pre-equalization, (b) channel information estimation, (c) pre-equalization coefficients, and (d) electric spectra of upconverted OFDM signals.

Fig. 9.4(b). The calculated pre-equalization coefficients w_n are illustrated in Fig. 9.4(c), and these coefficients are used for pre-equalization in Fig. 9.4(a). The electric spectra of the pre-equalized signals are given in Fig. 9.4(d). After employing pre-equalization, the available 3-dB bandwidth can be significantly increased.

9.3. Spectral-Efficient Modulation with Nyquist and Faster than Nyquist Shaping

As mentioned in the previous section, pre-equalization can be used for bandwidth-limitation compensation. However, the available 3-dB bandwidth can only be extended to about 100 MHz in white LED-based free-space communication system. How to increase the spectral efficiency (SE) is an interesting research topic. Generally speaking, there are two methods to increase the SE: one is utilizing the high-order modulation format. Figure 9.5 gives the constellations of various modulation formats, including quadrature phase shift keying (QPSK), 16-ary quadrature amplitude modulation (16QAM) and 64QAM. One QPSK symbol can carry 2 bits, while 64QAM can carry 6 bits. When the modulation formats are turned from spectral-inefficient to spectral-efficient, the SE can be largely enhanced.

The other method to increase the SE is by adopting the modulation formats with Nyquist or Faster than Nyquist shaping.

Assuming the symbol duration is Ts, the baud rate of the signal can be expressed as $B = 1/\text{Ts}$. The electric spectrum of the signals is depicted in Fig. 9.6. From this figure, we can find the main lobe occupied bandwidth is 2B. This main lobe can be compressed and the side lobes can be filtered out by passing through various low-pass filters (LPFs).

Figures 9.7(a)–(c) give the frequency responses of different low-pass filters, including Bessel filter with fourth order, Nyquist filter and faster than

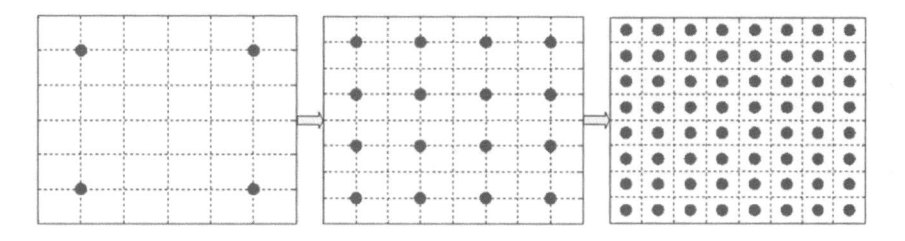

Fig. 9.5. Constellations of various modulation formats: QPSK, 16QAM and 64QAM.

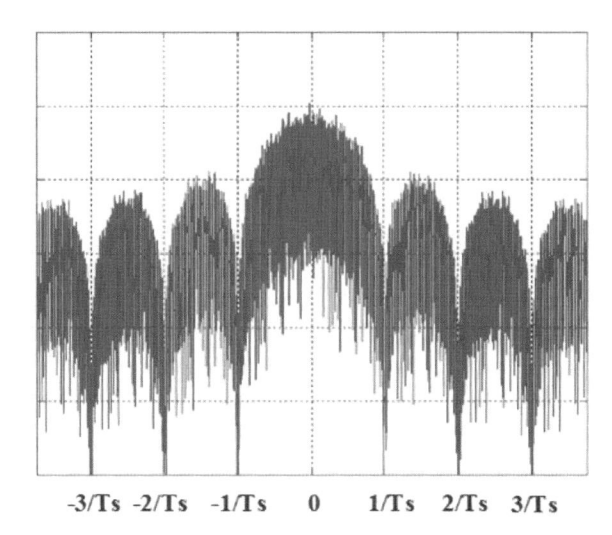

Fig. 9.6. Electric spectrum of conventional signals.

Nyquist filter. The 3-dB bandwidth of Bessel filter with fourth order is almost $2B$, and the 3-dB bandwidth of faster than Nyquist filter is smaller than B. The Nyquist filter used here is square root raised cosine filter, whose bandwidth is around B and related to the roll-off factor. Nyquist and faster than Nyquist modulation formats will be separately discussed in the remainder of this section.

9.3.1. *Nyquist modulation formats*

Nyquist modulation formats can be divided as single carrier modulation and multi-carrier modulation formats. Carrierless amplitude and phase modulation (CAP), half-cycle QAM, single carrier with frequency domain equalization (SC-FDE) are three popular types of single carrier Nyquist modulation. OFDM and its variance such as discrete multi-tone (DMT), DFT-Spread OFDM are multi-carrier Nyquist modulation formats. In this section, multi-carrier modulation OFDM and single-carrier modulation SC-FDE will be discussed and compared [3].

The block diagram of SC-FDE/OFDM free-space optical system is presented in Fig. 9.8. In this demonstration, a commercially available RGB-LED (Cree, PLCC) generating a luminous flux of about 6lm used as the transmitters (TX) and an avalanche photodiode (Hamamatsu APD) used as the receiver (RX) are adopted.

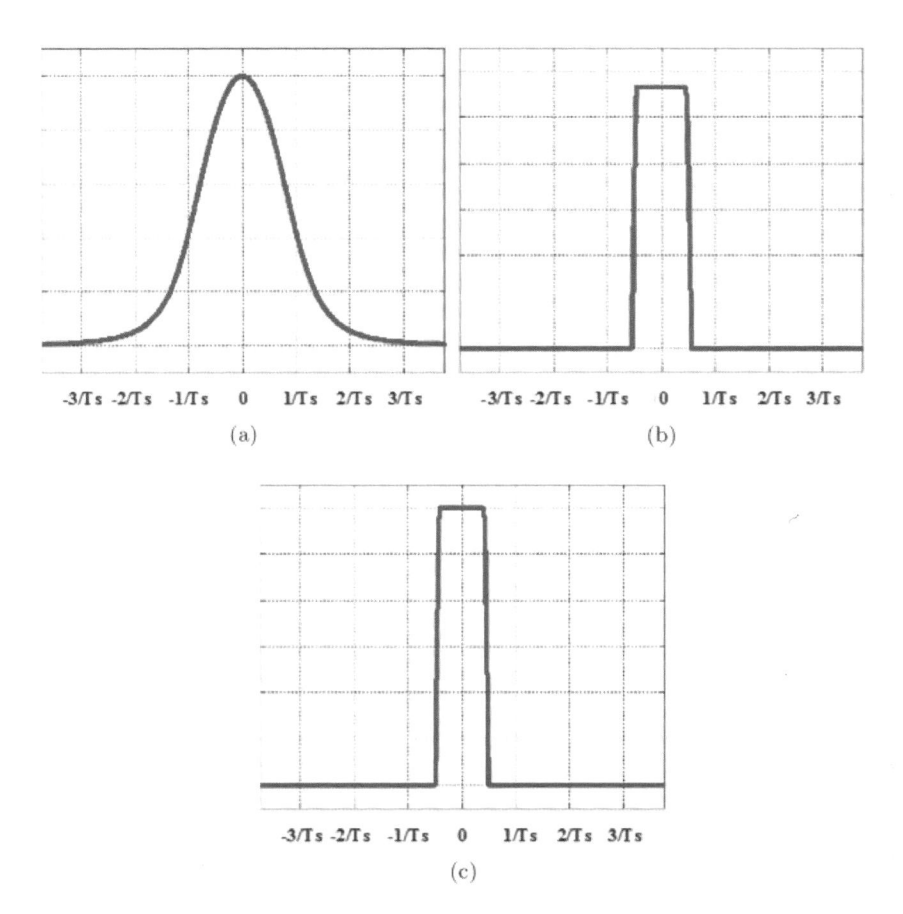

Fig. 9.7. Frequency responses of different low-pass filters: (a) Bessel filter with fourth order, (b) Nyquist filter, and (c) faster than Nyquist filter.

For the SC-FDE signals, the binary data would be firstly mapped into M-ary quadrature amplitude modulation (MQAM) format and then the training sequences (TS) are inserted into the signals. After making preequalization in frequency domain and upsampling, circle prefix (CP) is adding and low-pass filters are used to remove out-of-band radiation, and subsequently amplified by electrical amplifier (EA) (Minicircuits, 25-dB gain), combined with direct current (DC)-bias via bias tee, and then applied to these three different color chips. Passing through free-space transmission, lens (50-mm diameter) and optical R/G/B filter, the signals are recorded by a commercial high-speed digital oscilloscope (OSC)

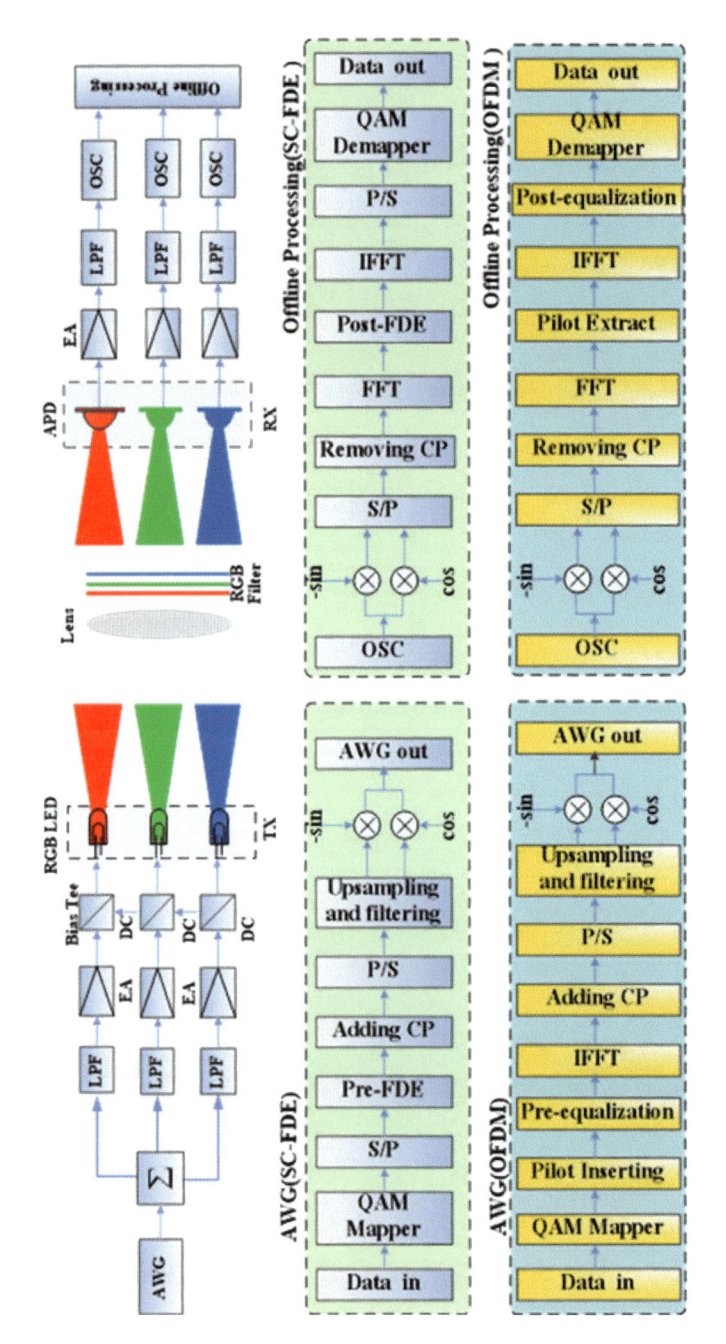

Fig. 9.8. RGB-LED-based free-space optical communication with SC-FDE and OFDM Nyquist modulation.

and sent for demodulation. The demodulation is performed with the offline MATLAB®DSP programs.

For OFDM demodulation, after downconversion to baseband and removal of CP, the time-domain OFDM signals are transformed into frequency-domain signals by DFT with fast Fourier transform (FFT) size of 128 to implement frequency-domain equalization. For SC-FDE demodulation, firstly, the SC-FDE signals are downconverted to baseband and subsequently filtered by a square filter. Second, mirroring the demodulation process of OFDM, frequency-domain equalization is implemented after removing CP and executing DFT of FFT size of 128. The frequency-domain SC-FDE signals are transformed into time-domain signals again by IDFT. The OFDM and SC-FDE schemes have many similarities in block diagram as shown in Fig. 9.8. For instance, both OFDM and SC-FDE schemes require CP to overcome inter-symbol interference (ISI) issue and discrete Fourier transform (DFT) for frequency-domain equalization. The only one dissimilarity between the OFDM and SC-FDE schemes is that the IFFT block is moved from the transmitter to the receiver as shown in Fig. 9.8.

Except the limited modulation bandwidth, one of the other drawbacks is the quite small modulation index caused by the significant nonlinearity of LED. The current–voltage characteristic is shown in Fig. 9.9. The C point represents the bias point and the current at this point can be noted as I_C, while the A and B points represent the highest current and lowest current

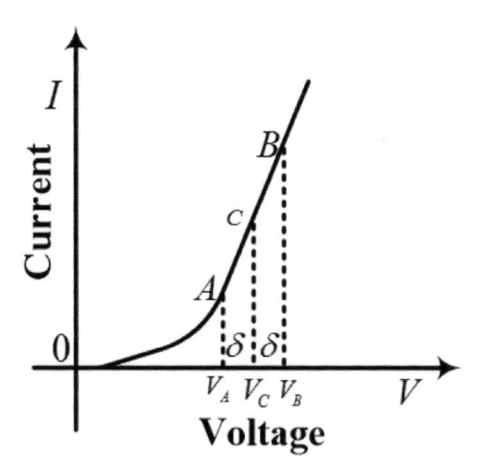

Fig. 9.9. Voltage–current curve.

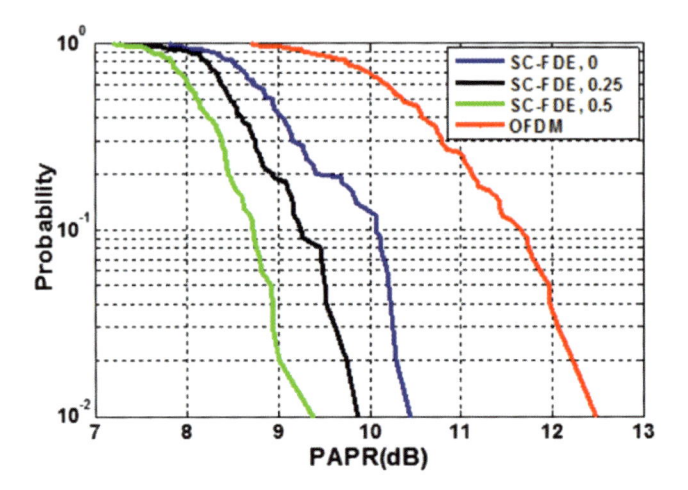

Fig. 9.10. PAPR comparison between OFDM and SC-FDE.

during the quasi-linear working area. So the modulation index m can be expressed as

$$m = \delta/I_C \tag{9.4}$$

As the modulation index of LED is very small, the dynamic range of the transmitted signals can also be limited. Therefore, the high peak to average power ratio (PAPR) of the signal can severely affect the performance. Figure 9.10 shows the complementary cumulative distribution function (CCDF) curves of PAPR of the OFDM and SC-FDE signals. The CCDF curves of the SC-FDE signals are obtained at the cases of roll-off factor with 0, 0.25 and 0.5. The FFT sizes of these two schemes are both 128. The PAPR of the OFDM is about 2 dB better than that of the SC-FDE signals with roll-off factor of 0. Hence, SC-FDE is more suitable for LED-based free-space optical communication system at this point.

A comparison of system performance between OFDM and SC-FDE has also been made experimentally. The BER performance with the different modulation formats is carried out. For SC-FDE, the modulation orders are ranging from 6 to 9, and for OFDM, the modulation orders are ranging from 5 to 9. The experimental results are shown in Figs. 9.11(a)–(c). It can be seen that the maximum modulation orders of SC-FDE in red, green and blue chips are 8 (256QAM), while for OFDM, the maximum modulation orders

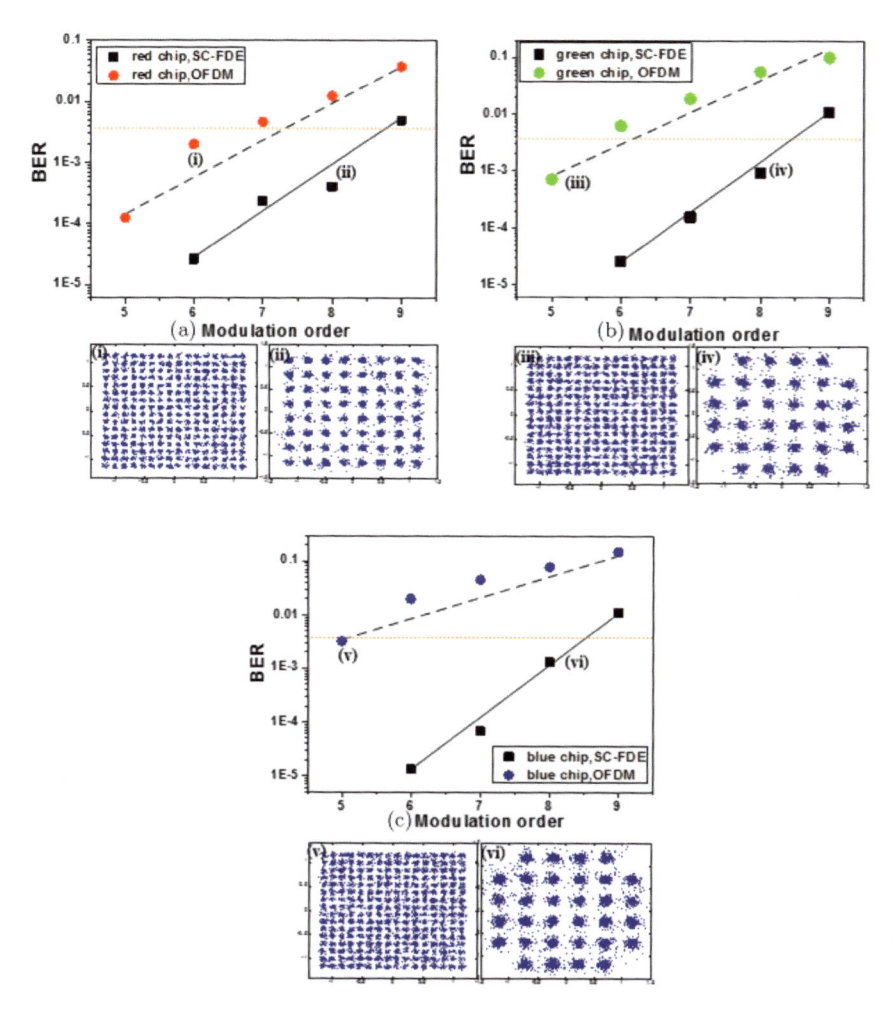

Fig. 9.11. The measured BER performance versus different modulation orders of SC-FDE and OFDM: (a) red LED chip, (b) green LED chip, and (c) blue LED chip.

in red, green and blue chips are 5 (32QAM), 5 (32QAM) and 6 (64QAM), respectively. Therefore, the achieved data rates of SC-FDE and OFDM are 3.75-Gb/s and 2.5-Gb/s, with the BERs for all wavelength channels under pre-FEC threshold of 3.8×10^{-3}. The constellations in the individual LED chips of different modulation schemes are also inserted in Fig. 9.11. Hence, we can find that the SC-FDE outperforms OFDM.

9.3.2. Faster than Nyquist modulation formats

Compared to conventional modulation, the Nyquist modulation formats can provide high SE, and the SE can be even higher by adopting faster than Nyquist (FTN) modulation formats. These FTN signals can be generated at electrical domain or optical domain, with the aid of faster than Nyquist filters, whose 3-dB bandwidth is smaller than Nyquist bandwidth. After passing through the FTN filter, the binary signals will become duobinary signals, and these constellation changes have been illustrated in Fig. 9.12.

From this figure, we can find the QPSK signals will become 9QAM, and the 16QAM will become 49QAM. Figures 9.12(a)–(d) show the constellations of QPSK, 9QAM, 16QAM and 49QAM. Additionally, the 9QAM are distributed on three circles, while the 49QAM are distributed on 10 circles with different radii. These distributions require high resolution and high signal-to-noise ratio (SNR).

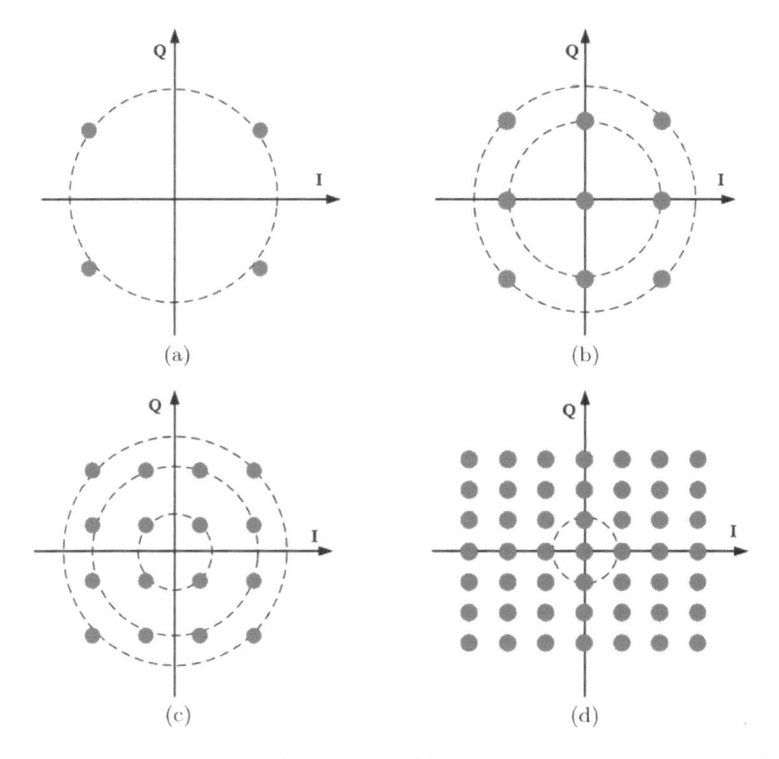

Fig. 9.12. The constellations of (a) QPSK, (b) 9QAM, (c) 16QAM, and (d) 49QAM signals.

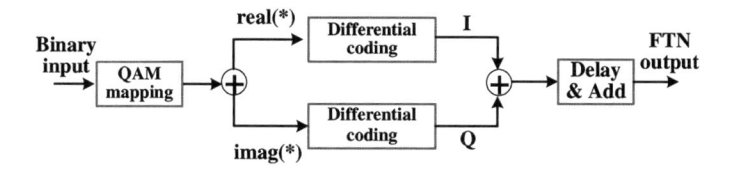

Fig. 9.13. The flowchart of faster than Nyquist signal generation.

In this section, the FTN signals are realized by duobinary delay-and-add filter. The response of the filter in z-domain can be expressed as [4]

$$H(z) = 1 + z^{-1}. \tag{9.5}$$

In practical implementation, the filter can be realized by one symbol delay and adding operation. It is also suitable for vector signals. Additionally, differential coding is needed at the transmitter side in order to help accurate signal recovery at the receiver side. The flowchart of FTN vector signal generation is depicted in Fig. 9.13, including 4/16QAM mapping, I/Q separation, differential coding, I/Q combination, delay-and-add filter.

Assuming the real and imaginary parts of the kth symbol before and after differential coding can be denoted as i_k, q_k, I_k and Q_k, respectively, the differential coding can be expressed as

$$I_k = i_k - I_{k-1} \bmod M \tag{9.6}$$

$$Q_k = q_k - Q_{k-1} \bmod M \tag{9.7}$$

$$d_k = I_k + jQ_k \tag{9.8}$$

where d_k represents the differential coded vector signals. After passing through the delay-and-add filter, the generated faster than Nyquist signals D_k can be written as

$$D_k = d_k + d_{k-1} \tag{9.9}$$

At the receiver side, differential decoding should be implemented after hard decision. It can be easily completed by module M operation. In this way, the generation and reception of faster than Nyquist signals can be realized.

In this section, a simple comparison between Nyquist 16QAM SC-FDE and faster than Nyquist 49QAM SC-FDE is carried out. Figure 9.14 gives the principle of generation and reception operations of Nyquist SC-FDE

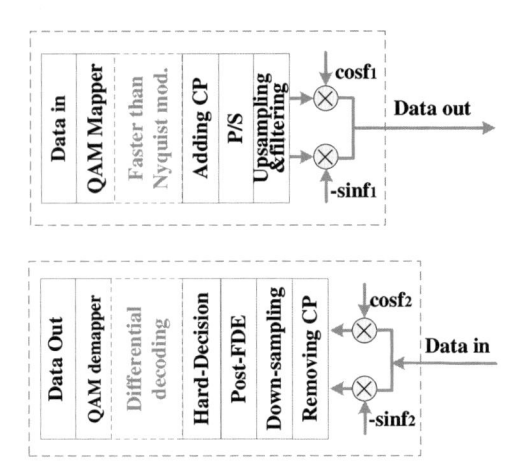

Fig. 9.14. The principle of generation and reception operations of Nyquist SC-FDE and faster than Nyquist SC-FDE signals.

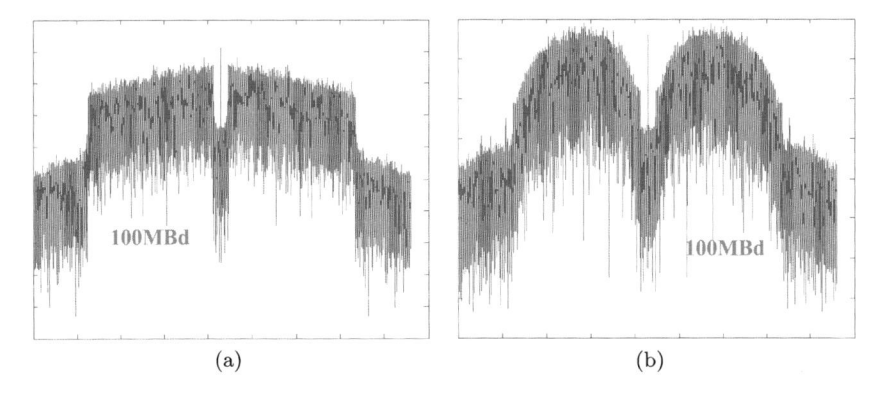

Fig. 9.15. The electrical spectrum of (a) Nyquist 16QAM SC-FDE signals and (b) faster than Nyquist 49QAM SC-FDE signals.

and faster than Nyquist SC-FDE signals. Compared to Nyquist SC-FDE, the faster than Nyquist SC-FDE needs additional differential coding and decoding.

Here, the baud rates of the Nyquist and faster than Nyquist SC-FDE signals are both 100 MBd, and the electrical spectra are illustrated in Fig. 9.15. From this figure, we can find that the 3-dB bandwidth of the FTN SC-FDE signals is below 100 MHz, which means the spectral efficiency can be largely enhanced.

9.4. Nonlinearity Compensation in the FSO System

Nonlinearity is another drawback in free-space optical communication, which mainly comes from the nonlinear characteristics of opto-electrical devices and square-law detection. The nonlinear effect will seriously affect the system performance. The current nonlinearity mitigation methods can be classified as low peak-to-average power ratio (PAPR) modulation scheme and adaptive nonlinearity compensation. In this section, Volterra series-based adaptive nonlinearity compensation method is proposed.

The voltage–current curve of LED is illustrated in Fig. 9.16. From this figure, we can find that the quasi-linear range is very narrow, thus leading to low modulation index and low signal-to-noise ratio (SNR). After passing through the nonlinear system, the output signals $r(t)$ can be represented as a function of input $x(t)$ by using a pth-order Volterra series expansion and can be written as

$$r(t) = h_0 + \sum_{k=0}^{M-1} h_1(k)x(t-k) + \sum_{k_1=0}^{M-1}\sum_{k_2=k_1}^{M-1} h_2(k_1,k_2)x(t-k_1)x(t-k_2)$$

$$+ \cdots + \sum_{k_1=0}^{M-1} \cdots \sum_{k_p=k_{p-1}}^{M-1} h_p(k_1,\ldots,k_p)x(t-k_1)\cdots x(t-k_p) \quad (9.10)$$

where M is the memory length and p is the nonlinear order. Here h_0 is a constant, and $h_p(\cdot)$ is the pth-order Volterra kernel coefficients. However, it is not practical to use the full Volterra series as the number of parameters increases exponentially with the increase of nonlinearity and memory length.

The nonlinear effect can be divided into memory and memoryless effects, based on whether we take into account the memory effects of nonlinearities. For the memoryless effect, the last equation can be simplified as

$$r(t) = h_0 + h_1 x(t) + h_2 x^2(t) + \cdots + h_p x^p(t). \quad (9.11)$$

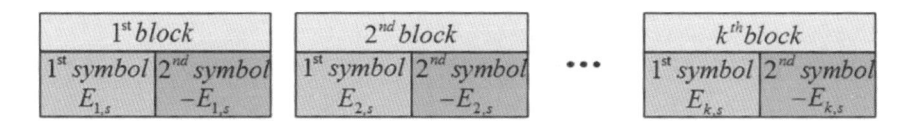

1st block	2nd block		k^{th} block
1st symbol | 2nd symbol	1st symbol | 2nd symbol	...	1st symbol | 2nd symbol
$E_{1,s}$ | $-E_{1,s}$	$E_{2,s}$ | $-E_{2,s}$		$E_{k,s}$ | $-E_{k,s}$

Fig. 9.16. Conceptual diagram of QBD-OFDM.

This memory nonlinear effect can be easily compensated by a post-adaptive equalizer, whose response can be expressed as

$$y(t) = w_0 + w_1 x(t) + w_2 x^2(t) + \cdots + w_L x^L(t) \quad (9.12)$$

where $x(t)$ is the captured signal at the receiver side and is also used as the input of the equalizer. $(w_0, w_1, w_2, \ldots w_L)$ are the coefficients of the adaptive filters, and L is the tap number.

As for the memory nonlinear system, the process is much more complicated. It can also be compensated by the adaptive equalizer filter, and the equalized signals can be written as

$$y(t) = w_0 + \sum_{k=0}^{N-1} w_1(k) x(t-k) + \sum_{k_1=0}^{N-1} \sum_{k_2=k_1}^{N-1} w_2(k_1, k_2) x(t-k_1) x(t-k_2)$$

$$+ \cdots + \sum_{k_1=0}^{N-1} \cdots \sum_{k_p=k_{p-1}}^{N-1} w_L(k_1, \ldots, k_p) x(t-k_1) \cdots x(t-k_p). \quad (9.13)$$

Generally speaking, the nonlinear order can be selected as 2 or 3.

9.5. Quasi-balanced Coding and Detection for Signal-to-Signal Beating Noise Elimination

The second-order nonlinearity distortion in OFDM introduced by direct detection (DD) is another drawback in free-space optical communication [5]. Several approaches have been reported to reduce the distortion [6–9]. In [6], Lowery has proposed an offset single sideband (SSB) OFDM scheme to allocate sufficient guard band such that the signals and the intermodulation distortion (IMD) are not overlapping. In [7], a baseband SSB-OFDM scheme is proposed. However, SSB-OFDM cannot be realized in the IM/DD FSO system. Moreover, the SE in [6] and receiver sensitivity in [7] are sacrificed. In [8], iterated distortion reduction is proposed. It has better SE, but with a burden of computational complexity. In [9], asymmetrically clipped optical (ACO) OFDM is proposed. In this scheme, only odd subcarriers are filled so that the IMD will be located at even subcarriers. But it will lose efficacy in the case of frequency deviation.

Balanced detection (BD) is an efficient solution to this problem, but the LED is an incoherent source that BD cannot be employed in the optical domain. In this section, a novel quasi-balanced detection (QBD) technique in the OFDM FSO system is proposed. By employing opposite signals to odd and even consecutive symbols, BD with one single detector can be realized in electrical domain. This scheme has both advantages of BD and direct detection, and can also be used in other similar cases such as intensity modulation (IM)/DD OFDM short-range, low-cost optical fiber system. Using this scheme, the second-order intermodulation distortion and direct current (DC) can be eliminated, and the sensitivity of receiver can be improved. In the remainder of this section, this scheme will be theoretically and experimentally investigated.

The QBD-OFDM is mainly realized in the electrical domain. The block diagram of this scheme is depicted in Fig. 9.16. The OFDM signals are divided into several blocks, and in each block, there are two symbols. The signals in the second symbol are opposite to the first one in the same block.

In this scheme, the baseband OFDM signals in the kth block are given by

$$E_{2k-1}(t) = \sum_{m=1}^{N} c_m e^{j2\pi f_m t} \quad (9.14)$$

$$E_{2k}(t) = -\sum_{m=1}^{N} c_m e^{j2\pi f_m t} \quad (9.15)$$

where N is the number of subcarriers, c_m is the information symbol at the mth subcarrier, and f_m is the frequency of the subcarrier. The OFDM signals which modulated on the LED light through bias tee can be expressed as

$$s_{2k-1}(t) = (V_0 - V_a)e^{j2\pi f_0 t} + \alpha e^{j2\pi(f_0+f_1)t} E_{2k-1}(t) \quad (9.16)$$

where V_0 and V_a are respectively the bias voltage and reversal voltage of LED. The coefficient α is used to describe the ratio of OFDM band strength related to the main carrier. f_1 is the upconverted frequency of the OFDM signals. After transmitting through the free-space environment, the OFDM signals considered frequency deviation and phase noise can be approximated as

$$r_{2k-1}(t) = (V_0 - V_a)e^{j(2\pi(f_0+\Delta f)t + \phi(t))}$$
$$+ \alpha e^{j(2\pi(f_0+f_1+\Delta f)t+\phi(t))} E_{2k-1}(t) + n_{0,2k-1}(t) \quad (9.17)$$

where Δf is the deviation frequency, $\phi(t)$ is the phase noise, and $n_0(t)$ is the AWGN. After detection by the square-law detector, the photocurrents for the kth block can be approximated as

$$I_{2k-1}(t) = |r_{2k-1}(t)|^2 + n_{r,2k-1}(t) = I_{s,2k-1} + I_{b,2k-1}$$
$$+ I_{nl,2k-1} + I_{n,2k-1} \qquad (9.18)$$

$$I_{n,2k-1} = 2\alpha \, \mathrm{Re} \left\{ e^{-j(2\pi(f_0+\Delta f)t+\phi(t))} n_{0,2k-1}(t) \sum_{m_1=1}^{N} c_i e^{-j2\pi f_m t} \right\}$$
$$+ |n_{0,2k-1}(t)|^2 + n_{r,2k-1}(t) + 2\alpha(V_0 - V_a)$$
$$\times \mathrm{Re}\{e^{-j(2\pi f_0 t+\phi(t))} n_{0,2k-1}(t)\} \qquad (9.19)$$

$$I_{s,2k-1} = 2\alpha(V_0 - V_a) \mathrm{Re} \left\{ e^{j2\pi \Delta f t} \sum_{m_1=1}^{N} c_i e^{j2\pi f_m t} \right\} \qquad (9.20)$$

$$I_{b,2k-1} = |V_0 - V_a|^2 \qquad (9.21)$$

$$I_{nl,2k-1} = |\alpha|^2 \sum_{m_1=1}^{N} \sum_{m_2=1}^{N} m_2^* m_1 e^{(j2\pi(f_{m_1}-f_{m_2})t)} \qquad (9.22)$$

$$I_{2k}(t) = |r_{2k}(t)|^2 + n_{r,2k}(t) = -I_{s,2k-1} + I_{b,2k-1} + I_{nl,2k-1} + I_{n,2k} \qquad (9.23)$$

where $n_r(t)$ is the noise introduced by the receiver. In Eq. (9.18), the first term is proportional to the original OFDM signal, the second term is DC, the third term is the signal-to-signal beating noise (SSBN), and the fourth term is the other noises listed in the right-hand side of Eq. (9.19). In Eq. (9.19), the first term is the signal-to-AWGN beating noise, the second term is the AWGN-to-AWGN noise, and the fourth term is proportional to the AWGN noise. The photocurrents of second in the kth block can be expressed in the same way in Eq. (9.23). By subtracted Eqs. (9.19) and (9.23), we can obtain the photocurrents of the kth block.

$$I_k(t) = I_{2k-1}(t) - I_{2k}(t) = 2I_{s,2k-1}(t) + \{I_{n,2k} - I_{n,2k-1}\} \qquad (9.24)$$

where the first term is the signal and the second term is the noise. From Eq. (9.24), we can find the second-order intermodulation distortion, DC can be totally eliminated, and the sensitivity of receiver can be improved by 3 dB, thus the signal-to-noise ratio can be improved.

The block diagram of QBD-OFDM FSO system is shown in Fig. 9.17. The flowchart of the OFDM signal generation and reception is given.

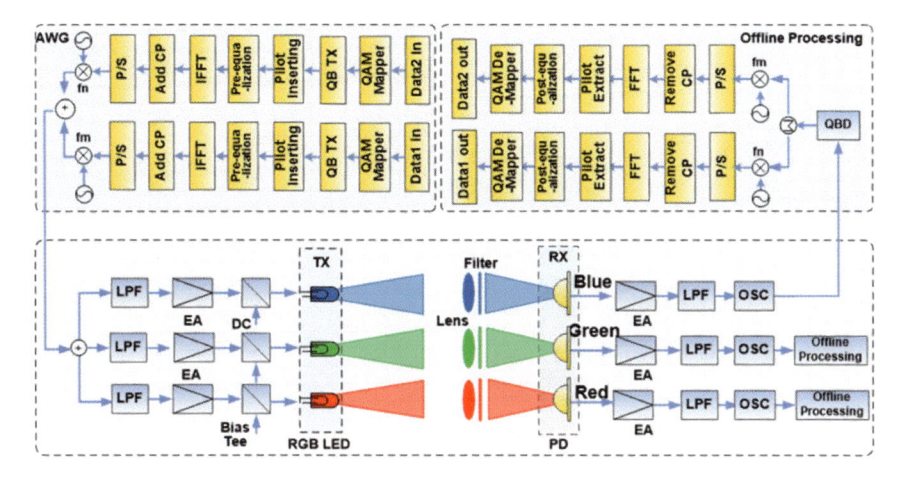

Advanced DSP for Free-Space Optical Communication 215

Fig. 9.17. Block diagrams of proposed QBD-OFDM FSO system (CP: cylix prefix; P/S: parallel to serial).

Compared to conventional OFDM generation, additional quasi-balanced coding and decoding are needed. The electrical QAM-OFDM signals and DC-bias voltage are combined via bias tee and applied to different LED chips. Through free-space transmission, lens (100-mm focus length) and R/G/B filter, the data are recorded by a commercial high-speed digital oscilloscope (OSC).

A comparison of DDO-OFDM, ACO-OFDM, QBD-OFDM and QBD-ACO-OFDM is carried out. For simplicity, we just utilize green LED chip, and only one sub-channel is adopted to avoid the crosstalk from other subchannels. The modulation format is 256QAM-OFDM, and the distance between the transmitter and the receiver is varied from 0.5 m to 2.5 m. As shown in this figure, only the QBD-OFDM and QBD-ACO-OFDM can eliminate the intermodulation distortion. The effectiveness of ACO-OFDM is not significant. Compared with the unoccupied spectra of ACO-OFDM or DDO-OFDM ranging from 60 MHz to 100 MHz, we can find that the background noises of QBD-OFDM and QBD-ACO-OFDM are also reduced. The experiment results of BER performances versus transmission distance are shown in Fig. 9.18. It can be seen the QBD-OFDM and QBD-ACO-OFDM can achieve error-free performance with 7% FEC threshold after 2.5 m indoor environment delivery. At the distance of 0.5 m, the BER performance of QBD-OFDM can be enhanced by 22.2 dB and 20.8 dB, compared with DDO-OFDM and ACO-OFDM, respectively. If we

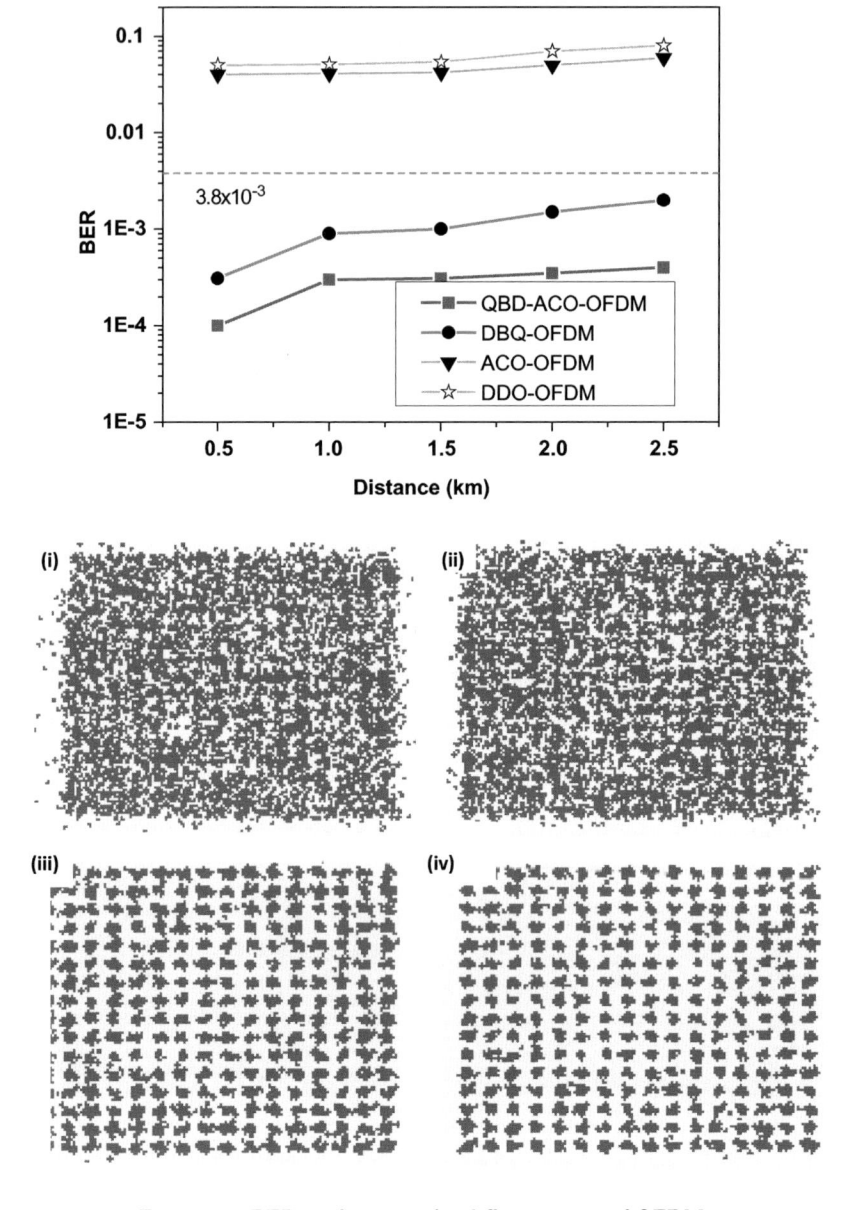

Fig. 9.18. BER performance for different types of OFDM.

combine ACO- and QBD-OFDM together (i.e., QBD-ACO-OFDM), the BER performance can be enhanced by 27.0 dB and 25.6 dB, compared with DDO-OFDM and ACO-OFDM, respectively. The received OFDM signal constellations inserted in Fig. 9.19 are shown for the four types of OFDM, respectively. It can be easily seen that the constellations of QBD-OFDM and QBD-ACO-OFDM can be clearly recognized.

9.6. Hybrid Time and Frequency Equalization

In the previous section, an alternative FSO modulation scheme using Nyquist single carrier frequency-domain equalization (N-SC-FDE) has been mentioned. This scheme has the similarity of spectral efficiency performance to the aforementioned OFDM technology, but with a reduced PAPR. In the SC system, the adaptive equalization is critical. The equalization method can be performed either in time domain such as cascaded multi-modulus algorithm (CMMA) [10] or in frequency domain such as pre-FDE and post-FDE. Generally speaking, time-domain equalization typically requires a number of multiplications per symbol that is proportional to the maximum channel impulse response length. FDE appears to offer a better complexity trade-off than time-domain equalization when large taps are needed [11].

In this section, a hybrid time-frequency adaptive algorithm in an N-SC FSO system based on a combination of FDE and decision-directed least mean square (DD-LMS) is proposed and experimentally investigated [12]. The non-flat frequency response of the FSO system can be first mitigated by FDE, and the system performance can be further improved by DD-LMS via symbol decision.

The architecture and principle of the FSO system with the proposed hybrid time-frequency equalizer is shown in Fig. 9.19. The concept of Nyquist SC-FDE is very similar to that of OFDM. If no pre-FDE is employed, the only difference is that, in SC-FDE, the inverse fast Fourier transform (IFFT) block is moved from the transmitter to the receiver. The binary data would be firstly mapped into 512-QAM format and then the training sequences (TSs) are inserted into the signals. After making pre-equalization in frequency domain and upsampling, cyclic prefix (CP) is added. CP is used to mitigate the multipath distortion. Then the real and imaginary components of signals are multiplied with sine and cosine

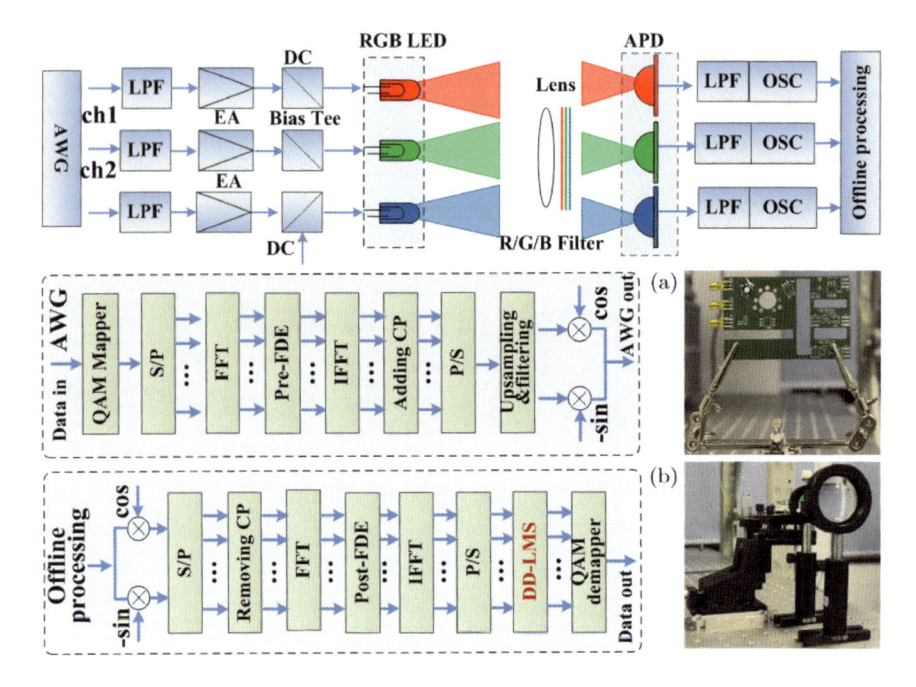

Fig. 9.19. The architecture of the proposed FSO system based on hybrid time-frequency adaptive equalization algorithm (AWG: arbitrary waveform generator, P/S: parallel to serial, EA: electrical amplifier, LPF: low-pass filter, DC: direct current, OSC: real-time oscilloscope).

functions, respectively. Then these signals are modulated on light waves. Passing through free-space transmission, lens (50-mm diameter) and optical R/G/B filter, the signals are recorded by a commercial high-speed digital oscilloscope and sent for offline processing.

At the receiver, after synchronization, resampling and removing CP, a two-fold hybrid time-frequency equalization method jointly employing FDE and DD-LMS is carried out. First, the non-flat frequency response is compensated by FDE via zero forcing (ZF) algorithms, then to switch to DD-LMS equalizer once the bit error rate (BER) has dropped to a sufficiently low level around 10^{-1} to 10^{-2} [13]. The signals in frequency domain are transformed to time domain and pass through the DD-LMS equalizer. DD-LMS is a stochastic gradient descent algorithm and does not depend on the statistics of symbols but rely on the symbol decisions. The skeleton structure of DD-LMS equalizer is illustrated in Fig. 9.20.

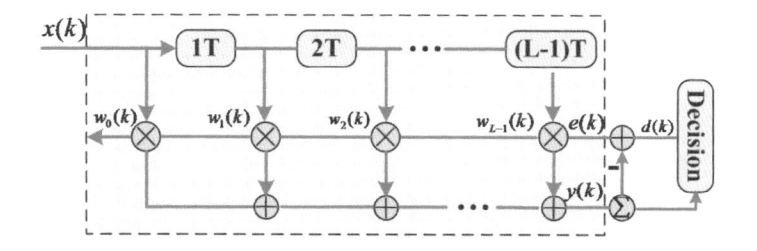

Fig. 9.20. The skeleton structure of DD-LMS equalizer.

The output $y(k)$ of DD-LMS equalizer with L taps is shown as

$$y(k) = w^H(k)X(k) \tag{9.25}$$

$$w(k) = [w_0(k), w_1(k), w_2(k), \ldots, w_L(k)]^T \tag{9.26}$$

$$X(k) = [x(k), x(k-1), x(k-2), \ldots, x(k-L+1)]^T \tag{9.27}$$

where $X(k)$ and $w(k)$ represent the input signal and weight vectors, respectively, of the kth DD-LMS section. $(\cdot)^H$ denotes the Hermitian matrix of $(\cdot)$.

The error signal $e(k)$ and weight vector for adaptive updating DD-LMS at the kth iteration are given by

$$e(k) = d(k) - y(k) \tag{9.28}$$

$$w(k+1) = w(k) + \mu e^*(k)X(k) \tag{9.29}$$

where $d(k)$ is expected output, μ is the step size, and $(\cdot)^*$ denotes the complex conjugate matrix of $(\cdot)$. The DD-LMS error term in Eq. (9.28) assumes zero values at the symbol points and hence the excess mean-squared error (EMSE) is greatly reduced [13].

The tap number of DD-LMS equalizer is a key parameter in this hybrid time and frequency equalization scheme. The optimal number of taps is investigated. In this investigation, two TSs are employed and the modulation format is 512-QAM. The number of taps ranges from 3 to 53. The results are shown in Fig. 9.21. The system performance can be improved with the increasing tap number of the equalizer. Considering the system performance and computational complexity, the proper number of taps can be set at 33. The tap number is related to the distortions and system.

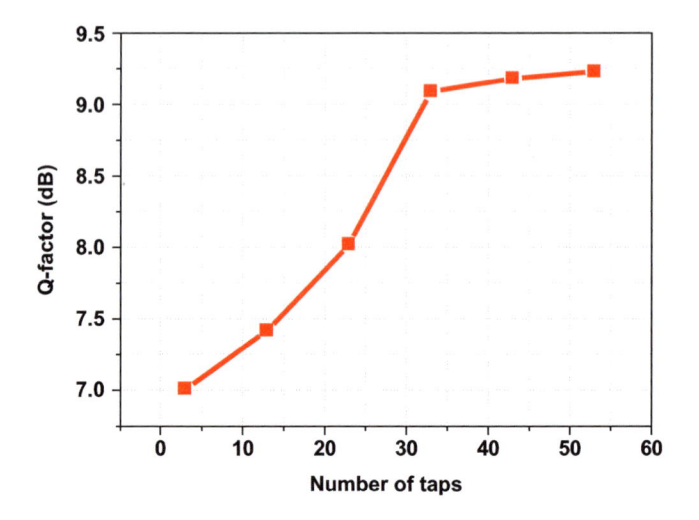

Fig. 9.21. The Q-factor performance versus number of taps.

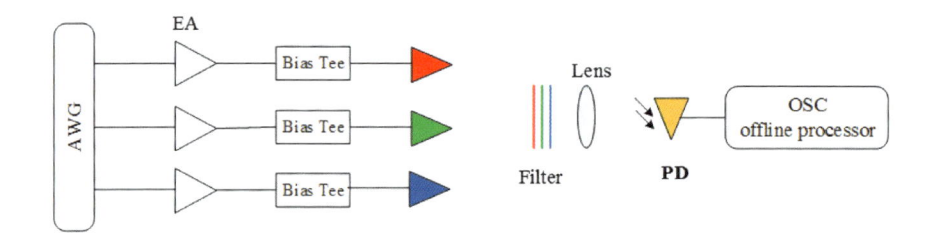

Fig. 9.22. Schematic diagram of WDM FSO system.

9.7. Crosstalk Elimination in Multi-dimensional Multiplexed FSO System

Multi-dimensional multiplexing is another method that can largely increase the system capacity by 'multiple' times. In the FSO system, wavelength division multiplexing (WDM), frequency division multiplexing (FDM), polarization division multiplexing (PDM) [14], and space division multiplexing (SDM) [15] are conventional multiplexing methods. Figures 9.22–9.24 give the system diagram of WDM, PDM and SDM-based FSO system.

By modulated-independent signals on multiple channels (wavelength channel, polarization channel, and space channel), the transmission capacity can be largely enhanced. However, crosstalk from different channels will

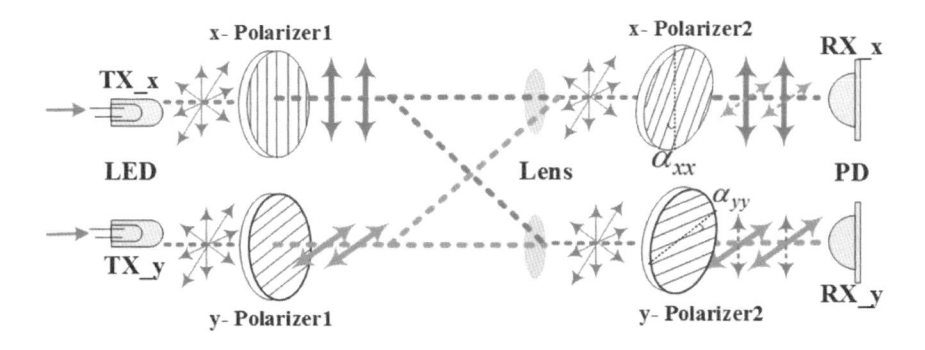

Advanced DSP for Free-Space Optical Communication 221

Fig. 9.23. Schematic diagram of PDM FSO system.

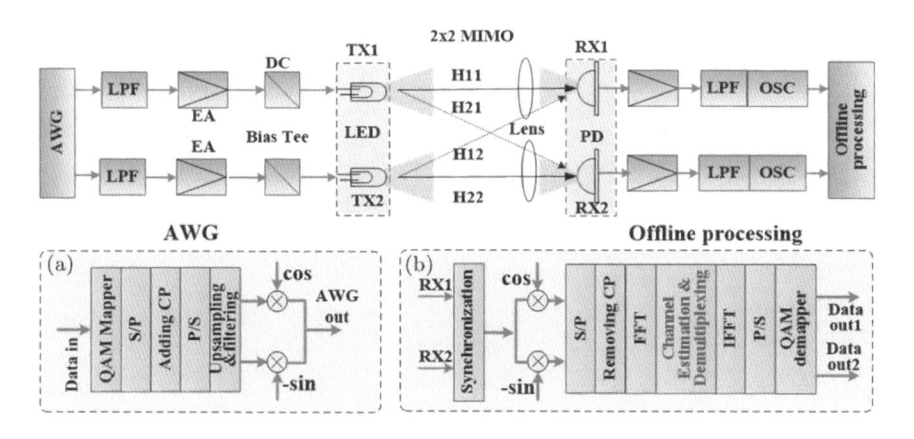

Fig. 9.24. Schematic diagram of SDM FSO system.

be induced and the system performance will be degraded. Therefore, optical and electrical devices or algorithms are needed for the crosstalk elimination. In this section, the distortions can be regarded as linear crosstalk, and the proposed crosstalk elimination algorithm is suitable for the three above-mentioned multiplexing systems. For simplicity, SDM-FSO system is taken for example.

Figure 9.25 shows the block diagram of this proposed SDM-FSO system with SC-FDE modulation. The free-space link can be regarded as MIMO model, which can be expressed as

$$\begin{pmatrix} Y_1 \\ Y_2 \end{pmatrix} = \begin{pmatrix} H_{11} & H_{12} \\ H_{21} & H_{22} \end{pmatrix} \cdot \begin{pmatrix} X_1 \\ X_2 \end{pmatrix} + \begin{pmatrix} N_1 \\ N_2 \end{pmatrix} \qquad (9.30)$$

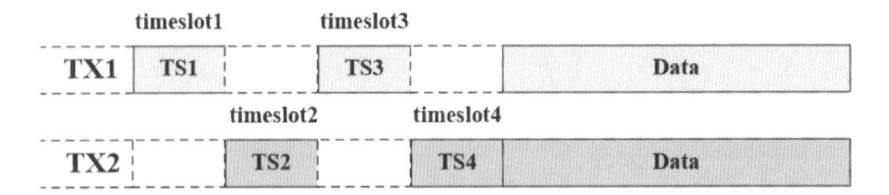

Fig. 9.25. Time-multiplexed training symbols for MIMO demultiplexing and post-equalization (TS: training sequence, TX: transmitter).

where $(Y_1 \ Y_2)^T$ represent two received SC-FDE signals after free-space transmission, and $(X_1 \ X_2)^T$ are the two independent SC-FDE signals at the transmitter, while $(N_1 \ N_2)^T$ denote the system noise. The channel matrix elements $H_{i,j}$ (i = 1, 2; j = 1, 2) represent the gain from jth transmitter to ith receiver. All the signals are processed at frequency domain.

The demultiplexing can be easily realized once the channel matrix H is known [16]. Unlike the method that an extra MIMO training run used in, we adopt the time-multiplexed TS-based frequency-domain equalization as shown in Fig. 9.25. Two pairs of TSs are transmitted in front of signal to obtain the matrix for channel estimation and they can be expressed as

$$T_1 = \begin{pmatrix} TS_1 \\ 0 \end{pmatrix}, \quad T_2 = \begin{pmatrix} 0 \\ TS_2 \end{pmatrix} \quad (9.31)$$

where TS_1 and TS_2 are training sequences which are made up with binary phase shift keying (BPSK) signals inserted in front of two independent streams. Zero-forcing (ZF) and minimum-mean-square-error (MMSE) can be applied for MIMO processing. In this demonstration, ZF is adopted due to the low algorithm complexity and the system noise can be ignored. The obtained channel matrix can be expressed as

$$H = \begin{pmatrix} H_{11} & H_{12} \\ H_{21} & H_{22} \end{pmatrix} = \begin{pmatrix} Y_{1,1}/TS_1 & Y_{1,2}/TS_2 \\ Y_{2,1}/TS_1 & Y_{2,2}/TS_2 \end{pmatrix} \quad (9.32)$$

where $Y_{1,1}$ and $Y_{2,1}$ represent the received training sequence of the first symbol of RX1 and RX2, while $Y_{1,2}$ and $Y_{2,2}$ represent the received training sequence of the second symbol of RX1 and RX2, respectively. After obtaining the channel matrix H, the transmitted signal would be recovered by using Eqs. (9.33) and (9.34).

$$X_1 = (H_{22}*Y_1 - H_{12}*Y_2)/(H_{22}*H_{11} - H_{12}*H_{21}) \quad (9.33)$$
$$X_2 = (H_{11}*Y_2 - H_{21}*Y_1)/(H_{22}*H_{11} - H_{12}*H_{21}) \quad (9.34)$$

By using this method, the demultiplexing and post-equalization can be simultaneously realized, and no extra phase recovery process of 4-QAM SC-FDE signals is needed.

The channel matrix obtained in Eq. (9.33) is critical to the recovery of the received signals. In order to improve the accuracy of channel estimation in the presence of noise, the time-domain averaging and frequency-domain averaging are needed [17, 18].

In this system, channel transfer function is usually highly correlated with adjacent frequencies. After obtaining channel matrix H at frequency domain in Eq. (9.28), frequency-domain averaging process can be applied. The frequency response at w_k can be smoothed by averaging the estimate for itself and its multiple adjacent frequencies in the ith training sequence pair. Typically, for w_k, the averaging can be performed over w_k and its m left neighbors and/or m right neighbors, or totally up to $(2m+1)$ adjacent frequencies. It should be noted that the window size has to be narrowed when w_k is near two edges of H. The improved channel matrix at frequency w_k after the FDA process can be expressed as

$$H_i(w_k) = \frac{1}{2m+1} \sum_{n=k-m}^{k+m} H_i(w_n) \quad (9.35)$$

As the indoor channel is a time slowly varying channel, the adjacent training symbols can be considered to experience the same channel effect, therefore, TDA can be performed by combining different pairs of TSs. A more accurate estimation of the channel matrix can be obtained through

$$H(w_k) = \frac{1}{N} \sum_{i=1}^{N} H_i(w_k) \quad (9.36)$$

where $H_i(w_k)$ is the channel response at w_k estimated by using the ith pair of TSs calculated in Eq. (9.35). After the two averaging strategies both in time and frequency domain, the channel matrix is much more precise.

Figure 9.27 shows the amplitudes of channel matrix coefficients estimated without and with the frequency-domain equalization process as a

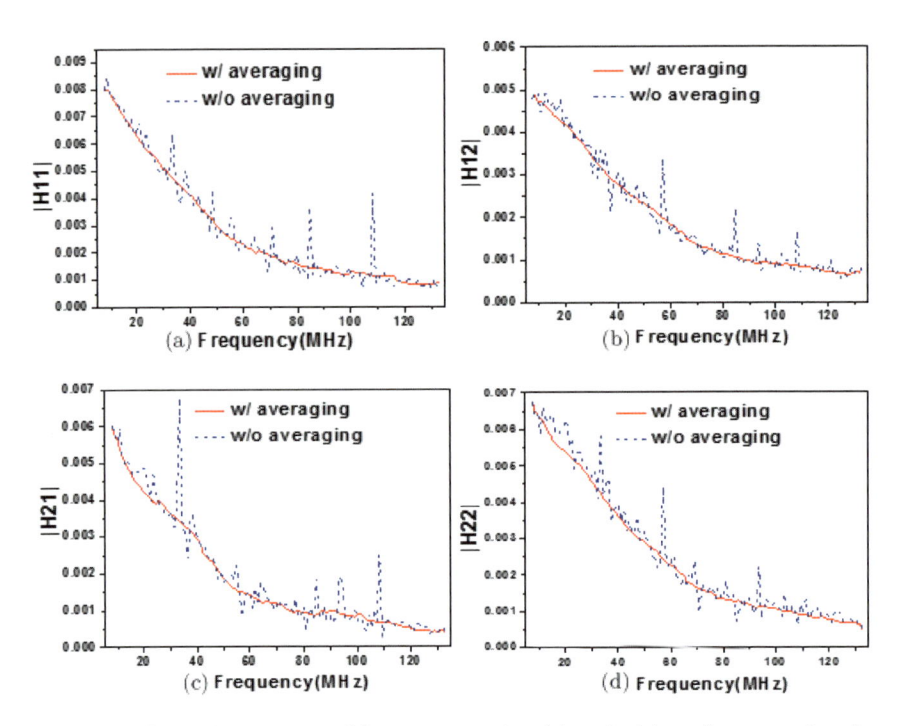

Fig. 9.26. Channel estimation of frequency matrix with and without frequency domain averaging: (a) H11, (b) H12, (c) H21, and (d) H22.

function of the frequency, under 40 cm free-space transmission of blue LED chip. The original elements of the estimated channel matrix are displayed in Fig. 9.26 as blue line. The estimated channel coefficient without the FDA exhibits high-frequency fluctuations due to the presence of the optical noise, and with FDA, the high-frequency fluctuations can be removed shown as red line.

Figure 9.27 shows constellations of the RX1 and RX2 N-SC-FDE signals after 40-cm free-space transmission of blue LED chips at a data rate of 500 Mb/s in different procedures of the offline DSP. The transmitted SC-FDE signals in two different transmitters both appear in RX1 and RX2, which causes the constellation to be distorted before demultiplexing as shown in insets (a) and (b) of Fig. 9.27. The channel matrix obtained through the channel estimation with FDA and TDA is applied to realize the demultiplexing, and after the demultiplexing, the signals can be recovered. The constellations of recovered signals are depicted in insets (c) and (d) of Fig. 9.27.

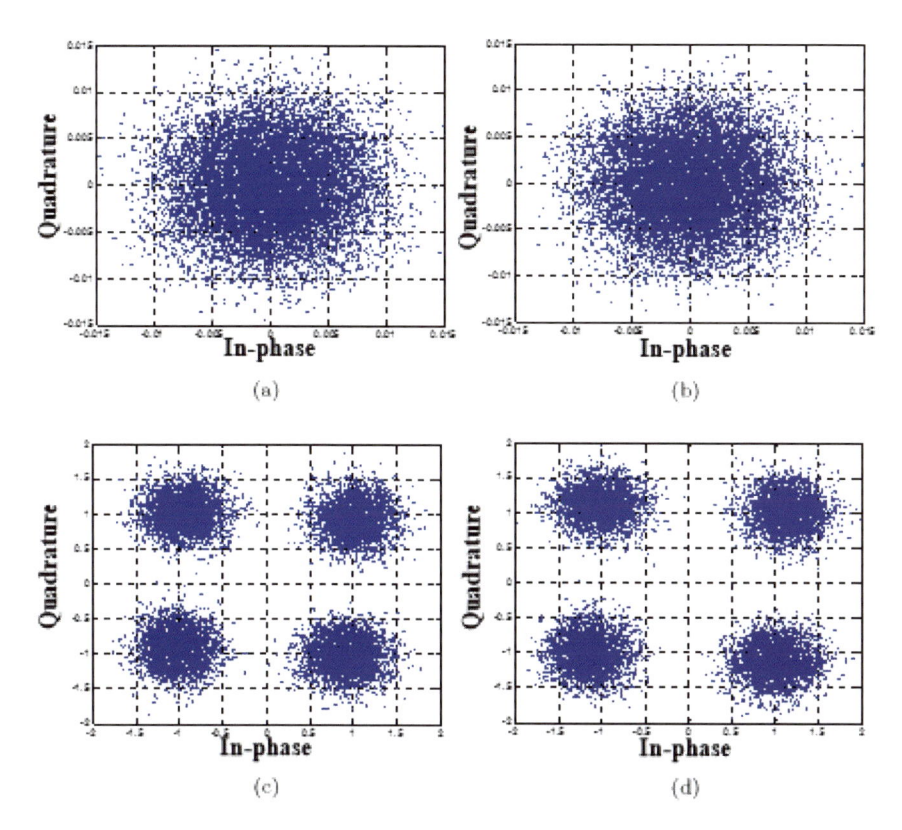

Fig. 9.27. Constellations of the RX1 and RX2 N-SC-FDE signals after 40-cm free-space transmission of blue LEDs at different procedures in the offline DSP: (a) RX1 before demultiplexing and post-equalization, (b) RX2 before demultiplexing and post-equalization, (c) RX1 after demultiplexing and post-equalization, and (d) RX2 after demultiplexing and post-equalization.

References

[1] Y. Wang, Y. Wang, N. Chi, J. Yu and H. Shang, Demonstration of 575-Mb/s downlink and 225-Mb/s uplink bi-directional SCM-WDM visible light communication using RGB LED and phosphor-based LED, *Opt. Express* **21**(1) (2013) 1203–1208.
[2] Y. Wang and N. Chi, Asynchronous multiple access using flexible bandwidth allocation scheme in SCM-based 32/64QAM-OFDM VLC system, *Photonic Network Communications* **27**(2) (2013) 57–64.
[3] N. Chi, Y. Wang, Y. Wang, X. Huang and X. Lu, Ultra-high speed single RGB LED based visible light communication system utilizing the advanced modulation formats, *Chinese Opt. Lett.* **12**(1) 010605.

[4] J. Zhang, J. Yu and N. Chi, Generation and transmission of 512-Gb/s quad-carrier digital super-Nyquist spectral shaped signal, *Opt. Express* **21**(25) (2013) 31212–31217.

[5] Y. Wang, N. Chi, Y. Wang, R. Li, X. Huang, C. Yang and Z. Zhang, High-speed quasi-balanced detection OFDM in visible light communication, *Opt. Express* **21**(23) (2013) 27558–27564.

[6] A. J. Lowery, L. Du and J. Armstrong, Orthogonal frequency division multiplexing for adaptive dispersion compensation in long haul WDM systems, in *Opt. Fiber Commun. Conf. (OFC)*, Anaheim, CA, (2006), PDP39.

[7] I. B. Djordjevic and B. Vasic, Orthogonal frequency division multiplexing for high speed optical transmission, *Opt. Express* **14**(9) (2006) 3767–3775.

[8] W. Peng, X. Wu, V. R. Arbab, B. Shamee, J. Yang, L. C. Christen, K. Feng, A. E. Willner and S. Chi, Experimental demonstration of 340 km SSMF transmission using a virtual single sideband OFDM signal that employs carrier suppressed and iterative detection techniques, in *Opt. Fiber Commun. Conf. (OFC)*, San Diego, (2008), OMU1.

[9] W. Peng, X. Wu, V. R. Arbab, B. Shamee, L. C. Christen, J. Yang, K. Feng, A. E. Willner and S. Chi, Experimental demonstration of a coherently modulated and directly detected optical OFDM system using an RF-tone insertion, in *Opt. Fiber Commun. Conf. (OFC)*, San Diego, (2008), OMU2.

[10] D. Falconer, S. L. Ariyavisitakul, A. Benyamin-Seeyar and B. Eidson, Frequency domain equalization for single-carrier broadband wireless systems, *IEEE Commun. Mag.* **40**(4) (2002) 58–66.

[11] Z. Zheng, R. Ding, F. Zhang and Z. Chen, 1.76Tb/s Nyquist PDM 16QAM signal transmission over 714km SSMF with the modified SCFDE technique, *Opt. Express* **21**(15) (2013) 17505–17511.

[12] Y. Wang, X. Huang, J. Zhang, Y. Wang and N. Chi, Enhanced performance of visible light communication employing 512-QAM N-SC-FDE and DD-LMS, *Opt. Express* **22**(13) (2014) 15328–15334.

[13] L. R. Litwin, Jr., M. D. Zoltowski, T. J. Endres and S. N. Hulyalkar, Blended CMA: smooth, adaptive transfer from CMA to DD-LMS, in *Wireless Communications and Networking Conf. (WCNC)*, New Orleans, LA, (1999), pp. 797–800.

[14] Y. Wang, C. Yang, Y. Wang and N. Chi, Gigabit polarization division multiplexing in visible light communication. *Optics Lett.* **39**(7) (2014) 1823–1826.

[15] Y. Wang and N. Chi, Demonstration of high-speed 2×2 non-imaging MIMO Nyquist single carrier visible light communication with frequency domain equalization, *J. Lightwave Technology* **32**(11) (2014) 2087–2093.

[16] F. Li et al., Fiber-wireless transmission system of PDM-MIMO-OFDM at 100 GHz frequency, *J. Lightwave Technol.* **31**(14) (2013) 2394–2399.

[17] Q. Yang, N. Kaneda, X. Liu and W. Shieh, Demonstration of frequency-domain averaging based channel estimation for 40 Gb/s CO-OFDM with high PMD, *IEEE Photon. Technol. Lett.* **21**(20) (2009) 1544–1546.

[18] X. Liu and F. Buchali, Intra-symbol frequency-domain averaging based channel estimation for coherent optical OFDM, *Opt. Express* **16**(26) (2008) 21944–21957.

Chapter 10

DSP Precoding for Photonic Vector Signal Generation

10.1. Introduction

Photonic micro-wave/millimeter-wave (mm-wave) generation can effectively overcome the bandwidth-insufficiency problem of commercially available electrical components and meanwhile seamlessly integrate fiber-optic and wireless networks. Simple and cost-effective photonic micro-wave/mm-wave generation techniques play a vital role in the potential commercial deployment of radio-over-fiber (RoF) systems, which nowadays are showing a tendency to employ higher-frequency micro-wave/mm-wave carrier and higher-spectral-efficiency vector signal modulation to provide large-capacity mobile data communication [1–10]. Therefore, it is interesting to investigate simple and cost-effective photonic vector micro-wave/mm-wave signal generation techniques.

As one kind of widespread photonic mm-wave generation techniques, external optical modulation combined with photonic frequency multiplication can realize high-stability high-frequency high-purity mm-wave generation with significantly reduced bandwidth requirement for transmitter components (both optical and electrical ones) [11]. Recently, a lot of researches have been carried out on photonic vector signal generation based on external optical modulation, which generally falls into two categories: (1) photonic vector signal generation based on optical carrier suppression (OCS) modulation with a multiplication factor of 2 and without optical filter [12–18], and (2) photonic vector signal generation based on an optical comb with an adaptive multiplication factor (2, 3, 4, 5, 6, ...)

and with optical filter [19–26]. These photonic vector signal generation schemes can employ both constant-amplitude quadrature-amplitude modulation (QAM), such as quadrature-phase-shift-keying (QPSK), and multi-amplitude QAM, such as 8QAM and 16QAM, and they can also employ both optical single-carrier modulation and optical orthogonal-frequency-division-modulation (OFDM) modulation. However, in these schemes, the change of both amplitude and phase information after square-law photodiode (PD) detection requires the employment of digital-signal-processing-based (DSP-based) amplitude and phase precoding at the transmitter.

As a result, in this chapter, we will introduce in detail the precoding principle for the photonic vector signal generation schemes both based on OCS modulation and on an optical comb. First, in the scenario of optical single-carrier modulation, we will introduce the precoding principle for photonic vector signal generation based on OCS modulation and on an optical comb in Sections 10.2 and 10.3, respectively. Then, we will introduce in Section 10.4 the precoding principle in the counterpart scenario of optical OFDM modulation. Section 10.5 provides the summary.

10.2. Precoding for Photonic Vector Signal Generation Based on Optical Carrier Suppression Modulation Employing a Single MZM

Photonic vector signal generation based on OCS modulation can realize frequency doubling with the aid of a single Mach–Zehnder modulator (MZM) biased at its OCS point. Precoding is needed at the transmitter to offset the change of both amplitude and phase information incurred by frequency doubling in order to ensure that the generated vector signal after square-law PD detection displays regular vector signal modulation [12–18].

10.2.1. *Principle for photonic vector signal generation based on OCS modulation employing a single MZM*

Figure 10.1(a) shows the principle of our proposed photonic multi-amplitude QAM vector signal generation, using OCS modulation enabled by a single MZM [12]. As shown in Fig. 10.1(a), the continuous wave (CW) output, at frequency f_c, from a laser, is modulated by a radio-frequency

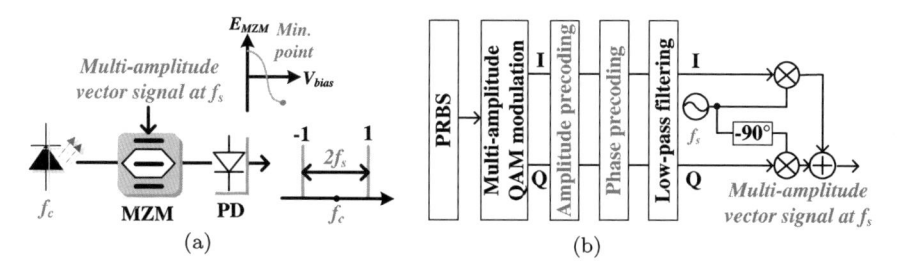

Fig. 10.1. (a) Principle of photonic vector signal generation based on OCS modulation employing a single MZM. (b) Precoded RF signal generation for the driver of the MZM. MZM: Mach–Zehnder modulator, PD: photodiode.

(RF) carrier at frequency f_s, which carries a multi-amplitude QAM data and drives the MZM. Assume that the CW output at frequency f_c and the driving RF signal at frequency f_s can be respectively expressed as

$$E_{CW}(t) = K_1 \exp(j2\pi f_c t) \tag{10.1}$$

$$E_{RF}(t) = K_2(t) \sin[2\pi f_s t + \varphi(t)] \tag{10.2}$$

where K_1 is constant and denotes the amplitude of the CW output. K_2 and φ denote the amplitude and phase of the driving RF signal, respectively. K_2 is invariant with time when the transmitter data adopt constant-amplitude vector modulation, such as QPSK, and has several different values when the transmitter data adopt multi-amplitude vector modulation, such as 8QAM and 16QAM. Here, no channel distortions are considered for simplification.

Thus, when the MZM is biased at its minimum transmission point to realize OCS modulation, its output is expressed as

$$E_{\text{MZM}}(t) \approx 2jK_1\{J_{-1}(\kappa) \exp[j2\pi(f_c - f_s)t - j\varphi(t)]$$
$$+ J_{+1}(\kappa) \exp[j2\pi(f_c + f_s)t + j\varphi(t)]\} \tag{10.3}$$

where J_n is the Bessel function of the first kind and order n. κ is equal to $\pi V_{\text{drive}} K_2(t)/V_\pi$, while V_{drive} and V_π denote driving voltage and half-wave voltage of the MZM, respectively. We can see from Eq. (10.3) that the MZM generates two first-order subcarriers spaced by $2f_s$ as shown by the inset of Fig. 10.1(a). When the two generated first-order subcarriers are heterodyne mixed in a PD, the leading term of the generated RF current after the PD is given by

$$i_{\text{RF}}(t) = \frac{1}{2} R J_1^2(\kappa) \cos[2\pi \cdot 2f_s t + 2\varphi(t)] \tag{10.4}$$

where R denotes the PD sensitivity. We can see from Eq. (10.4) that the frequency $2f_s$ of the obtained RF signal is double of the driving RF signal (f_s).

Therefore, we can realize photonic frequency doubling of the driving RF signal based on our proposed scheme as shown in Fig. 10.1(a), which, in the meantime, can reduce the bandwidth requirement for photonic and electronic components at the transmitter end. However, it is realized that in our proposed scheme, after square-law PD conversion, frequency doubling also simultaneously leads to phase doubling. Moreover, the amplitude information of the driving RF signal is carried by the term of the square of $J_1(\kappa)$, which depends on the ratio of V_{drive} to V_π. In order to directly attain the amplitude information and phase information of the multi-amplitude QAM transmitter data after PD conversion, the amplitude K_2 and phase φ of the driving RF signal should satisfy

$$K_{\text{data}} = J_1^2(\pi K_2 V_{\text{drive}}/V_\pi)$$
$$\varphi_{\text{data}} = 2\varphi - 2m\pi, \quad m = 0, 1 \qquad (10.5)$$

where K_{data} and φ_{data} denote the amplitude and phase of the original transmitter data, respectively. Therefore, the amplitude and phase of the driving RF signal needs to be precoded at the transmitter end. For a known multi-amplitude QAM transmitter data, the values of K_2 and φ obtained by resolving Eq. (10.5) are just the precoded amplitude and phase which can be assigned to the driving RF signal. Figure 10.1(b) shows the generation procedure of driving precoded RF signal at frequency f_s carrying multi-amplitude QAM data, which can be implemented by MATLAB programming. Here, the pseudorandom binary sequence (PRBS) is first multi-amplitude QAM modulated, then amplitude- and phase-precoded, and finally upconverted into RF band by simultaneous cosine and sine functions. Note that when the transmitter data adopt constant-amplitude vector modulation, such as QPSK, only phase precoding is needed. However, when the transmitter data adopt multi-amplitude vector modulation, such as 8QAM and 16QAM, both phase and amplitude precoding are required.

10.2.2. *Principle for imbalanced precoding*

Due to the periodicity of the cosine function in Eq. (10.4), the factor m in Eq. (10.5) can be selected as 0 or 1. When the factor m is fixed at 0, the precoded constellations are situated only in the first and second quadrants, which we define as imbalanced phase precoding [12].

Figure 10.2 gives the calculated original constellations as well as the calculated constellations after only imbalanced phase precoding, after only amplitude precoding, and after both amplitude and imbalanced phase precoding for the 1-Gbaud 6-GHz QPSK/8QAM/16QAM vector signals, respectively. The 1-Gbaud 6-GHz QPSK/8QAM/16QAM vector signals are generated by MATLAB programming. According to Eq. (10.5), the phase of the 1-Gbaud 6-GHz QPSK/8QAM/16QAM vector signals is 1/2 of that of the regular QPSK/8QAM/16QAM symbols, while their amplitude depends on the amplitude of the regular QPSK/8QAM/16QAM symbols and the ratio of V_{drive} to V_π. The ratio of V_{drive} to V_π, which should be equal to the practical ratio of V_{drive} to V_π of the adopted MZM in the experiment, is set at 3 for the MATLAB-based amplitude precoding in Fig. 10.2. It is worth noting that, for both 8QAM and 16QAM cases, the order of amplitude precoding and phase precoding can be exchanged. Figures 10.3(a) and 10.3(b) show the transmitter spectra before and after 6-GHz digital upconversion for the QPSK cases with 64-GSa/s sampling rate, and both the 8QAM and 16QAM cases have quite similar transmitter spectra, which are not shown here.

Correspondingly, Fig. 10.4 gives the signal spectra and constellations for the generated 1-Gbaud 12-GHz QPSK/8QAM/16QAM vector signals after 20-km single-mode fiber-28 (SMF-28) transmission and PD detection, respectively [12]. The bit-error rates (BERs) are all zero for these three cases. For each case, the constellations from left to right in Fig. 10.4 correspond to those before clock extraction, after clock extraction, after constant modulus algorithm (CMA) equalization for QPSK modulation or cascaded multi-modulus algorithm (CMMA) equalization for 8QAM/16QAM modulation, after frequency offset estimation (FOE), and after carrier phase estimation (CPE), respectively. It is worth noting that, in the experiment, for the 8QAM/16QAM cases, the practical ratio of V_{drive} to V_π of the MZM may have a certain deviation from that set in the MATLAB-based amplitude precoding and thus affects the detected amplitude information after the PD, which, however, can be compensated by the CMMA equalization.

10.2.3. *Principle for balanced precoding*

The imbalanced-precoded transmitter constellations shown in Fig. 10.2 have unequal Euclidean distances, which will lead to an asymmetrical distribution of the constellation points of the received constellations and thus

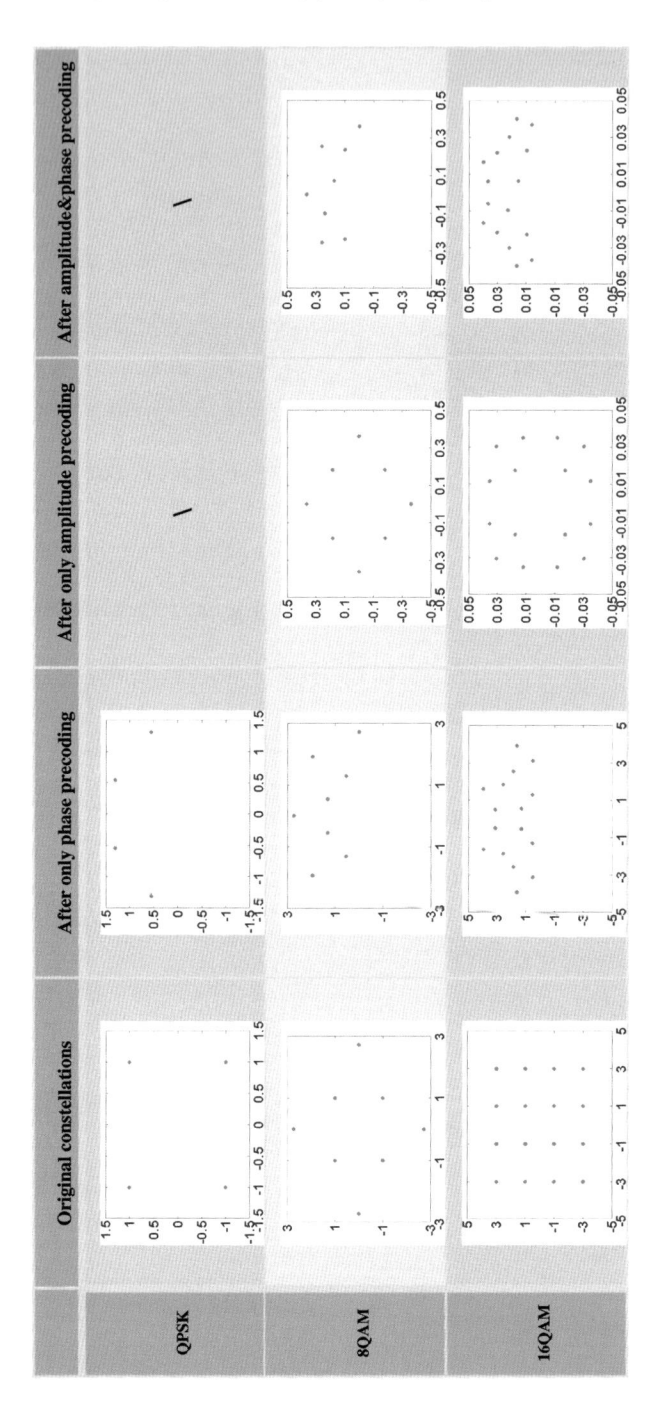

234   *Digital Signal Processing for High-Speed Optical Communication*

Fig. 10.2. Original and imbalanced-precoded constellations for QPSK/8QAM/16QAM vector signals at the transmitter.

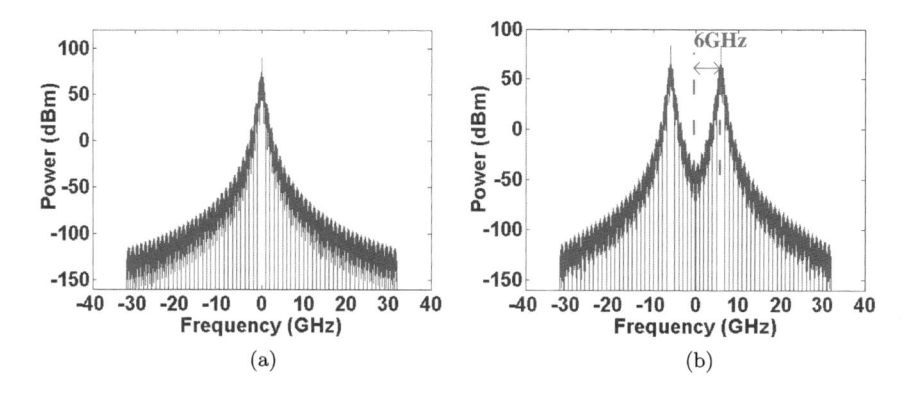

Fig. 10.3. Transmitter spectra (a) before and (b) after upconversion.

the degradation of the overall system performance [12]. In order to solve this problem, balanced precoding [17] is proposed. To implement balanced precoding, the factor m in Eq. (10.5) is randomly selected as 0 or 1, which means the balanced-precoded phase can be half of the desired phase ($\varphi_{data}/2$) or plus a constant value $\pi(\varphi_{data}/2+\pi)$. For example, if the original phase is $\pi/4$, the balanced-precoded phase can be $\pi/8$ or $9\pi/8$. The balanced-precoded constellations are located at four quadrants with a balanced distribution, and the Euclidean distances between adjacent points are equal. Balanced precoding is suitable for both constant-amplitude QAM modulation, such as QPSK, and multi-amplitude QAM modulation, such as 8QAM and 16QAM. Figure 10.5 gives the calculated constellations of original, imbalanced-precoded, and balanced-precoded vector signals in the scenario of QPSK and 8QAM modulation, respectively.

Figure 10.6 gives the calculated transmitter spectra of an 8-Gbaud 8-GHz vector driving signal adopting QPSK modulation with imbalanced precoding and balanced precoding, respectively. We can see that an 8-GHz RF component is displayed in Fig. 10.6(a) due to the imbalanced constellation distribution after imbalanced precoding.

Figure 10.7 shows the recovered constellations for the generated 8-Gbaud 16-GHz QPSK vector signals with imbalanced precoding and balanced precoding after employing digital signal processing at the receiver, respectively [17]. Figures 10.7(a) and 10.7(c) depict the recovered constellations with imbalanced precoding in the cases of BTB and after 25-km SMF-28 transmission, respectively. Figures 10.7(b) and 10.7(d) show the recovered constellations with balanced precoding in the cases of BTB and after

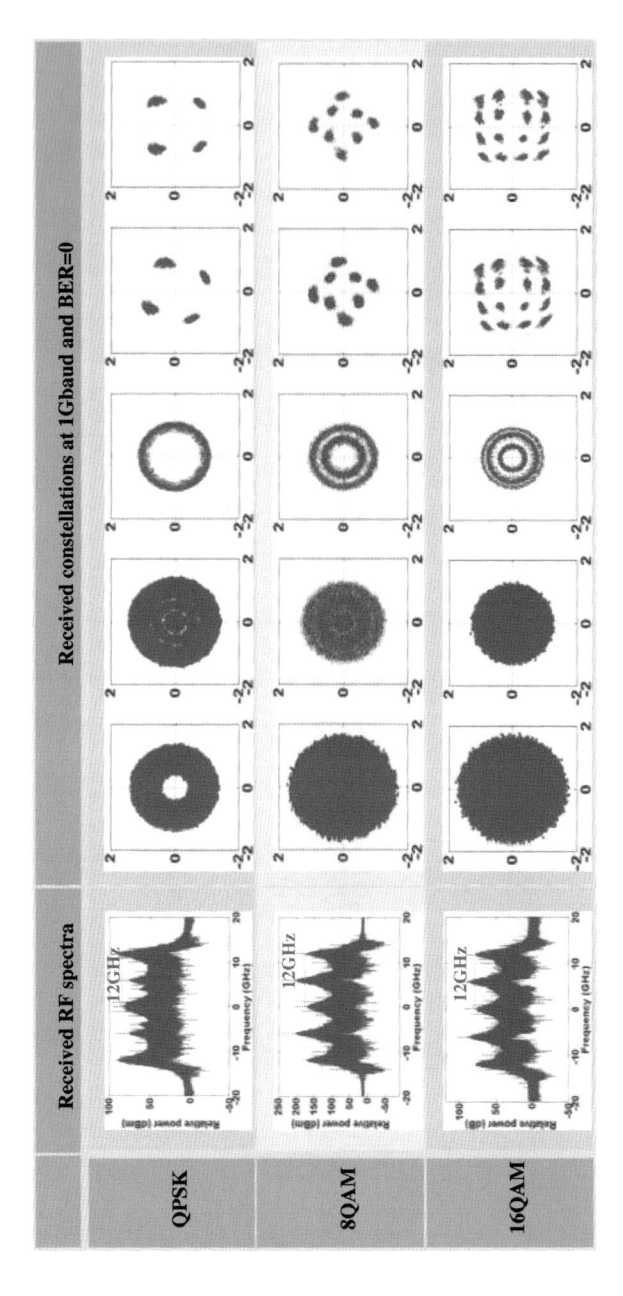

Fig. 10.4. Received signal spectra and constellations for QPSK/8QAM/16QAM vector signals at the receiver.

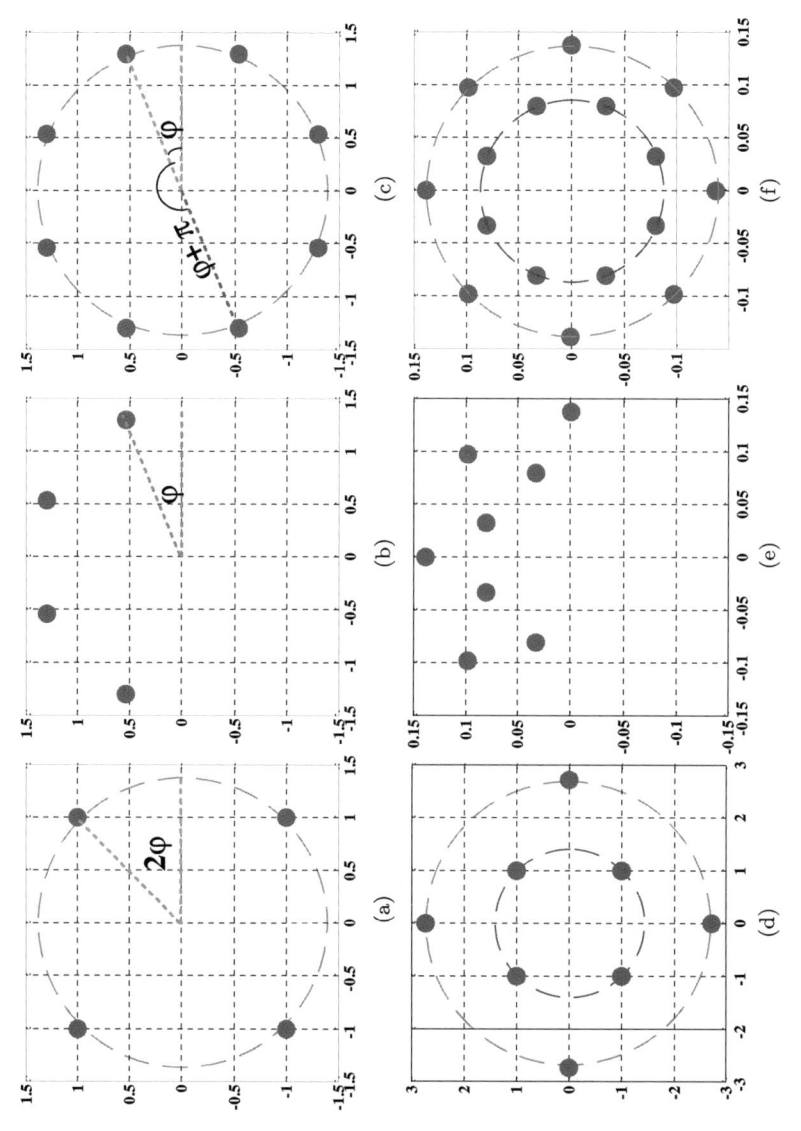

Fig. 10.5. Constellations of (a) original QPSK, (b) imbalanced-precoded QPSK, (c) balanced-precoded QPSK, (d) original 8QAM, (e) imbalanced-precoded 8QAM, and (f) balanced-precoded 8QAM vector signals.

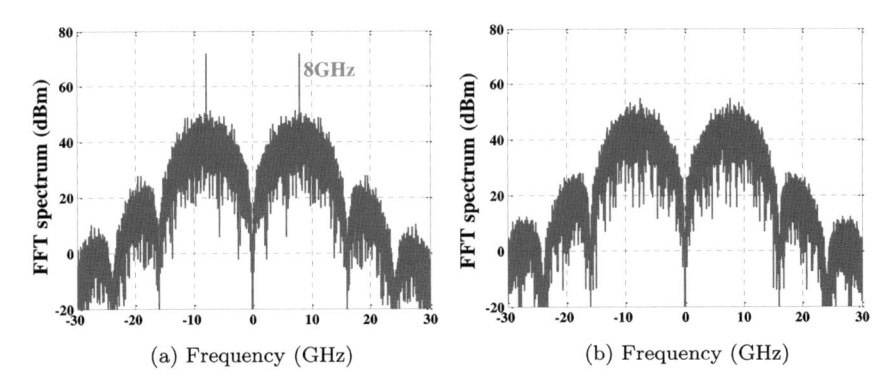

Fig. 10.6. Calculated transmitter spectra for an 8-Gbaud 8-GHz QPSK vector driving signal with (a) imbalanced precoding and (b) balanced precoding.

25-km SMF-28 transmission, respectively. We can see that in Fig. 10.7(a), the distance between points A and B is larger than the other adjacent points, while in Fig. 10.7(c), the distance between points C and D is larger than the other adjacent points. It can be obviously found that the constellations in Figs. 10.7(b) and 10.7(d) are much more symmetrical than those in Figs. 10.7(a) and 10.7(c). The symmetry of the constellations shows in-phase (I) and quadrature (Q) components have quite similar performance, and this conclusion can also be validated from the BER values of I and Q signals. The BERs of I and Q signals in Fig. 10.7(c) are 0 and 6.2×10^{-3}, respectively, while those in Fig. 10.7(d) are 6.5×10^{-4} and 5.8×10^{-4}, respectively. In the scenario of frequency-doubling photonic QPSK vector signal generation, balanced precoding can significantly improve the receiver sensitivity by 2 dB compared to imbalanced precoding [17].

10.3. Precoding for Photonic Vector Signal Generation Based on an Optical Comb

As mentioned in Section 10.2, frequency-doubling photonic vector signal generation can be realized based on OCS modulation employing a single MZM. However, because of the very limited multiplication factor of only 2 and the limited bandwidth of commercially available digital-to-analog converters (DACs) (typically less than 40 GHz), it is difficult to generate vector signal with a carrier frequency over 40 GHz simply based on the OCS scheme [13, 14]. In order to solve this problem, photonic vector signal generation based on an optical comb [19–26] is proposed with an adaptive

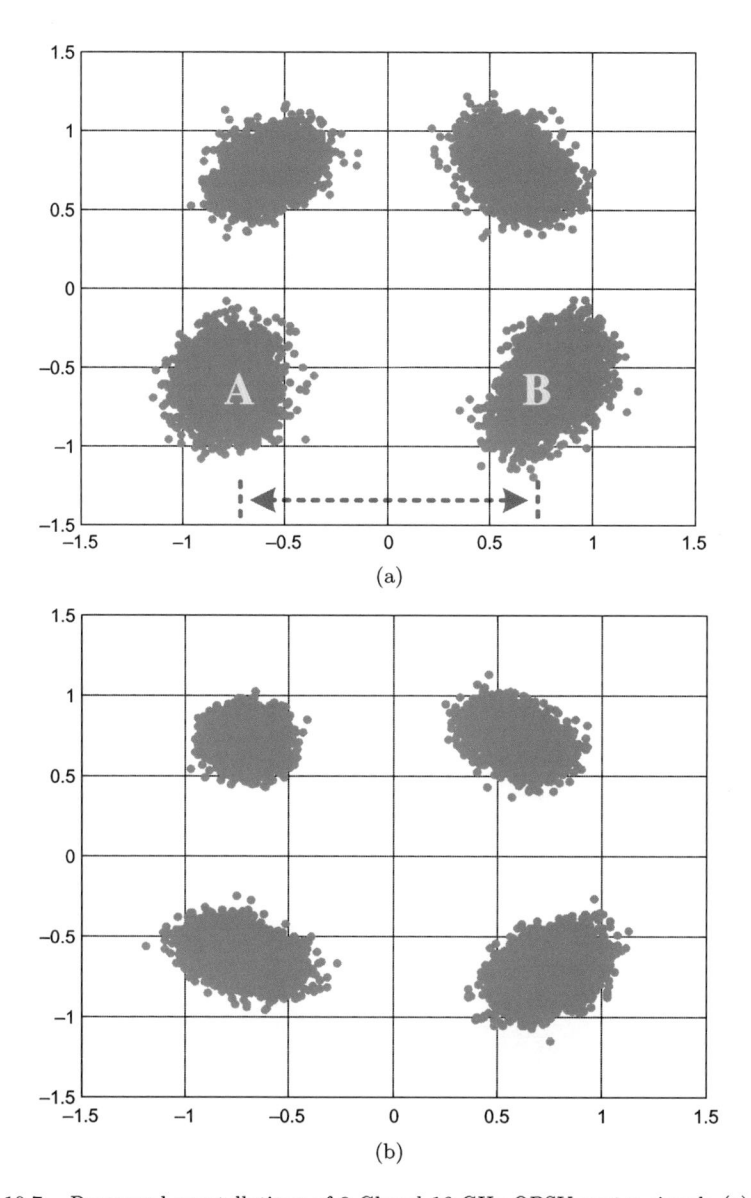

Fig. 10.7. Recovered constellations of 8-Gbaud 16-GHz QPSK vector signals (a) with imbalanced precoding in the case of BTB, (b) with balanced precoding in the case of BTB, (c) with imbalanced precoding in the case of after 25-km SMF-28 transmission, and (d) with balanced precoding in the case of after 25-km SMF-28 transmission.

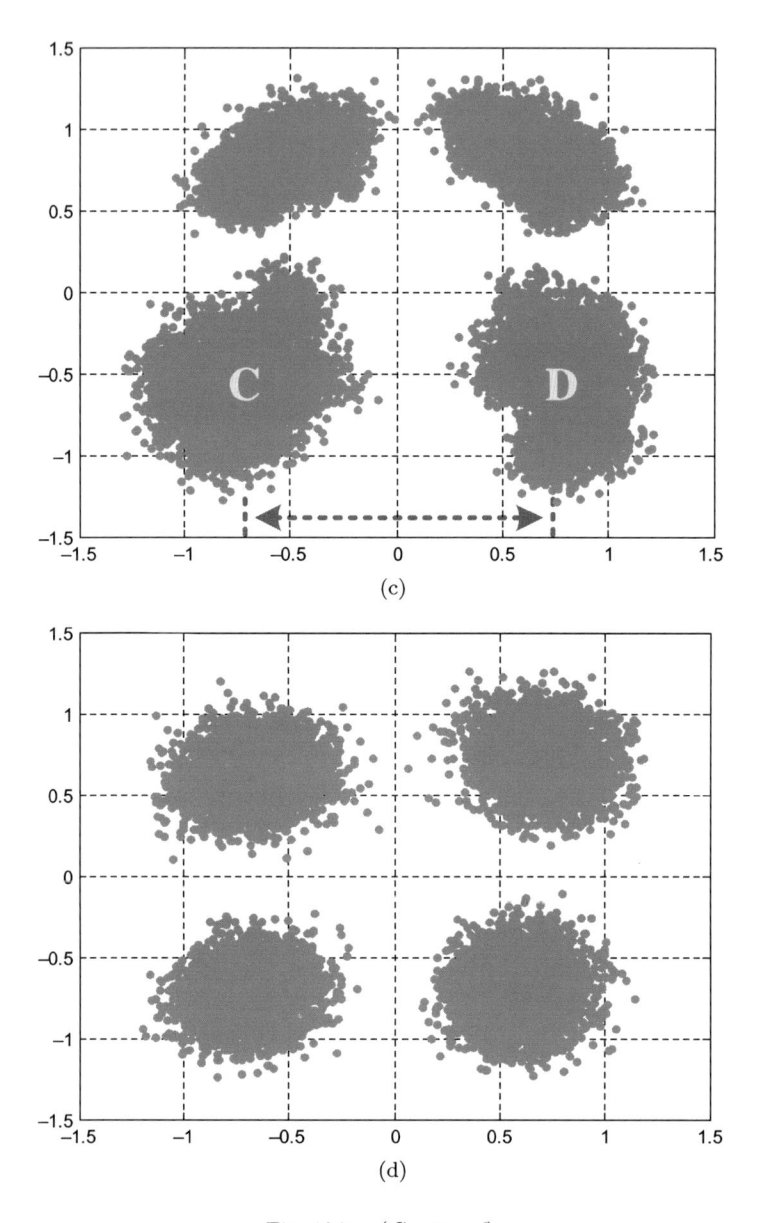

Fig. 10.7. (*Continued*)

multiplication factor over 2. Here, the optical comb can be generated by a CW laser cascaded with a single MZM [19–22], a CW laser cascaded with a single phase modulator [23–25], a directly modulated laser (DML) [26], or an electro-absorption modulated laser (EML). In the following, we will take the case of a CW laser cascaded with a single MZM as an example to explain the principle of precoding for photonic vector signal generation based on an optical comb.

10.3.1. *Principle for photonic vector signal generation based on an optical comb employing a CW laser and a single MZM*

Figure 10.8 shows the schematic diagram of our proposed photonic multi-amplitude QAM vector signal generation based on an optical comb

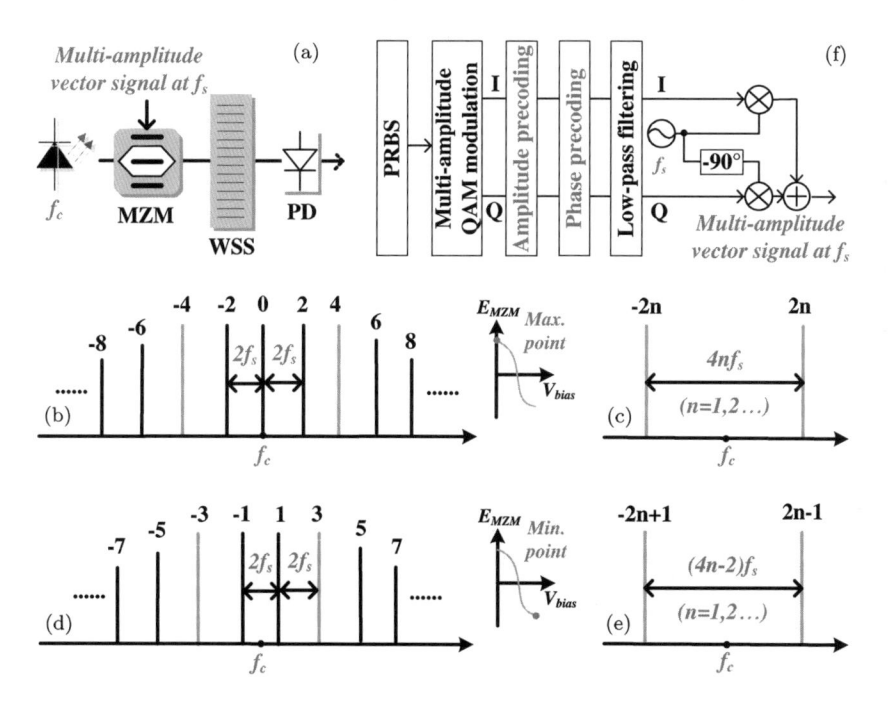

Fig. 10.8. (a) Principle of photonic vector signal generation based on an optical comb employing a CW laser and a single MZM; (b, c) output optical spectra of MZM and WSS when MZM biased at the maximum transmission point; (d, e) output optical spectra of MZM and WSS when MZM biased at the minimum transmission point; and (f) precoded RF signal generation for the driver of the MZM. MZM: Mach–Zehnder modulator, WSS: wavelength selective switch, PD: photodiode.

employing a CW laser and a single MZM [22]. Identical to Fig. 10.1(a), the CW output, at frequency f_c, from a laser, and an RF driving carrier, at frequency f_s, carrying a vector-modulated multi-amplitude QAM data, can be expressed by Eqs. (10.1) and (10.2), respectively.

Thus, when the MZM is biased at its maximum transmission point, its output can be expressed as

$$\begin{aligned} E_{\text{MZM}}(t) &= K_1 \exp(j2\pi f_c t) \exp\{j\kappa \sin[2\pi f_s t + \varphi(t)]\} \\ &\quad + K_1 \exp(j2\pi f_c t) \exp\{-j\kappa \sin[2\pi f_s t + \varphi(t)]\} \\ &= K_1 \sum_{n=-\infty}^{\infty} J_n(\kappa) \exp[j2\pi(f_c + nf_s)t + jn\varphi(t)] \\ &\quad + K_1 \sum_{n=-\infty}^{\infty} J_n(-\kappa) \exp[j2\pi(f_c + nf_s)t + jn\varphi(t)] \\ &= 2K_1 \sum_{n=-\infty}^{\infty} J_{2n}(\kappa) \exp[j2\pi(f_c + 2nf_s)t + j2n\varphi(t)] \quad (10.6) \end{aligned}$$

We can see from Eq. (10.6) that only even-order optical subcarriers spaced by $2f_s$ are generated by the MZM biased at its maximum transmission point, as shown in Fig. 10.8(b). A wavelength selective switch (WSS) is used to select two optical subcarriers with the same order $2n$ and a frequency spacing $4nf_s$ ($n = 1, 2, \ldots$), as shown in Fig. 10.8(c). The WSS output can be expressed as

$$\begin{aligned} E_{WSS}(t) &= 2K_1 J_{2n}(\kappa)\{\exp[j2\pi(f_c + 2nf_s)t + j2n\varphi(t)] \\ &\quad + \exp[j2\pi(f_c - 2nf_s)t - j2n\varphi(t)]\} \quad (n = 1, 2, \ldots) \quad (10.7) \end{aligned}$$

obtained directly from Eq. (10.6). Upon heterodyne mixing in a PD, the leading term of the generated RF current is given by

$$i_{\text{RF}}(t) = \frac{1}{2} R J_{2n}^2(\kappa) \cos[2\pi \cdot 4nf_s t + 4n\varphi(t)] \quad (n = 1, 2, \ldots) \quad (10.8)$$

We can see from Eq. (10.8) that the frequency $4nf_s$ of the obtained RF signal is $4n$ times that of the driving RF signal (f_s).

Similarly, the generated optical vector signal from the MZM, biased at its minimum transmission point, is given by

$$E_{\text{MZM}'}(t) = K_1 \exp(j2\pi f_c t) \exp\left\{j\kappa \sin[2\pi f_s t + \varphi(t)] + j\frac{\pi}{2}\right\}$$

$$+ K_1 \exp(j2\pi f_c t) \exp\left\{-j\kappa \sin[2\pi f_s t + \varphi(t)] - j\frac{\pi}{2}\right\}$$

$$= jK_1 \sum_{n=-\infty}^{\infty} J_n(\kappa) \exp[j2\pi(f_c + nf_s)t + jn\varphi(t)]$$

$$- K_1 \sum_{n=-\infty}^{\infty} J_n(-\kappa) \exp[j2\pi(f_c + nf_s)t + jn\varphi(t)]$$

$$= 2jK_1 \sum_{n=-\infty}^{\infty} J_{2n-1}(\kappa) \exp\{j2\pi[f_c + (2n-1)f_s]t$$

$$+ j(2n-1)\varphi(t)\} \tag{10.9}$$

We can see from Eq. (10.9) that only odd-order optical subcarriers spaced by $2f_s$ are generated by the MZM biased at its minimum transmission point, as shown in Fig. 10.8(d). Thus, the WSS selects two optical subcarriers with the same order $2n-1$ and a frequency spacing $(4n-2)f_s$ ($n = 1, 2, \ldots$), as shown in Fig. 10.8(e). The WSS output is given by

$$E_{\text{WSS}'}(t) = 2jK_1 J_{2n-1}(\kappa)\{\exp[j2\pi(f_c + (2n-1)f_s)t + j(2n-1)\varphi(t)]$$

$$- \exp[j2\pi(f_c - (2n-1)f_s)t - j(2n-1)\varphi(t)]\}, \quad (n = 1, 2, \ldots) \tag{10.10}$$

After square-law PD conversion, we can obtain an electrical RF signal expressed as

$$i_{\text{RF}'}(t) = \frac{1}{2} R J_{2n-1}^2(\kappa) \cos[2\pi \cdot (4n-2)f_s t + (4n-2)\varphi(t)], \quad (n = 1, 2, \ldots) \tag{10.11}$$

in analogy with Eq. (10.8). The frequency $(4n-2)f_s$ of the obtained RF signal is $(4n-2)$ times that of the driving RF signal (f_s).

Therefore, based on our proposed scheme, we can realize adaptive photonic frequency multiplication of the driving RF signal, and thus a lower-frequency microwave signal can be upconverted into a higher-frequency mm-wave signal by employing lower-bandwidth photonic and electronic components at the transmitter end. However, it is realized that in our proposed scheme, after square-law PD conversion, frequency multiplication also simultaneously leads to phase multiplication with the same multiplicative factor by reference to the frequency and phase of the driving RF signal. Moreover, the amplitude information of the driving RF signal is carried by the term of the square of $J_n(\kappa)$, which depends on the order n of the selected optical subcarriers as well as the ratio of V_{drive} to V_π. In order to directly attain the amplitude information and phase information of the multi-amplitude QAM transmitter data after PD conversion, the amplitude K_2 and phase φ of the driving RF signal should satisfy

$$K_{\text{data}} = J_n^2(\pi K_2 V_{\text{drive}}/V_\pi)$$
$$\varphi_{\text{data}} = 2n\varphi \quad (n = 1, 2, 3, 4, \ldots) \tag{10.12}$$

where n is the order of the selected optical subcarriers. Similar to the aforementioned OCS case, for a known multi-amplitude QAM transmitter data, the obtained values of K_2 and φ by resolving Eq. (10.12) are just the precoded amplitude and phase which can be assigned to the driving RF signal. Also, similar to the aforementioned OCS case, when the transmitter data adopt constant-amplitude vector modulation, such as QPSK, only phase precoding is needed. However, when the transmitter data adopt multi-amplitude vector modulation, such as 8QAM, both phase and amplitude precoding are required. Figure 10.8(f) shows the corresponding generation procedure of driving precoded RF signal at frequency f_s carrying vector-modulated multi-amplitude QAM data, which can be implemented by MATLAB programming.

10.3.2. Precoding for photonic constant-amplitude QPSK vector signal generation based on an optical comb

Assume that the precoded vector driving signal for photonic constant-amplitude QPSK vector signal generation based on an optical comb carries 2-Gbaud transmitter data and has a 12-GHz center frequency. Figure 10.9 gives the calculated precoded phase for photonic frequency doubling (×2), quadrupling (×4), sextupling (×6) and octupling (×8) of the 12-GHz precoded vector driving signal, respectively. We can see that the precoded phase for photonic frequency doubling (×2), quadrupling (×4), sextupling (×6) and octupling (×8) is 1/2, 1/4, 1/6 and 1/8 of the phase of the regular QPSK signal, respectively. For the case of frequency octupling (×8), Figures 10.10(a) and 10.10(b) show the transmitter constellations before

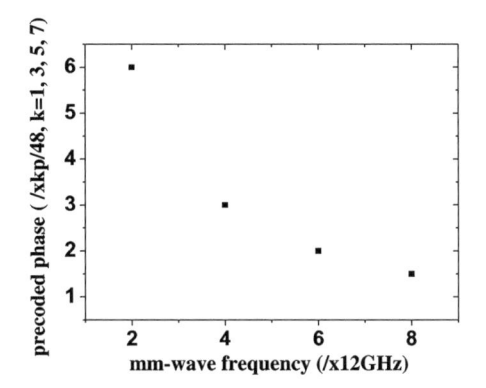

Fig. 10.9. Calculated precoded phase for photonic frequency doubling (×2), quadrupling (×4), sextupling (×6) and octupling (×8) of the 12-GHz precoded vector driving signal.

and after phase-precoding, while Figs. 10.10(c) and 10.10(d) show the transmitter spectra before and after upconversion.

Here, balanced precoding discussed in Section 10.2.3 can be expanded to be applied to frequency-quadrupling photonic constant-amplitude QPSK vector signal generation based on an optical comb [27], which can be expressed by

$$\varphi_{\text{data}} = 4\varphi_{\text{balanced}} - m\pi/2 \quad (m = 0, 1, 2, 3) \tag{10.13}$$

where m is randomly assigned as 0, 1, 2, or 3. This means the frequency-quadrupling balanced-precoded phase is 1/4 of that of the regular QPSK symbol, or plus $\pi/2, \pi$, or $3\pi/2$. Figure 10.11 gives the original QPSK constellation and the constellation after frequency-quadrupling balanced precoding, respectively.

However, different from frequency-doubling balanced precoding [17], which can significantly improve the receiver sensitivity by 2dB compared to frequency-doubling imbalanced precoding, frequency-quadrupling balanced precoding can only slightly improve the receiver sensitivity by ∼0.2 dB compared to frequency-quadrupling imbalanced precoding. We think this is mainly because of the constellation points further doubled from 8 to 16. Due to the imperfect characteristics of the adopted transmitter components in the experiment, the four counterpart constellation points after frequency-quadrupling balanced precoding, which correspond to a certain regular QPSK constellation point, may have a different phase rotation. The superimposed phase rotation of the four counterpart constellation points after

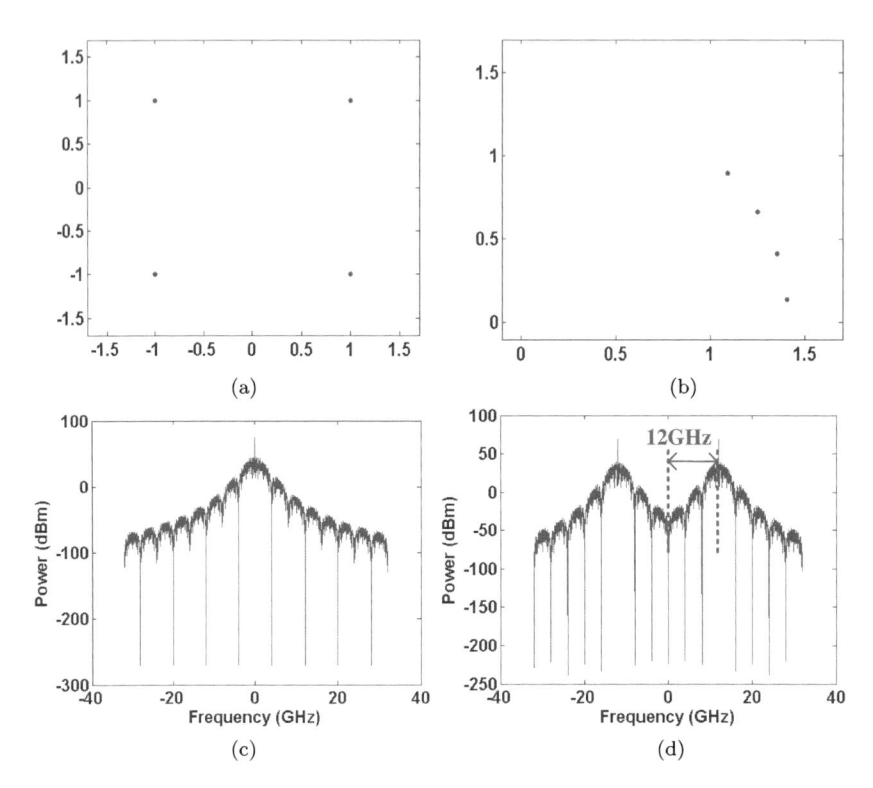

Fig. 10.10. For the case of frequency octupling (×8), (a) and (b) show the transmitter constellations before and after phase-precoding, while (c) and (d) show the transmitter spectra before and after upconversion.

PD detection may affect the effectiveness of receiver-based DSP algorithms and thus degrade the system performance.

10.3.3. *Precoding for photonic multi-amplitude QAM vector signal generation based on an optical comb*

Assume that the precoded vector driving signal for photonic multi-amplitude QAM vector signal generation based on an optical comb carries 2-Gbaud 8QAM transmitter data and has a 12-GHz center frequency. For the case of frequency octupling (×8), Fig. 10.12(a) shows the transmitter 8QAM constellation, while Figs. 10.12(b)–(d) show the constellations after only amplitude precoding, after only phase precoding, and after both amplitude and phase precoding. Figures 10.12(e) and 10.12(f) show the transmitter spectra before and after upconversion with 64-GSa/s

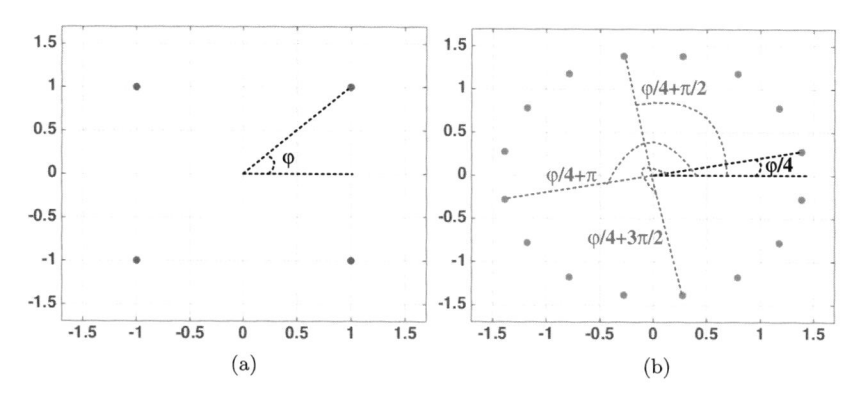

Fig. 10.11. (a) Original QPSK constellation and (b) the constellation after frequency-quadrupling balanced precoding.

sampling rate. Here, the ratio of V_{drive} to V_π is set at 3 for the transmitter MATLAB-based amplitude precoding.

Here, it is worth noting that balanced precoding discussed in Section 10.2.3 cannot be expanded to be applied to photonic multi-amplitude QAM vector signal generation based on an optical comb with a multiplication factor over 2, since balanced phase precoding becomes ineffectiveness because of more constellation points of multi-amplitude QAM symbols and a higher multiplication factor over 2 [18].

10.4. Precoding for Photonic Vector Signal Generation Employing OFDM Modulation

The aforementioned photonic vector signal generation schemes based on OCS modulation (in Section 10.2) and an optical comb (in Section 10.3) both employ optical single-carrier modulation. As we know, optical OFDM modulation has some advantages relative to optical signal-carrier modulation [28]. For example, optical OFDM modulation brings the benefits of robustness to multipath fading and high-frequency efficiency, as well as elimination of inter-symbol interference (ISI), inter-carrier interference (ICI), chromatic dispersion (CD) and polarization mode dispersion (PMD) in optical transmission. As a result, optical OFDM modulation is introduced to photonic vector signal generation schemes both based on OCS modulation [29] and an optical comb [30], in which precoding is also needed at the transmitter.

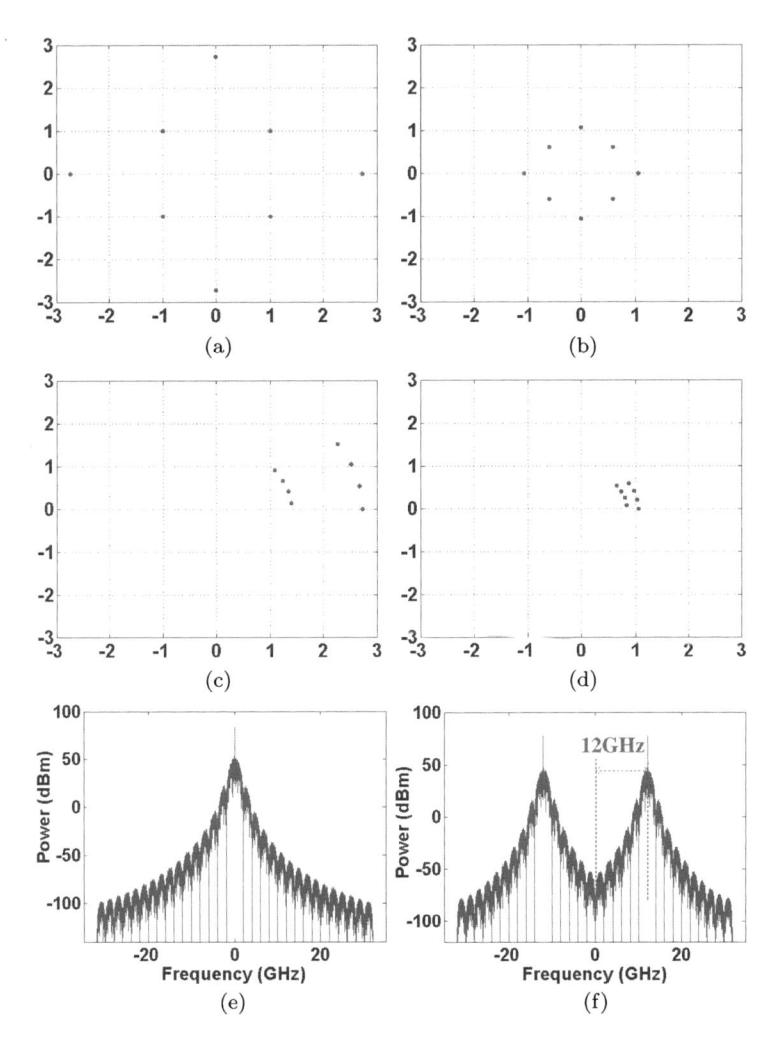

Fig. 10.12. Transmitter 8QAM constellations, (a) before precoding, (b) after only amplitude precoding, (c) after only phase precoding, (d) after both amplitude and phase precoding. Transmitter spectra (e) before and (f) after upconversion.

10.4.1. *Precoding for photonic OFDM vector signal generation based on OCS modulation*

For photonic OFDM vector signal generation based on OCS modulation employing a single MZM, the precoded vector driving signal carries OFDM data and its generation procedure is given by Fig. 10.13 [29]. The input

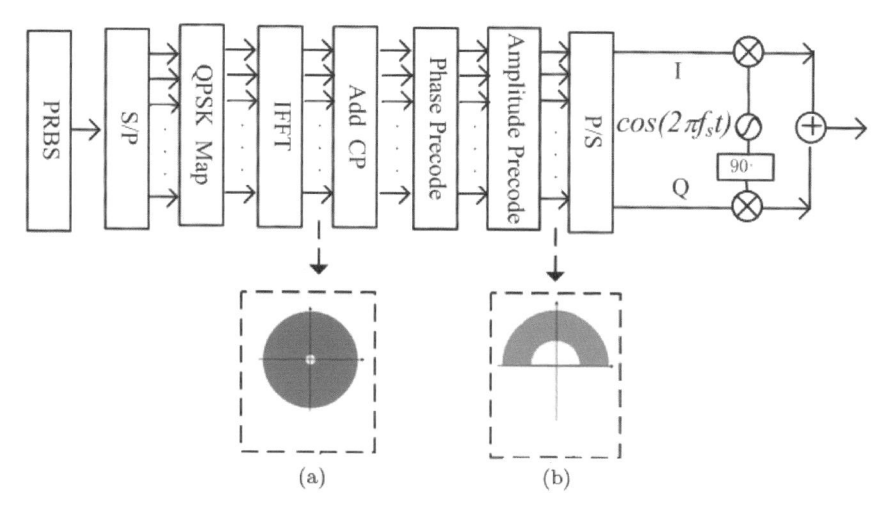

Fig. 10.13. Precoded RF signal generation for the driver of the MZM: (a) constellation of regular QPSK OFDM signal after IFFT and (b) constellation of precoded QPSK OFDM signal.

PRBS is first converted into many parallel data pipes (S/P), and then the parallel data is QPSK mapped. The digital time-domain signal is obtained by using inverse faster Fourier transform (IFFT), and then guard interval is inserted to prevent ISI due to channel dispersion. The baseband QPSK OFDM signal is precoded firstly based on phase information and then amplitude information, and then the I and Q branches of the QPSK OFDM precoded signal are upconverted into intermediate-frequency (IF) signals at f_s by mixing with two sinusoidal RF signals with quadrature phase at f_s, respectively. The summation of the two IF signals is the desired electrical OFDM vector driving signal.

The baseband electrical OFDM signal can be described as

$$S_{BB_{OFDM}}(t) = \frac{1}{\sqrt{N}} \sum_{n=0}^{N} d_n \exp\left(j2\pi \frac{n}{T} t\right) \quad (10.14)$$

where t is the discrete time index, N is the number of the subcarrier, and T is the symbol duration. d_n is the data symbol modulated on the nth subcarrier and can be expressed as

$$d_n = a_n \exp(\varphi_n) \quad (10.15)$$

where a_n and φ_n are the amplitude and phase of data symbol modulated on the nth subcarrier, respectively. The baseband electrical OFDM signal is subcarrier-modulated with RF carrier at f_s. The modulated electrical OFDM vector signal at f_s can be expressed as

$$S_{\text{OFDM\_drive}}(t) = S_{\text{BB\_OFDM}}(t) * \exp(j2\pi f_s t)$$
$$= \frac{1}{\sqrt{N}} \sum_{n=0}^{N} a_n \exp\left(j2\pi \left(\frac{n}{T} + f_s\right) t + \varphi_n\right) \quad (10.16)$$

Thus, the output optical signal of the MZM biased at its OCS point can be written as

$$E_{\text{MZM}}(t) \approx -2K_1 J_1(\kappa) \sum_{n=0}^{N} \cos\left(j2\pi \left(\frac{n}{T} + f_s\right) t + j\varphi_n\right) \quad (10.17)$$

where κ is equal to $a_n \pi/(\text{sqrt}(N) V_\pi)$. The output current of the PD can be expressed as

$$i_{\text{RF}}(t) = \frac{1}{2} R J_1^2(\kappa) \sum_{n=0}^{N} \cos\left(j2 \times 2\pi \left(\frac{n}{T} + f_s\right) t + j2\varphi_n\right) \quad (10.18)$$

We can see from Eq. (10.18) that the phase and amplitude of OFDM signal on each subcarrier after the PD is 2 and $J_1^2(a_n A)$ times that of the electrical OFDM vector signal on corresponding subcarrier for the driver of the MZM, respectively. In order to ensure that the output signal of the PD is the original OFDM signal, the amplitude and phase of the driving RF OFDM signal on each subcarrier need to be precoded. The precoded phase on each subcarrier after IFFT should be 1/2 of that of the original OFDM signal on corresponding subcarrier after IFFT, and the precoded amplitude on each subcarrier after IFFT should be $1/J_1^2(a_n A)$ of that of the original OFDM signal on corresponding subcarrier after IFFT, which ensures that the OFDM signal after the PD can be restored to the original OFDM signal. The schematic constellations of the original OFDM signal after IFFT and the precoded OFDM signal are shown in Figs. 10.13(a) and 10.13(b), respectively.

It is worth noting that the precoding scheme for frequency-doubling photonic OFDM vector signal generation adopting multi-amplitude QAM modulation is identical to that for the aforementioned frequency-doubling photonic OFDM vector signal generation adopting constant-amplitude QPSK modulation, and it can be expanded to be applied to photonic OFDM vector signal generation based on an optical comb with a multiplication factor over 2 [30]. The balanced precoding mentioned in Section 10.2.3 can also be expanded to be applied to this frequency-doubling photonic OFDM vector signal generation.

10.5. Summary

In this chapter, we introduce in detail the precoding principle for the photonic vector signal generation schemes both based on OCS modulation and on an optical comb. These photonic vector signal generation schemes can employ both constant-amplitude QAM modulation, such as QPSK, and multi-amplitude QAM modulation, such as 8QAM and 16QAM, and they can also employ both optical single-carrier modulation and optical OFDM modulation. We introduce the precoding principle in these different scenarios.

References

[1] X. Li, J. Xiao and J. Yu, Long-distance wireless mm-wave signal delivery at W-band, *J. Lightwave Technol.* **34**(2) (2016) 661–668.
[2] X. Li, J. Yu and J. Xiao, Demonstration of ultra-capacity wireless signal delivery at W-band, *J. Lightwave Technol.* **34**(1) (2016) 180–187.
[3] T. P. Mckenna, J. A. Nanzer and T. R. Clark, Experimental demonstration of photonic millimeter-wave system for high capacity point-to-point wireless communication, *J. Lightwave Technol.* **32**(20) (2014) 3588–3594.
[4] K. Kitayama, A. Maruta and Y. Yoshida, Digital coherent technology for optical fiber and radio-over-fiber transmission systems, *J. Lightwave Technol.* **32**(20) (2014) 3411–3420.
[5] J. Yu, X. Li, J. Zhang and J. Xiao, 432-Gb/s PDM-16QAM signal wireless delivery at W-band using optical and antenna polarization multiplexing, in *Proc. ECOC 2014*, Cannes, France (2014), We.3.6.6.
[6] A. Kanno, K. Inagaki, I. Morohashi, T. Sakamoto, T. Kuri, I. Hosako, T. Kawanishi, Y. Yoshida and K. I. Kitayama, 40 Gb/s W-band (75-110 GHz) 16-QAM radio-over-fiber signal generation and its wireless transmission, in *Proc. ECOC 2011*, Geneva, Switzerland (2011), We.10.P1.112.
[7] A. Kanno, T. Kuri, I. Hosako, T. Kawanishi, Y. Yasumura, Y. Yoshida and K. Kitayama, Optical and radio seamless MIMO transmission with 20-Gbaud QPSK, in *Proc. ECOC 2012*, Amsterdam, The Netherlands (2012), We.3.B.2.

[8] D. Zibar, R. Sambaraju, A. Caballero, J. Herrera, U. Westergren, A. Walber, J. B. Jensen, J. Marti and I. T. Monroy, High-capacity wireless signal generation and demodulation in 75- to 110-GHz band employing all optical OFDM, *IEEE Photon. Technol. Lett.* **23**(12) (2011) 810–812.

[9] C. W. Chow, F. M. Kuo, J. W. Shi, C. H. Yeh, Y. F. Wu, C. H. Wang, Y. T. Li and C. L. Pan, 100 GHz ultra-wideband (UWB) fiber-to-the-antenna (FTTA) system for in-building and in-home networks, *Opt. Express* **18**(2) (2010) 473–478.

[10] Y. Yang, C. Lim and A. Nirmalathas, Investigation on transport schemes for efficient high-frequency broadband OFDM transmission in fibre-wireless links, *J. Lightwave Technol.* **32**(2) (2014) 267–274.

[11] J. Yu, Z. Jia, L. Yi, Y. Su, G. K. Chang and T. Wang, Optical millimeter-wave generation or up-conversion using external modulators, *IEEE Photon. Technol. Lett.* **18**(1) (2006) 265–267.

[12] X. Li, J. Yu, J. Zhang, J. Xiao, Z. Zhang, Y. Xu and L. Chen, QAM vector signal generation by optical carrier suppression and precoding techniques, *IEEE Photon. Technol. Lett.* **27**(18) (2015) 1977–1980.

[13] X. Li, J. Xiao, Y. Xu and J. Yu, QPSK vector signal generation based on photonic heterodyne beating and optical carrier suppression, *IEEE Photon. J.* **7**(5) (2015) 7102606.

[14] X. Li, J. Xiao, Y. Wang, Y. Xu, L. Chen and J. Yu, W-band QPSK vector signal generation based on photonic heterodyne beating and optical carrier suppression, in *OFC 2016*, Anaheim, CA (2016) Th2A.15.

[15] X. Li, J. Yu, J. Xiao, N. Chi, Y. Xu and L. Chen, PDM-QPSK vector signal generation by MZM-based optical carrier suppression and direct detection, *Opt. Commun.* **355** (2015) 538–542.

[16] L. Chen, J. Yu and X. Li, PDM-16QAM vector signal generation and detection based on intensity modulation and direct detection, *Opt. Commun.* **371** (2016) 15–18.

[17] Y. Wang, Y. Xu, X. Li, J. Yu and N. Chi, Balanced precoding technique for vector signal generation based on OCS, *IEEE Photon. Technol. Lett.* **27**(13) (2015) 2469–2472.

[18] C. Qin, X. Li, N. Chi and J. Yu, Comparison between balanced and unbalanced precoding technique in high-order QAM vector mm-wave signal generation based on intensity modulator with photonic frequency doubling, *Opt. Express* **24**(5) (2016) 4399–4404.

[19] X. Li, J. Yu, Z. Zhang, J. Xiao and G. K. Chang, Photonic vector signal generation at W-band employing an optical frequency octupling scheme enabled by a single MZM, *Opt. Commun.* **349** (2015) 6–10.

[20] X. Li, J. Yu, J. Xiao, N. Chi and Y. Xu, W-band PDM-QPSK vector signal generation by MZM-based photonic frequency octupling and precoding, *IEEE Photon. J.* **7**(4) (2015) 7101906.

[21] X. Li, J. Yu, J. Xiao, F. Li, Y. Xu, N. Chi and G. K. Chang, Mm-wave vector signal generation and transport for W-band MIMO system with intensity modulation and direct detection, in *OFC 2016*, Anaheim, CA (2016), M3B.2.

[22] X. Li, J. Zhang, J. Xiao, Z. Zhang, Y. Xu and J. Yu, W-band 8QAM vector signal generation by MZM-based photonic frequency octupling, *IEEE Photon. Technol. Lett.* **27**(12) (2015) 1257–1260.

[23] J. Xiao, Z. Zhang, X. Li, Y. Xu, L. Chen and J. Yu, High-frequency photonic vector signal generation employing a single phase modulator, *IEEE Photon. J.* **7**(2) (2015) 7101206.

[24] L. Zhao, J. Yu, L. Chen, P. Min, J. Li and R. Wang, 16QAM vector millimeter-wave signal generation based on phase modulator with photonic frequency doubling and precoding, *IEEE Photon. J.* **8**(2) (2016) 5500708.

[25] X. Li, Y. Xu, J. Xiao and J. Yu, W-band mm-wave vector signal generation based on precoding-assisted random photonic frequency tripling scheme enabled by phase modulator, *IEEE Photon. J.* **8**(2) (2016) 5500410.

[26] X. Li, J. Xiao, Y. Xu, L. Chen and J. Yu, Frequency-doubling photonic vector millimeter-wave signal generation from one DML, *IEEE Photon. J.* **7**(6) (2015) 5501207.

[27] X. Li, J. Yu and G. K. Chang, Frequency-quadrupling vector mm-wave signal generation by only one single-drive MZM, *IEEE Photon. Technol. Lett.* **28**(12) (2016) 1302–1305.

[28] W. Peng, X. Wu, V. Arbab, K. Feng, B. Shamee, L. Christen, J. Yang, A. Willner and S. Chi, Theoretical and experimental investigations of direct-detected RF-tone-assisted optical OFDM systems, *J. Lightwave Technol.* **27**(10) (2009) 1332–1339.

[29] J. Xiao, Z. Zhang, X. Li, Y. Xu, L. Chen and J. Yu, OFDM vector signal generation based on optical carrier suppression, *IEEE Photon. Technol. Lett.* **27**(23) (2015) 2449–2452.

[30] J. Xiao, X. Li, Y. Xu, Z. Zhang, L. Chen and J. Yu, W-band OFDM photonic vector signal generation employing a single Mach–Zehnder modulator and precoding, *Opt. Express* **23**(18) (2015) 24029–24034.

Index

A

above analysis, 11
access network, 60, 80, 90
access optical communication, 72
ACO-OFDM, 215, 217
adaptive 9-tap FIR filter, 39
adaptive equalizations, 70
adaptive equalizer filter, 15, 212
ADCs, 184
adding weights, 65
additive white Gaussian noise, 9
advanced DSP, 25, 27
advanced modulation formats, 1
aggressive spectral filtering, 14
algorithms, 4, 221
amplitude modulation, 72
amplitude response, 112
analog TSs and conventional OFDM, 103
analog-to-digital conversion, 82
analog-to-digital converters, 26
ASE noise, 38, 64
autocorrelation, 178
AWGN-to-AWGN noise, 214

B

back-to-back, 65, 167
balanced detection, 212
balanced precoding, 233, 235, 237–239, 245, 247

bandwidth, 120, 122
bandwidth limitation, 51, 110
baseband QPSK OFDM signal, 249
baud rate, 201
BER counting, 149
BER curve, 59
BER performance, 71, 94, 119
bit, 184
bit error ratio (BER), 36, 129, 156, 171, 190–191, 216
blind equalization, 134–135, 141, 143, 145, 152, 160
blind polarization, 32
butterfly equalizers, 6

C

C-band CW, 125
calculations, 188
cancellation, 89
CAP filters, 62
CAP signal, 58
CAP-64QAM, 60
CAP-64QAM signal generation, 59
CAP-QAM signal, 56
carrier phase drift, 38
carrier phase estimation, 30
carrier recovery, 31
carrierless amplitude, 202

carrierless amplitude phase (CAP) modulation, 85
CCDF calculation, 102
CCDF curve, 102
CCDF probability, 117
CD compensation, 172
CD-induced ICI, 96
CD-induced serious ICI, 95
channel crosstalk, 36
channel equalization, 10
channel estimation, 181, 184
channel estimation accuracy, 112
channel information estimation, 200
channel matrix, 222, 224
channel matrix coefficients, 223
channel multiplexing, 26
channel response, 9, 92
channel spacing, 129
channel transfer function, 223
chromatic dispersion, 3, 90, 187
chromatic dispersion compensation, 156
Clock recovery, 4
CMA + post-filter algorithms, 41
CMEQ-based DSP with post-filter, 29
CMMA, 57
CMMA blind equalization, 159
CMMA equalization method, 144, 154–155
CO-OFDM, 114, 193
coherent detection, 1, 69, 71, 79, 107, 181
coherent optical communication, 1
coherent receiver, 30
commercial DD-MZM, 174
complementary cumulative distribution function, 97, 116
complex conjugates, 33
complex symbols, 55
computational complexity, 181
constant modulus algorithm, 6, 15
constant or multi-modulus algorithms (CMA, MMA), 53
constant step-size SSFM, 18
constant-amplitude, 235

constellation, 38, 66, 94, 137, 143, 150–151, 157–158, 207, 225, 233, 236–239, 245, 247, 250
constellation diagrams, 190
constellation points, 30
constellation recovery process, 8
conventional constant modulus algorithm, 28
conventional OFDM, 98, 115, 122–123
conventional OFDM and DFT-spread OFDM, 102
conventional OFDM signal, 117, 121, 124
conventional SSB, 167, 171–173
CP removal, 140
crosstalk elimination, 221
cyclic prefix, 81, 170

D

DAC and ADC, 188
DD-LMS, 58, 218
DD-LMS equalizer, 219
DD-LMS post-equalizer, 130
DD-MZM, 166
DD-MZM modulator, 164
DDO-16QAM-DMT, 183
DDO-OFDM, 84, 87, 215
DDO-OFDM link, 94
DDO-OFDM signal, 97
decision-directed least mean square, 6, 15, 217
decision-directed least radius distance (DD-LRD) algorithm, 37, 41
delay-and-add filter, 209
demultiplexing, 32, 225
detection methods, 80
detection OFDM system, 77
DFT-spread, 97, 107, 112, 130
DFT-spread OFDM (DFT-S OFDM), 85, 98, 100, 103, 108, 163, 173, 110, 115–117, 119–121, 124
DFT-spread technique, 96, 109
different matched-filters, 64
differential decoding, 209
digital and analog, 100

digital backward propagation (DBP), 18
digital coherent receiver with DSP, 2
digital coherent system, 2
digital equalizations and compensations, 51
digital FIR filter, 56
digital pre-equalization, 9, 21
digital QDB filtering, 44
digital QPSK TS and analog TS, 103
digital signal processing (DSP), 1, 2, 17, 21, 32, 40, 51, 53, 77, 133, 149, 158, 163, 175, 188, 197
digital-to-analog converter, 61, 142, 178
direct detection, 77, 79
direct detection (IM/DD)-based FSO system, 197
directly modulated laser, 53, 78, 198, 241
distortions, 221
distributed feedback (DFB)-based DML, 91
distributions, 208
DMT modulation, 184
DMT transmission, 176
downstream, 61
DP 16QAM-CO-OFDM, 190
DSP algorithms, 246
DSP precoding, 229
DSP processing, 189
dual optical carrier, 182
dual-polarization, 134, 136
dual-polarization traditional 256-subcarrier OFDM signal, 143
dual-subcarrier 16QAM OFDM, 154, 159
dual-subcarrier QPSK-OFDM, 153

E

each pairs of filters, 63
EDC, 192
electrical and optical impairments, 51
electrical dispersion compensation, 192
electrical domain, 150, 213
electrical pre-filtering, 27
electrical spectra, 138, 210
electronic pre-equalization, 8
equalization schemes, 58
equalize ISI impairment, 30
equalized signals, 200
equalizer, 11, 33
error floor, 95
ETDM, 66
Euclidean distance, 45
experimental setup, 114, 125, 139, 147–148, 155, 169–170, 182–183, 185
external cavity laser, 68
external modulators, 84
extracting the training sequence, 82

F

4-point FFT, 150
4-subcarrier, 145
9-QAM signal, 137, 141
fast Fourier transform, 9, 31, 82, 205
fast inverse Fourier transform, 81
faster than Nyquist modulation formats, 208
faster-than-Nyquist-WDM signals, 29
few subcarriers CO-OFDM system, 160
FFT, 92, 170
FFT and IFFT-based digital pre-equalizations, 12
FFT size, 94
FFT spectra analysis, 7
fiber distance, 192
fiber loss, 115
fiber-wireless transmission, 54
filter or the WSS, 28
filter pair, 55
finite-impulse-response, 16
FIR filters, 6, 56
first-order subcarriers, 231
FOE, 150
forward-error-correction, 83
fourth-order Gaussian filter, 38

fourth-order Gaussian type, 37
FPGA implantation, 180
FPGAs, 184, 192, 193
FPGAs modules, 188
Free-space optical communication, 197
free-space optical communication system, 206
free-space transmission, 215
frequency division multiplexing, 220
frequency domain, 181, 199
frequency domain and upsampling, 203
frequency domain channel estimation, 175
frequency domain equalization, 127, 202
frequency equalization, 219
frequency multiplication, 243–244
frequency octupling, 246
frequency offset, 35, 118, 180–181
frequency offset estimation, 31
frequency response, 11, 16, 32, 166, 201, 203
frequency subcarriers, 126
frequency symmetrization, 16
frequency synchronization, 180
frequency-domain, 4, 17
frequency-domain averaging process, 223
frequency-domain equalization, 135, 205, 223
frequency-domain pre-equalization, 63
frequency-doubling, 245
frequency-doubling photonic OFDM vector signal generation, 250
frequency-doubling photonic QPSK vector signal generation, 238
frequency-offset, 72
frequency-offset estimation, 6
FSO system, 211
FTN, 208
FTN signals, 209
FWM, 19

G

Gardner-timing recovery, 5
Gaussian distribution, 70
Gaussian noise, 199
Godard scheme, 5
Gram–Schmidt Orthogonalization Process, 3

H

half cycle, 89
half-cycle 16QAM, 85
half-cycle QAM, 202
half-cycled DDO-OFDM, 87
half-cycled OFDM, 86
Hermitian conjugate symmetry, 176
Hermitian symmetric, 98
heterodyne, 242
high output power, 84
high PAPR, 133
high SE, 53
high spectral efficiency, 20
high-end modulation format, 83
high-frequency attenuation, 82
high-frequency fluctuation, 103, 224
high-frequency power attenuation, 111
high-level QAM, 93
high-order modulation formats, 107
high-order QAM-OFDM Signals, 90
high-speed and long-haul optical coherent transmission system, 21
high-speed DAC, 8, 62
high-speed SERDES interfaces, 189
higher modulation formats, 17
higher spectrum efficiency, 36, 54
Hilbert pair, 55
Hilbert transform, 167
hybrid time and frequency equalization, 217

I

I/Q modulator, 113, 186
IFFT, 136–137, 145, 178, 250
IFFT size, 86

IFFT/FFT, 138
imbalance, 3
imbalanced phase precoding, 233
imbalanced precoding, 232, 239, 245
imbalanced precoding and balanced precoding, 235
include, 32
individual symbol ISI, 93
indoor channel, 223
inherent inter-symbol memory, 39
intensity modulation and direct detection, 60
inter-channel crosstalk, 26
inter-channel crosstalk enhancement, 8
inter-channel nonlinear impairments, 20
inter-symbol interference, 26, 81
interference and nonlinearity penalty, 164
interleaved OFDM, 88
inverse fast Fourier transform, 186, 249
IQ modulator, 13, 118
ISFA, 92–93, 189
ISFA technique, 184
ISI, 40, 249
ISI and crosstalk, 57

J

joint-polarization QPSK partitioning algorithm, 33–34
JTAG interface, 189

L

Lagrange interpolation, 178
laser phase noise, 14
least-mean-square (LMS) algorithm, 53
LED, 207, 225
LED chips, 224
LED light, 213
light-emitting diode, 198
linear equalization algorithm, 28
linear interpolation, 39

linear term and nonlinear terms, 168
LMS equalization, 59
local oscillator, 69
low pass filter, 152
low power consumption, 84
low-capital expenditures, 84
lower-bandwidth photonic and electronic components, 243

M

Mach–Zehnder modulator, 67, 68, 113, 139, 186
middle ring symbols, 34
millimeter-wave, 229
MIMO, 172
MIMO OFDM Signal, 163
MIMO training run, 222
MIMO-Volterra algorithm, 174
MIMO-Volterra equalization algorithm, 168
MIMO-Volterra equalizer, 169
minimum, 45
minimum mean square error, 10
minimum MSE, 11–12
minimum symbol spacing, 58
minimum transmission point, 243
MMEQ, 32, 41
modified carrier recovery scheme, 32
modified logarithmic step-size distribution NLC, 21
modulation bandwidth, 205
modulation format, 25, 54, 77, 100, 201
modulation index, 206
modulation modes, 78
modulator, 78
moduli, 32, 33
MSE, 40
MSE-criterion equalizers, 10
multi-amplitude QAM data, 231–232
multi-amplitude QAM modulation, 235

multi-amplitude vector modulation, 244
multi-band CAP, 60–62
multi-carrier generation, 124, 127
multi-filter pairs, 62
multi-modulus algorithm, 134–135, 144
multiplexed signal, 68
multiplication factor, 251

N

N-point FFT, 177
narrowband optical filtering, 27
narrowband interference, 117
NLC, 18
non-constant logarithmic, 19
non-return-to-zero (NRZ), 52
nonlinear compensation, 17
nonlinear effect, 142, 212
nonlinear impairments, 19
nonlinear noise tolerance, 108
nonlinear propagation impairments, 14
nonlinearity compensation, 211
nonlinearity equalization, 167–168
nonlinearity impairment compensation, 21
nonlinearity noise, 167
normalized symbols, 35
Nyquist and faster than Nyquist shaping, 201
Nyquist bandwidth, 11
Nyquist filtered, 29, 44
Nyquist limit, 28
Nyquist SC-FDE signals, 210
Nyquist spectral shaping, 28
Nyquist WDM multi-channel, 20
Nyquist modulation formats, 202

O

OCS modulation, 230–231, 248
OFDM, 81, 107, 138, 142, 176, 207, 216
OFDM band, 213
OFDM coherent detection, 133

OFDM data, 248
OFDM modulation, 93–94, 136, 186, 247
OFDM Nyquist modulation, 204
OFDM optical communication system, 80
OFDM signal, 82, 85, 91, 96, 101, 170, 214, 250
OFDM symbol, 88–89, 126, 136–137, 187
OFDM transmission, 173
OFDM vector signal, 248
offline DSP, 92
offline processing, 64
OOK, 61
optical 16QAM, 13
optical carriers generated, 125
optical coherent communication system, 15
optical comb, 238, 241, 246, 251
optical comb carries, 244
optical communication system, 51, 77, 199
optical domain, 166
optical eye diagram, 141
optical homodyne coherent detection, 38
optical interfaces, 52
optical narrowband filtering, 27
optical OFDM transmission, 175
optical orthogonal-frequency-division-modulation, 230
optical power, 191
optical pre-equalization, 69, 128, 130
optical signal-to-noise ratio, 28, 149
optical spectra, 155, 171
optical transmission, 163
optical vector signal, 243
optical-to-electrical, 138
optimal filter, 10
optimal filter length, 59
optimal ISFA tap number, 93
original and imbalanced-precoded constellations, 234

original constellations, 233
original OFDM, 110
orthogonal filter pair, 52, 62
orthogonal frequency division multiplexing, 107
OSNR, 128, 191
OSNR penalty, 120, 122, 129

P

pairs, 222
PAM or OFDM, 54
PAM-4 for high-speed Short-Haul transmission, 66
PAM-4 signal, 68, 69, 73
PAPR, 97, 116–117, 142, 146–147, 206
PAPR reduction, 96
partial response signaling, 29
PDM dual-subcarrier, 158
PDM dual-subcarrier 16QAM OFDM, 156
PDM dual-subcarrier 16QAM-OFDM signal, 159
PDM dual-subcarrier coherent 16QAM-OFDM, 153
PDM-PAM-4 signal, 68, 70
peak-to-average power ratio, 108, 211
peer-to-peer data center, 80
phase angle estimation, 34
phase modulation, 126, 202
phase noise, 72, 112
phase offset, 179
phase recovery, 152
phase rotation, 57
photocurrents, 214
photodiode, 198
photonic constant-amplitude, 244
photonic frequency doubling, 232
photonic mm-wave generation techniques, 229
photonic multi-amplitude QAM vector signal generation, 241
photonic vector signal generation schemes, 229–231
pilot tone, 140
polarization crosstalk, 111
polarization demultiplexing, 5, 16, 20, 182
polarization division multiplexed, 25
polarization maintaining optical coupler, 113
polarization multiplexed coherent system, 33
polarization multiplexer, 68, 115, 126, 139, 187
polarization–division–multiplexing, 177
post-equalization, 12, 128, 225
post-filter, 29, 39
power fading effect, 66
power spectrum, 44
pre-compensated, 119
pre-convergence, 16
pre-emphasized process, 14
pre-equalization, 12, 14, 17, 65, 70–71, 108–109, 115–116, 121, 123, 128, 130, 198, 200
pre-equalization coefficients, 201
pre-equalization OFDM, 123
pre-equalization technique, 109, 112, 199
pre-equalized OFDM, 122
pre-equalized OFDM signal, 124
precoding, 230
precoding principle, 251
probability, 116
probability density function, 101
probability distribution, 97, 100
procedure, 92
progress of DSP, 45
pseudo-random binary sequence, 140

Q

Q-factor, 123, 220
QBD-ACO-OFDM, 215
QBD-OFDM, 213, 215, 217
QBD-OFDM FSO system, 214
QDB 9-QAM signal, 45
QDB spectrum shaped signals, 35

QDB spectrum shaping, 36
QPSK format, 177
QPSK partitioning scheme, 35
QPSK symbol, 201
QPSK/8QAM/16QAM vector signals at the receiver, 236
quad-carrier QPSK-OFDM, 145, 153
quad-carrier QPSK-OFDM signal, 144, 147, 149, 151
quadrature amplitude modulation, 203
quadrature duobinary (QDB) delay and add filter, 42
quadrature-amplitude modulation, 176
quadrature-phase-shift-keying, 230
quasi-balanced coding and detection, 212

R

R/G/B filter, 215
Raman amplifier, 127
real and imaginary parts, 179, 209
real-time, 183, 189
real-time OFDM signal, 177
real-time oscilloscope, 151
received optical power, 65
received signal spectra, 236
receiver sensitivity, 83
recirculating loop, 139
record, 11
required OSNR, 71, 121
resampling and removing CP, 218
residual sideband DMT, 182
RGB-LED, 204
right-band SSB, 172
ROADMs, 41

S

16-Gbaud PDM dual-subcarrier 16QAM-OFDM signal transmission, 157
16QAM-OFDM signal, 157
sample rate, 55
SC-FDE, 204, 206, 207
SC-FDE signals, 203, 224
SDM-FSO system, 221
SE, 135, 208
second stage DD-LMS, 70
self-phase modulation, 17
series parallel conversion, 82
short-distance system, 72
signal resampling, 178
signal switch software, 8
signal-to-noise ratio, 80
signal-to-noise ratio (OSNR) degradation, 155
signal-to-signal beating noise elimination, 212
signals, 153
simulated electrical spectrum, 165
simulation results, 7, 57
single optical carrier, 185
software algorithm, 198
space division multiplexing, 220
spectral distribution, 146
spectral efficiency, 85, 174
spectrum efficiency, 175
split-step method (SSM), 19
square-law PD conversion, 232, 244
square-timing recovery, 5
SSB DFT-S OFDM signal, 171
SSB signal, 167
SSMI, 83
SSMI cancellation, 86
structure, 169
sub-band CAP signal, 63
subcarriers, 88, 109, 110
subcarriers CO-OFDM system, 160
super-Nyquist filtering, 27
super-Nyquist signal, 44–45
super-Nyquist transmission, 25, 42
symbol hard decision, 177
symbol synchronization, 82
synchronization, 218

T

2-subcarrier, 136, 138
2-subcarrier OFDM signal, 140

2048QAM-OFDM, 91
256-point IFFT, 140
Taylor expansion, 166
template, 198, 251
tighter filter bandwidth, 40
time-division multiplexing (TDM), 52
time domain, 89, 141, 144, 154, 159–160
time-domain impulse response, 42
time-domain joint image cancellation, 168
time-domain MIMO-Volterra algorithm, 164
time-domain MIMO-Volterra equalization algorithm, 173
time-domain MIMO-Volterra equalizer, 168
time-domain OFDM signals, 205
time-domain truncation method, 4
time-interleaved, 182
timing metric and frequency offset metric, 193
timing synchronization, 177–178, 192
traditional OFDM signal, 89, 140
traditional QPSK-OFDM signal, 146–147
transfer function, 30, 42, 79
transmission fiber, 142
transmitter, 81
TS-based frequency-domain equalization, 222
tunable filter, 64
twin-SSB signal, 164
two parallel half-cycled OFDM symbols, 87
two-dimensional CAP, 55
two-fold hybrid time-frequency equalization, 218
two-stage algorithm, 37

U

upconversion, 246
upstream, 61

V

vector signal, 234, 237, 244
vector-modulated multi-amplitude QAM data, 242
Viterbi–Viterbi algorithm, 7–8, 31

W

wavelength-division multiplexing, 52
WDM-CAP-PON, 62
weight vector, 219
window shaping, 143